Mathematics and Visualization

For further volumes:
http://www.springer.com/series/4562

Lars Linsen
Hans Hagen
Bernd Hamann
Hans-Christian Hege
Editors

Visualization in Medicine and Life Sciences II

Progress and New Challenges

With 132 Figures, 116 in color

Springer

Editors

Lars Linsen
Jacobs University, Bremen
School of Engineering and Science
Postfach 75 05 61
28725 Bremen
Germany
l.linsen@jacobs-university.de

Bernd Hamann
Institute for Data Analysis and Visualization
Department of Computer Science
University of California, Davis
One Shields Avenue
Davis, CA 95616-8562
U.S.A.
hamann@cs.ucdavis.edu

Hans Hagen
Universität Kaiserslautern
FB Informatik
Postfach 30 49
67653 Kaiserslautern
Germany
hagen@informatik.uni-kl.de

Hans-Christian Hege
Zuse Institute Berlin (ZIB)
Visualization and Data Analysis
Takustr. 7
14195 Berlin
Germany
hege@zib.de

ISBN 978-3-642-21607-7 e-ISBN 978-3-642-21608-4
DOI 10.1007/978-3-642-21608-4
Springer Heidelberg Dordrecht London New York

Library of Congress Control Number: 2007935102

Printed on acid-free paper

Springer is part of Springer Science+Business Media (www.springer.com)

Preface

Medicine has for a long time been a major driver for the development of data processing and visualization techniques. Many exciting and challenging visualization problems are continually arising in medicine as a consequence of our ability to generate increasingly large and complicated data (image data, simulated data) that require us to devise effective tools for meaningful interpretation and utilization in medical practice. The first VMLS workshop, which led to the book entitled "Visualization in Medicine and Life Sciences (VMLS)," was driven by the fact that emerging technologies in the life sciences had produced significant data visualization challenges. One interesting question was: Can medical data visualization approaches be devised and/or improved to meet these challenges with the promise of ultimately being adopted by medical experts.

Life sciences are understood by us in a broad sense, including animal and human biology, biochemistry, bioinformatics, biomathematics, food sciences, environmental sciences, and pharmacology. Different data acquisition technologies lead to different types of data, including both spatial and non-spatial data. The aim of the second international VMLS workshop was to document and discuss the progress that had been made since the first workshop and to explore what novel solution approaches for data processing and visualization had been developed and what new challenges had come up.

Internationally leading experts from the visualization and driving medical application areas came together for this second workshop held in Bremerhaven, Germany, in July 2009. Research and survey papers were solicited and peer-reviewed, ultimately leading to the collection of papers included in this book.

The research topics covered by the papers in this book deal with these themes:

- Feature Extraction
- Classification
- Volumes and Shapes
- Tensor Visualization
- Visualizing Genes, Proteins, and Molecules

The workshop was supported, in part, by the Deutsche Forschungsgemeinschaft (DFG). We would like to thank Diana Babiac for her help with compiling this book.

Bremen, Germany *Lars Linsen*
Kaiserslautern, Germany *Hans Hagen*
Davis, California, U.S.A. *Bernd Hamann*
Berlin, Germany *Hans-Christian Hege*

Contents

Part I
Feature Extraction

Discrete Distortion for 3D Data Analysis

Leila De Floriani, Federico Iuricich, Paola Magillo, Mohammed Mostefa Mesmoudi, and Kenneth Weiss

Abstract We investigate a morphological approach to the analysis and understanding of three-dimensional scalar fields, and we consider applications to 3D medical and molecular images as examples. We consider a discrete model of the scalar field obtained by discretizing its 3D domain into a tetrahedral mesh. In particular, our meshes correspond to approximations at uniform or variable resolution extracted from a multi-resolution model of the 3D scalar field, that we call a *hierarchy of diamonds*. We analyze the images based on the concept of *discrete distortion*, that we have introduced in [26], and on segmentations based on Morse theory. Discrete distortion is defined by considering the graph of the discrete 3D field, which is a tetrahedral hypersurface in R^4, and measuring the distortion of the transformation which maps the tetrahedral mesh discretizing the scalar field domain into the mesh representing its graph in R^4. We describe a segmentation algorithm to produce Morse decompositions of a 3D scalar field which uses a watershed approach and we apply it to 3D images by using as scalar field both intensity and discrete distortion. We present experimental results by considering the influence of resolution on distortion computation. In particular, we show that the salient features of the distortion field appear prominently in lower resolution approximations to the dataset.

L. De Floriani (✉) · F. Iuricich · P. Magillo · M.M. Mesmoudi
Department of Computer Science and Information Science (DISI), University of Genova, Italy, Via Dodecaneso 35, 16146 Genova (Italy)
e-mail: deflo@disi.unige.it; federico.iuricich@gmail.com; magillo@disi.unige.it; mmesmoudi@creteil.fr

K. Weiss
Department of Computer Science, University of Maryland, College Park, MD 20742 (USA)
e-mail: kweiss@cs.umd.edu

M.M. Mesmoudi
Labo. Pure and Appl. Math. UMAB, University of Mostaganem, Route Belhacel, 27000 Mostaganem (Algeria)

L. Linsen et al. (eds.), *Visualization in Medicine and Life Sciences II*, Mathematics and Visualization, DOI 10.1007/978-3-642-21608-4_1,

1 Introduction

We consider a three-dimensional scalar field which is defined by a collection of function values, each given at a point in a 3D domain. Examples of 3D scalar fields of interest in biomedical applications are 3D images, where the intensity at each voxel defines the scalar field. A scalar field is known at a finite set of points in 3D space, and a digital model of the field is constructed based on such points. Models of 3D scalar fields are based on voxels, or on tetrahedral meshes with vertices at the points in the 3D space at which the field is known. In both cases, such models tend to be verbose and may not be immediately useful to understand the behavior of the field.

Here, we consider tetrahedral meshes extracted from a multi-resolution representation of 3D images provided by a regular tetrahedral hierarchy. We have developed an efficient representation of a regular tetrahedral hierarchy, called a hierarchy of diamonds, as discussed in [43].

The aim of morphological analysis is to provide a tool for understanding the structure of a scalar field through structural representations of the field so that its basic features can be easily recognized. Here, we use the notion of discrete distortion to support morphological analysis. In [26] we have introduced a discrete approach to curvature for three-dimensional tetrahedralized shapes embedded in 4D space, that we called *discrete distortion*. If we consider 3D scalar fields, we can view the values of the field as constraints on the vertices of a tetrahedral mesh. From this perspective, the values induce a distortion of the geometry of the mesh, seen as a hypersurface representing the graph of the scalar field in R^4. As for surface curvature, discrete distortion highlights the local curvature of the constrained shape (the graph of the 3D scalar field) which cannot be perceived in the three-dimensional domain. As curvature gives interesting insights in terrain analysis, we show that distortion provides additional information to analyze the behavior of the intensity field. A null distortion value highlights a linear behavior of the intensity field, while a constant distortion corresponds to a uniform non-linear behavior. We observe that directions in which distortion changes indicate interesting directions in which the intensity field varies its growth speed.

One way to perform morphological analysis is to automatically decompose the domain of the field into meaningful parts in order to support understanding and semantic annotation. Segmentation has been the basic tool to support reasoning on terrains and 3D shapes. Here, we propose segmentations for a 3D image based the intensity value or on discrete distortion, in a similar way as done for terrains where segmentations are computed based on elevations and/or on curvature values.

The segmentation of a scalar field is performed based on its critical points, and the steepest directions through which the scalar field increases or decreases. This leads to two dual decompositions. The *stable Morse decomposition* associates a 3D cell with each local minimum of the field, and two adjacent 3D cells touch at ridge surfaces (i.e., surfaces where the field decreases on both sides). The *unstable Morse decomposition* symmetrically associates a 3D cell to each local maximum, and the boundaries of the 3D cells are at valley surfaces. Here, we present an algorithm for computing Morse decompositions based on a watershed approach and we compute Morse decompositions based on the intensity field and on a new field induced by discrete distortion. This approach reveals relevant features, different from those that are generally extracted by studying the behavior of the gradient field (i.e., the critical points of the intensity field). For example, the extrema of distortion correspond to locations in which the intensity field has abrupt variations, which might not be perceptible from the intensity values.

We apply our approach to the analysis of the morphology of scalar fields through examples on synthetic, biological and medical datasets. We show color-coded visualization of the fields based on the intensity field and of distortion. We study the interaction between the resolution of the tetrahedral mesh approximating the field and the distortion values, showing that we can reasonably approximate the 3D image at fairly low resolutions. Finally, we show results on Morse segmentations based on the intensity and on the distortion values and we compare them.

The remainder of this paper is organized as follows. In Sect. 2, we provide background notions on concentrated curvature, and on Morse theory and Morse complexes. In Sect. 3, we review some related work. In Sect. 4, we briefly describe a mesh-based multi-resolution model that we use for representing the 3D image. In Sect. 5, we discuss the notion of discrete distortion for a tetrahedralized shape representing the graph of a 3D scalar field, and we present some of its properties. In Sect. 6, we present an algorithm to segment tetrahedral meshes endowed with discrete scalar fields, and produce Morse decompositions. In Sect. 7, we present experimental results on medical data set and we discuss the results. Finally, in Sect. 8, we draw some concluding remarks and discuss on-going and future work.

2 Background Notions

In this Section, we discuss first concentrated curvature, since the notion of discrete distortion generalizes concentrated curvature to hypersurfaces in R^4. Then, we briefly review some notions from Morse theory which is the basis for defining morphological segmentations for a scalar field.

2.1 *Concentrated Curvature*

Concentrated curvature is the discrete counterpart of Gaussian curvature for triangulated surfaces [1, 25, 40], and was introduced by Aleksandrov for 2D scalar fields represented as triangle meshes [1]. Given a triangulated surface in R^3 and a vertex p in the interior of the corresponding triangle mesh, the local neighborhood of p is the union of the angular sectors of the triangles incident at p. The *total*

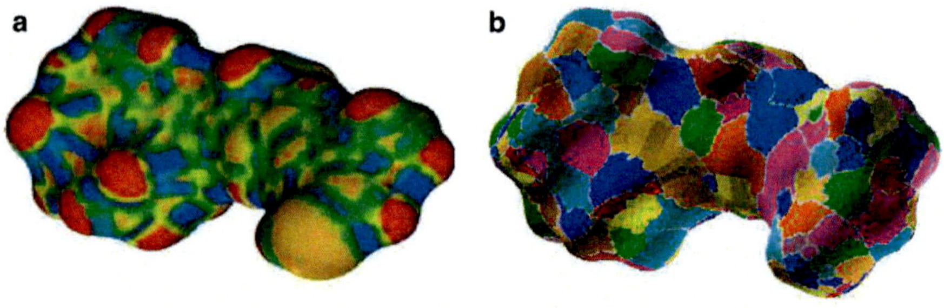

Fig. 1 In (**a**), surface of a molecule with associated concentrated curvature represented in a rainbow color scale (blue = low values, red = high values). In (**b**), Morse decomposition of the surface based on concentrated curvature

angle Θ_p at p is given by the sum of the angles at p of all the triangles incident in p. The *concentrated curvature* at a vertex p is defined as $K(p) = 2\pi - \Theta_p$, when p is an internal vertex [40]. If the sum of the angles of all these sectors is not equal to 2π, p is called a *singular conical point*. When the surface is defined by a scalar field, the concentrated curvature for boundary points can be defined as the defect between the angle at the vertex on the surface and its corresponding angle on the xy-projection domain.

Note that concentrated curvature involves only the sum of angles of the triangles incident at a vertex and does not take into account their geometric position. *Planar points* have null concentrated curvature. *Saddle points* have negative concentrated curvature, while *convex and concave points* have positive concentrated curvature. Figure 1a illustrates the behavior of concentrated curvature on the surface of a molecule: red areas correspond to high curvature, and blue areas to low curvature. The segmentation in Fig. 1b represents a Morse segmentation of the surface based on concentrated curvature.

2.2 *Morse Theory and Morse Complexes*

Let $f(x, y, z)$ be a scalar field defined on a domain D of R^3, and let function f be continuous and smooth in D. A point p of D is a *critical point* of f if the gradient of f at p is null. Points that are not critical are called *regular points*. A critical point p is *degenerate* if the Hessian matrix of the second partial derivatives of f at p is not singular. A scalar field f is a *Morse function* if and only if all its critical points are not degenerate. Morse [27] showed that the critical points of a Morse function are isolated. The number i of negative eigenvalues of the Hessian matrix is called the *index* of critical point p, and p is called an i*-saddle*. A 0-saddle is a *minimum* and a 3-saddle a *maximum*.

An *integral line* of f is a maximal path which is everywhere tangent to the gradient of f. Each integral line connects two critical points of f, called its *origin* and its *destination*. Integral lines that converge to a critical point p of index i form

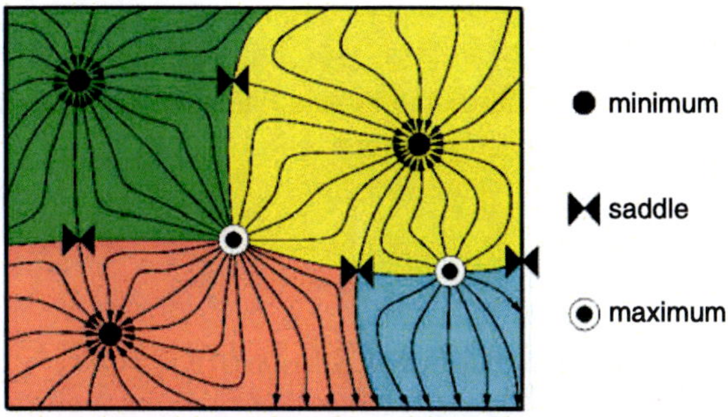

Fig. 2 Stable Morse complex for a 2D scalar field. Arrows denote the negative gradient field. Only a window over the domain of the field is shown; the stable cells of the minima are shown in different colors: each stable cell is the region covered by arrows converging to the same minimum

an i-cell, called a *stable* cell of p. Dually, integral lines that originate at p form its *unstable* $(n - i)$-cell. The stable and unstable cells decompose D into *stable* and *unstable Morse complexes*. Figure 2 illustrates the above concepts for the domain of a 2D scalar field.

A Morse function f is called a *Morse-Smale function* if each non-empty intersection of a stable and an unstable cell is transversal. The connected components of the intersection of the stable and unstable cells define a *Morse-Smale complex*. Note that the Morse-Smale complex for f can be obtained by the overlay of its stable and unstable complexes.

3 Related Work

In this Section, we briefly review related work on discrete curvature estimators, on multi-resolution modeling of 3D scalar fields, on algorithms for computing Morse decompositions of 3D scalar fields, and on features of interest in medical imaging.

3.1 Discrete Curvature Estimators

Curvature is an important notion in mathematics that found a great interest in the last century. Curvature is also used to study the local geometry and topology of surfaces from the metric point of view. With the development of discrete geometry, many authors tried to define a discrete counterpart of curvature based on the properties observed in the continuum [12, 18, 37, 38]. There is a rich literature dealing with the problem of defining and computing discrete curvature estimators for triangle meshes, and more recently for tetrahedral meshes (see [12,18,37,38] for a survey). Concentrated curvature [1,40] is a simple and efficient method to define a discrete curvature, as discussed in Sect. 2.1. Dyn et al. discuss how to optimize the triangulation of the boundary of a 3D object based on discrete curvature [9].

In the 3D case, the Ricci tensor is used to define the curvature notion for three-dimensional shapes [2], and, in the discrete case, the Laplace operator is generally used to define a discrete approach to curvature [33].

In the 4D case, curvature is one of the most important mathematical notions on which general relativity is based. Curvature of the space-time gave an important contribution to understand many phenomena in physics (black holes, gravitational lenses, light trajectories, interaction between planets, ...). Based on Aleksandrov's concentrated curvature, Regge introduced a discrete version of curvature for the four dimensional space-time [32].

3.2 Multi-Resolution Modeling

A very large class of multiresolution models of volumetric scalar fields is provided by nested meshes, in which all elements are defined by the uniform subdivision of a small set of primitive cells. Examples include octrees formed by cubes [36] and tetrahedral meshes generated by the so-called *Red/Green* tetrahedron refinement [5].

Nested tetrahedral meshes based on the *Longest Edge Bisection (LEB)* operation were originally introduced for domain decomposition in finite element analysis [20, 23, 34], and have since then been applied in many different contexts, including scientific computing [13, 14, 46], surface reconstruction [24] and volume segmentation [21]. A recent survey on nested simplicial meshes based on bisection can be found in [44].

The *LEB* operation is defined by bisecting a tetrahedron t along the plane defined by the midpoint of its longest edge e and the two vertices of t not incident to e. The containment relation among the tetrahedra generated by successive LEB operations naturally defines a binary tree, where the two tetrahedra generated by bisecting a *parent* tetrahedron t are the *children* of t. When a full binary tree is stored, this representation can be efficiently encoded as a linear array, and the parent-child relation can be implicitly determined from the array indices [13, 22, 46]. A forest of six such tetrahedral binary trees, whose roots share a common cube diagonal can thus decompose a cubic region of space.

We are often interested in generating crack-free, or *conforming*, tetrahedral meshes, since cracks in the mesh correspond to discontinuities in scalar fields discretized through it. Methods of ensuring continuity have been proposed based on a hierarchical monotonic error metric [29], symbolic *neighbor-finding* operations [20, 22] or an implicit clustering of tetrahedra sharing a common bisection edge into a *diamond* primitive [14, 42]. In this work, we utilize diamonds to extract conforming tetrahedral meshes from the multi-resolution model.

3.3 Algorithms for Morse Decompositions in 3D

Most of the algorithms proposed in the literature for extracting an approximation of the Morse complexes in the discrete case have been developed for terrains. The majority of them use a *boundary-based* approach, since they extract the decomposition by computing the critical points and then tracing the integral lines, or their approximations, starting from saddle points and converging to minima and maxima. Other algorithms use a *region-based* approach, in the sense that they compute an approximation of a Morse decomposition by growing a 2D region defined and started by the minima and the maxima of a Morse function f. Curvature has been applied to the segmentation of 3D shapes and terrains in combination with Morse decompositions (see, for instance, [28]). A comprehensive analysis of techniques for Morse decomposition can be found in [6].

Alternative region-based techniques for computing the stable and unstable Morse decompositions are those based on the discrete watershed transform (see [35] for a survey). For a C^2-differentiable function f, the watershed transform provides a decomposition of the domain of f into regions of influence of the minima, called *catchment basins*, which are bounded by watershed lines. If f is a Morse function, it can be shown that the catchment basins of the minima of f and the watershed lines correspond to the 2-cells and the 1-cells, respectively, in the stable Morse decomposition of f [6].

Much less work has been done on computing Morse decompositions for 3D scalar fields. In [10], an algorithm for extracting the Morse-Smale decomposition (i.e. the intersection of the stable and unstable Morse decompositions) from a tetrahedral mesh approximating a 3D scalar field is proposed. The algorithm computes the Morse-Smale decomposition by extracting the critical points, then the unstable Morse decomposition and finally the stable cells in pieces inside the unstable cells. The algorithm, while interesting from a theoretical point of view, has a large computation overhead, as discussed in [17]. In [17], a region growing method is proposed to compute the Morse-Smale decomposition inspired by the watershed approach. A procedural approach based on discrete Morse theory is described in [15] which also computes the Morse-Smale complex.

A major issue when computing Morse decompositions for 2D and 3D scalar fields is over-segmentation, which is due to the presence of noise in the data. To this aim, *generalization algorithms* have been developed in order to eliminate less significant features from a Morse or Morse-Smale decomposition, mainly for 2D scalar fields. Generalization is achieved by applying an operation, called *cancellation* of critical points. Cancellations of critical points for a 3D scalar field consist of collapsing a maximum and a 2-saddle into a single maximum, a minimum and 1-saddle into a single minimum, or a 1-saddle and a 2-saddle into either a 1-saddle or a 2-saddle [7, 16].

3.4 Features in Medical Images

Features of interest in medical images (also called *landmarks*) may correspond to points, lines, surfaces or volumes. Many techniques for landmark extraction are based on curvature. In [3, 31, 39] an algorithm that computes a polygonal approximation of the so-called *Gaussian frontier* is described. Points with large curvature values along a contour are selected following different scale-space Gaussian filters. In [4], isolines of extremal values of mean curvature are selected to segment MRI and CT-scans images. These isolines generally correspond to ravine and ridge lines of the surface. Based on contour detection and curvature, an automatic landmark extraction is proposed in [11]. Several techniques using first and second differentials of 3D operators (that extend the 2D case) are compared on MRI and CT-scan images in [19].

4 A Diamond-Based Multi-Resolution Model

In this Section, we describe a mesh-based multi-resolution model for 3D scalar fields called a *hierarchy of diamonds*, which we use to generate approximate representations of a scalar field at variable and uniform resolutions. We have studied the theory and the properties of hierarchy of diamonds in arbitrary dimensions in [42].

A multi-resolution model M of a shape Σ is typically defined by three components [8]: (a) A coarse base mesh Γ_0 that approximates Σ, (b) a set of modifications U, each of which replaces a set of cells γ_1 with a new set of cells γ_2 sharing the same combinatorial boundary and (c) a *direct dependency relation* R on the modifications U, where a modification u_2 depends on another modification u_1 if u_2 replaces a cell that was introduced in u_1.

As pointed out in Sect. 3.2, the LEB (longest edge bisection) operation is defined by bisecting a tetrahedron t along the plane defined by the midpoint of its longest edge e and the two vertices of t not incident to e.

A *Hierarchy of Diamonds* Δ is a multi-resolution representation of a regularly sampled 3D scalar field covering a cubic base domain D and generated through longest edge bisection. It is based on clusters of tetrahedra, called *diamonds*, sharing the same bisection edge, that we call the *spine* of the diamond. The diamond whose spine is a diagonal of D defines the base mesh Γ_0 of the model. Each diamond δ corresponds to a modification (γ_1, γ_2) in the multi-resolution model, where γ_1 consists of the tetrahedra of δ, and γ_2 consists of the tetrahedra generated by bisecting the tetrahedra in γ_1 along the spine of δ.

A diamond δ_p is said to be a *parent* of another diamond δ_c if one or more of the tetrahedra in δ_c is generated during the bisection of the tetrahedra in δ_p. This parent-child relation defines the *direct dependency* relation of the model, which can be encoded as a *Directed Acyclic Graph* (DAG) (see Fig. 4(a) for an illustration in 2D).

An explicit encoding of the hierarchy as a directed acyclic graph would require each modification to list the tetrahedra before and after bisection as well as the dependencies among these modifications. However, due to the regularity of the vertex distribution and the subdivision rule, this model generates only three *classes* of diamonds composed of six, four and eight *similar* tetrahedra [23], respectively (see Fig. 3). Furthermore, the diamond classes have three, two and four parents, and six, four and eight children, respectively.

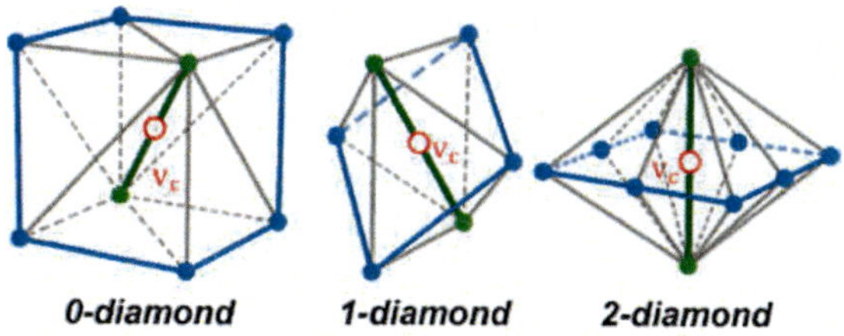

Fig. 3 The three classes of three-dimensional diamonds. The spine (internal green edge) of a diamond of class i is aligned with the diagonal of a $(d - i)$-cube. A diamond's central vertex v_c (hollow red circle) coincides with the midpoint of its spine

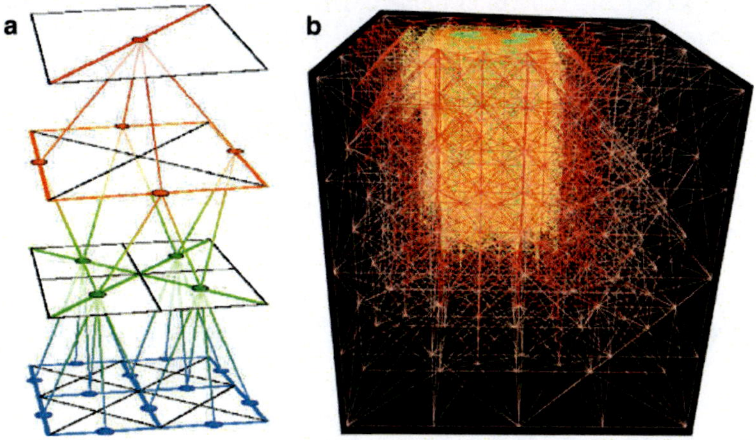

Fig. 4 (**a**) A *hierarchy of diamonds* (shown in 2D) is a rooted DAG whose nodes are the diamonds and whose arcs encode the dependency relation among the diamonds. (**b**) Example variable-resolution tetrahedral mesh extracted from a hierarchy of diamonds. A tetrahedron's color indicates its level of resolution

Thus, diamond hierarchies admit extremely compact encodings of the underlying multi-resolution model which exploit the implicit relationships among the modifications and their dependencies. Each diamond is completely defined by its spine, and all its tetrahedra are split by the diamond's *central vertex*, the unique midpoint of its spine. Thus, diamonds are in one-to-one correspondence with their central vertices, which, in turn, are in one-to-one correspondence with the samples of the dataset. From the coordinates of the central vertex, we use bit manipulations to extract the complete parent-child relations. A hierarchy of diamonds can therefore be encoded as the collection of the central vertices of its diamonds [42], from which all geometric and hierarchical relationships can be implicitly determined [14, 43].

A hierarchy of diamonds Δ is used to efficiently extract variable-resolution tetrahedral meshes Σ approximating a 3D image while satisfying an application-dependent *selection criterion*. The selection criterion can be defined on properties of the domain, such as proximity to a specified region of interest, or on properties of the range such as its degree of approximation to the underlying dataset. In contrast to octree-based approaches, such tetrahedral meshes have a higher degree of adaptability to the selection criterion [30], and are guaranteed to be free of cracks. Figure 4(b) shows a variable-resolution tetrahedral mesh extracted from a hierarchy of diamonds.

5 Discrete Distortion

In this Section, we briefly review the notion of distortion that we have introduced in [26] and we describe some of its properties. *Discrete distortion* is a generalization of concentrated curvature to tetrahedral meshes in 4D space [26]. In [25], we have investigated discrete distortion for triangulated terrains, and we have shown that it behaves as a discrete counterpart of mean curvature. Here, we discuss the notion of distortion for tetrahedral meshes endowed with a scalar field.

The graphical representation of a scalar field f defined on a tetrahedral mesh Σ is a hypersurface $(\Sigma; f)$ in R^4, namely, a tetrahedral mesh embedded in R^4. Hypersurface $(\Sigma; f)$ is generally curved due to the effects of the scalar field values. As for concentrated curvature, one may compare the defect solid angle at the vertices of Σ, when applying the scalar field.

The *distortion* at a vertex p of Σ is defined as the quantity

$$D(p) = 4\pi - \sum_{T_i \in T(p)} S_i, \tag{1}$$

where S_i is the solid angle, after applying the scalar field (i.e., within $(\Sigma; f)$), at vertex p of tetrahedron T_i, and $T(p)$ is the set of all tetrahedra incident at p. A similar formula holds for boundary vertices:

$$D(p) = \sum_{t_i \in t(p)} s_i - \sum_{T_i \in T(p)} S_i, \tag{2}$$

where s_i is the solid angle, within Σ, at vertex p of tetrahedron t_i, and $T(p)$ is the set of all tetrahedra incident at p.

Discrete distortion for 3D scalar fields has similar properties as concentrated curvature for 2D fields. Concentrated curvature gives positive values to locally convex, or concave, areas of the surface, negative values to saddles, and null values to flat areas. Similarly, positive values of distortion correspond to locally convex, or concave, portions of the hypersurface which is the graph of the field. Negative values correspond to saddle and degenerate saddle configurations.

Constant scalar fields are distortion-free (i.e., their distortion is null). This can easily be understood since, for a constant scalar field, mesh $(\Sigma; f)$ is only a translation, in the fourth dimension, of the mesh Σ decomposing the domain of the field. Hence, the Euclidean geometric structure of the mesh is preserved. More generally, affine scalar fields are distortion-free, since they combine rotations and translations of the whole mesh. Hence, the geometrical structure is not subject to any distortion.

As a consequence, piecewise linear scalar fields are distortion-free at the interior vertices of regions where the field is linear, as they act affinely within such regions. Another relevant property is that distortion is mesh-dependent. This means that the distortion value at a vertex depends on the way in which the neighborhood of such vertex is triangulated.

6 Computing Morse Complexes

We compute a discrete approximation of the unstable and the stable Morse complexes for a 3D scalar field f defined at the vertices of a tetrahedral mesh Σ by extending the watershed approach by simulated immersion developed for 2D images in [41]. We describe only the algorithm for the stable complex since the unstable complex can be built by considering field $-f$.

The watershed algorithm performs the following three steps:

1. Sort the vertices of the mesh by increasing values of field f.
2. Associate all mesh vertices with a local minimum. This is done starting from minima and proceeding based on increasing field values and on increasing distance from already discovered local minima.
3. Assign each tetrahedron to the stable 2-cell of a local minimum, based on the assignments of its vertices.

In the second step we process the vertices of the mesh according to their field values. Let h be the current field value (initially, h is the minimum field value over the mesh). We consider the set H of all the vertices whose field value is equal to h. A priority queue is used to ensure the processing vertices in H in increasing distance from an already assigned vertex. We iteratively pick the first vertex $v \in H$ from the priority queue, and we check if some of its neighboring vertices has already been assigned to a local minimum. If they are all either unassigned, or assigned to the same local minimum, then v is assigned to that local minimum. If two or more neighboring vertices are assigned to different local minima, then v is marked as a watershed vertex. After assigning v, the priorities of the unassigned neighbors of v are updated in the priority queue. The above process is repeated until no more assignments are possible (i.e., the priority queue is empty). Then, for each vertex $w \in H$ that is still unassigned, w is marked as a new local minimum, and all vertices with the same field values equal to h, which are connected to w, are assigned to w. Now, all vertices in H have been assigned to some local minimum (possibly equal to the vertex itself), and the algorithm proceeds with the next field value.

The third step examines each tetrahedron τ and assigns it to a local minimum based on its vertices. If all four vertices of τ are marked as watershed, then τ is marked as a watershed tetrahedron. Otherwise, among the local minima assigned to the vertices of τ, we choose the one having minimum field value and assign τ to such minimum.

The watershed algorithm described above often produces an over-segmentation of the graph of the scalar field. To overcome this problem, we perform an iterative merging process of the 3D cells of the stable (unstable) Morse complex, which correspond to a topological generalization operator, namely, to the cancellation of a maximum and a 2-saddle in the stable complex, or of a minimum and a 1-saddle in the unstable complex. We consider pairs of adjacent 3D cells in the stable (unstable) Morse complex. We associate with each cell γ_1 a *saliency* measure with respect to any adjacent 3D cell γ_2, which takes into account: (i) the maximum field difference between the local minimum [maximum] associated with cells γ_1 and γ_2 and the largest [smallest] field value of a point on the common boundary 2-cell of γ_1 and γ_2; (ii) the product of the sizes of γ_1 and γ_2; (iii) the extent of the common boundary of γ_1 and γ_2.

Thus, priority is given to merging cells with small differences in field values, with a small area and to cells with long common boundaries. The first measure is the persistence value, as defined in [10, 16].

7 Experimental Results

In this Section, we present some experimental results which show the behavior of discrete distortion as a tool for analysis of 3D images. Because of the large size of current data sets, it is also important to perform accurate analysis on low-resolution representations of the field. Here, we study the influence of mesh resolution on distortion by considering variable-resolution conforming tetrahedral meshes extracted from a hierarchy of diamonds according to a user-defined threshold on the approximation error. In this case, resolution can be coarsened locally in less interesting regions, without affecting the quality of the approximation. Finally, we show and compare segmentations of the 3D images obtained through Morse decompositions of the intensity and of the distortion fields. To this aim, we present results on a synthetic data set in which the intensity field is defined by an analytic function and on two real data sets.

7.1 3D Datasets and Distortion

Our first example is a synthetic dataset defined over a regularly sampled domain of 65^3 vertices. The intensity field is obtained by sampling the analytic function $f(x, y, z) = sin(x) + sin(y) + sin(z)$. We show here the tetrahedral mesh at full resolution extracted from a hierarchy of diamonds built on such data set. It is composed of 275 K vertices and 1.57 million tetrahedra. The relationship between

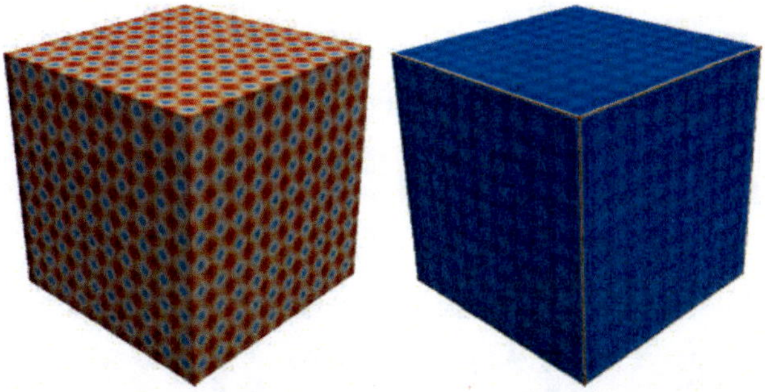

Fig. 5 Intensity field (left) and distortion field (right) for synthetic data set sampling function $f(x, y, z) = sin(x) + sin(y) + sin(z)$ over a 65^3 grid

the intensity field and the induced distortion field over this domain is illustrated in Fig. 5 along the boundary of the cubic domain using a blue-red color scale to indicate the low and high scalar and distortion values.

The second dataset, called Neghip, is a simulation of the spatial probability distribution of electrons in a high potential protein molecule. The knowledge of electron distribution within such molecules is important in pharmacology to understand the interactions between molecules and an organism. The inhibition of some protein molecules can reduce complications in diseases such as cataracts and neuropathies for diabetic subjects. The understanding of the catalytic mechanism and the electrostatic potential of the molecule plays a relevant role here. It may help to study, at the atomic scale, the transfer of electrons and protons in complex biological processes such as oxidation/reduction in relation to metallic ions by considering the reaction between hemoglobin (containing iron ions) and the oxygen molecule.

In Fig. 6, we show the intensity field and distortion field for the tetrahedral mesh extracted from the Neghip hierarchy at variable resolution corresponding to 0% approximation error. The mesh has 129 K vertices and 728 K tetrahedra. The range of colors used for visualization goes from blue for low values to red for high values, with gray indicating mean values. Discrete distortion highlights the growth behavior of the density scalar field, which is maximal around the atoms. We see that the density field grows quickly around atoms within small regions and then stabilizes its growth. Distortion becomes nearly constant in such case. We observe also that, within regions where the electron density has low values, many small regions have high distortion values. This indicates changes in the electron density and may be due to the interference between adjacent atoms or to some artifacts in the processing of the data. Regions in blue (for distortion) indicate that the scalar field grows differently in different directions. This corresponds to saddle regions where the convexity of the electron density field changes.

The third dataset, called CTA-Brain, is a CTA-scan of a human brain with an aneurysm. Computed Tomographic Angiography (CTA) is a minimally invasive technique that uses imaging technologies (e.g., X-rays) to explore the structure

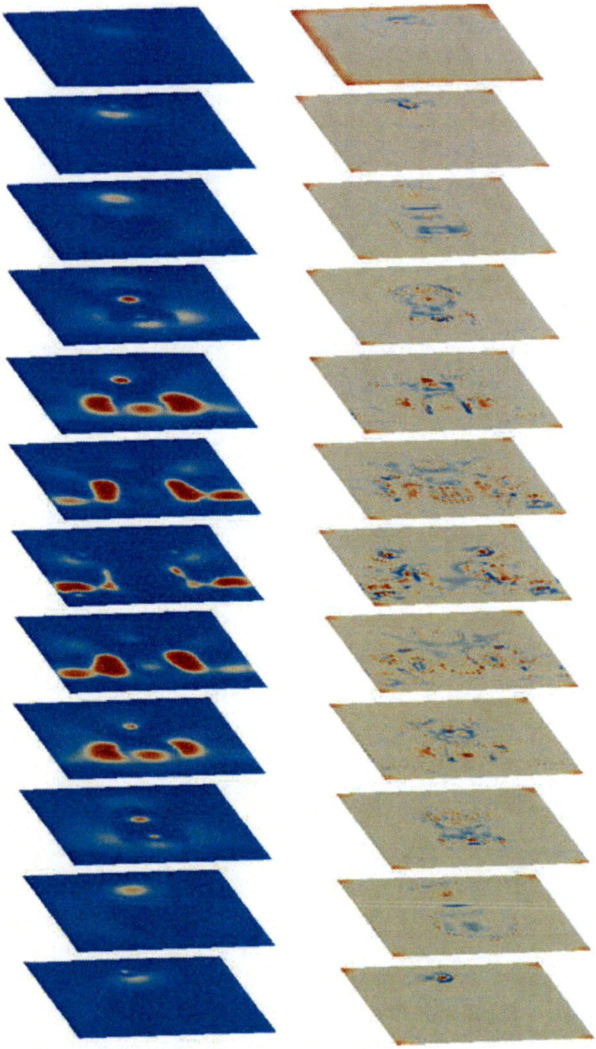

Fig. 6 Twelve equally spaced slices (along the z-axis) of the intensity field (left), and the distortion field (right) of the Neghip dataset at 0% error. The colors of the distortion field are scaled to highlight the extreme values

of vessels and tissues. A contrast agent is generally used to produce clear images. The original dataset has $512 \times 512 \times 120$ vertices and measure the intensity of the contrast agent. To show the behavior of the intensity field and of distortion, we have extracted a variable-resolution mesh from the diamond hierarchy, which has 1.74 million vertices and 9.52 million tetrahedra.

Figure 7 illustrates the dataset, where the scalar field corresponds to the intensity of the contrast agent, and its distortion, through equally spaced horizontal slices.

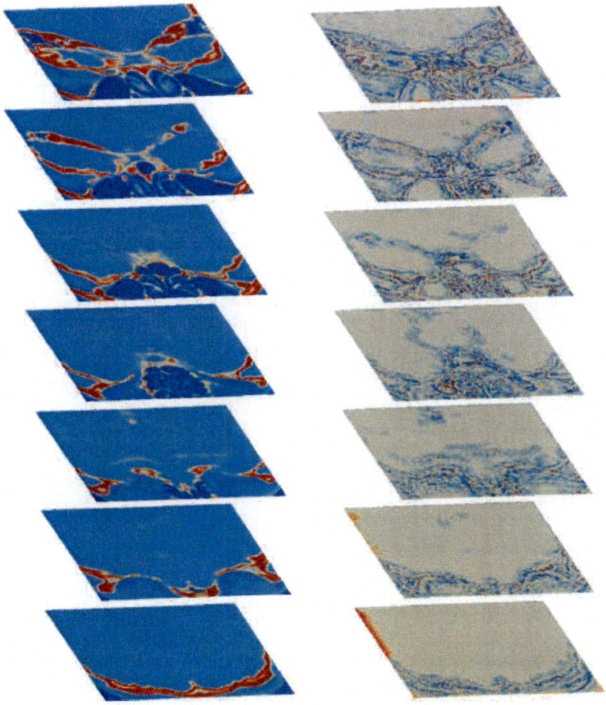

Fig. 7 Seven equally spaced slices (along the z-axis) of the CTA-brain dataset at 10% error illustrating the scalar field (left), and the distortion field (right). The colors of the distortion field are scaled to highlight the extreme values

The geometric structure of the scanned region is well represented by distortion. We see that most regions have gray or light blue color, which indicates a uniform distribution of the contrast agent within the brain. The regions with high distortion correspond to changes in the intensity of the contrast product.

7.2 Distortion and Mesh Resolution

We now demonstrate the validity of distortion analysis on lower resolution approximations by considering the distribution of distortion values over a set of extracted meshes with increasingly fine resolution. For brevity, we show results only on the two real data sets, Neghip and CT-Brain.

In the first case, we generate a diamond hierarchy Δ_H based on the intensity values of the 64^3 Neghip dataset, which contains 262 K vertices. The error of a diamond δ is computed as the maximum difference between the intensity values of all grid points within the domain of δ, and the value obtained by linear interpolation over the vertices of δ's tetrahedra. We extract a series of meshes

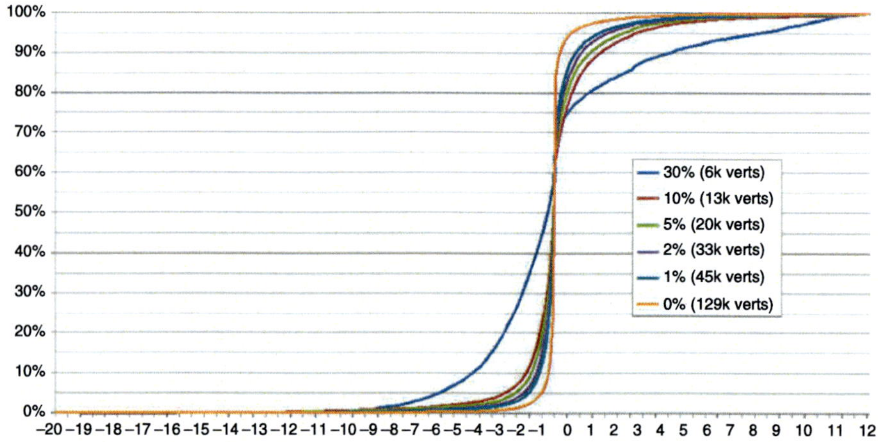

Fig. 8 Cumulative distribution functions of distortion values (horizontal axis) over increasingly fine meshes extracted from the Neghip dataset

Σ_{ϵ_i} of uniform approximation error ϵ_i from Δ_H, using threshold values of $\epsilon_i \in \{30\%, 10\%, 5\%, 2\%, 1\%, 0\%\}$ of the total error and then evaluate the distortion of the vertices of these meshes.

Figure 8 shows the *Cumulative Distribution Function (CDF)* of the discrete distortion (horizontal axis) of the vertices of each mesh. The sharp spike in the CDF of all datasets around a null distortion value indicates that the vast majority of vertices have (nearly) null distortion. As the resolution increases, this spike becomes steeper, indicating that the increased resolution is distributed among regions with nearly null distortion. Thus, the distortion is concentrated in relatively few vertices within the mesh, and appears prominently in lower resolution approximations. For example, when $\epsilon = 0$, more than 94% of the 129 K vertices in $\Sigma_{0\%}$ have distortion $D(v) \leq |1|$, and for $\epsilon = 2\%$, more than 83% of the 33 K vertices in $\Sigma_{2\%}$ have distortion $D(v) \leq |1|$.

Similarly, Fig. 9 shows the CDF of meshes Σ_{ϵ_i} using threshold values of $\epsilon_i \in \{99\%, 75\%, 50\%, 30\%, 10\%, 5\%\}$ extracted from the CTA-Brain dataset. These meshes illustrate the same general trend as the Neghip approximations, although they are a bit noisier since they are scanned images.

We have obtained similar results for several other datasets in other application domains [45]. These experiments indicate that we can obtain a fairly accurate understanding of the image via its discrete distortion even at lower resolutions, without the need to compute the distortion on the full image.

7.3 *Morse Decompositions*

In this Subsection, we show Morse decompositions of the synthetic and real data sets computed using the intensity and the distortion fields.

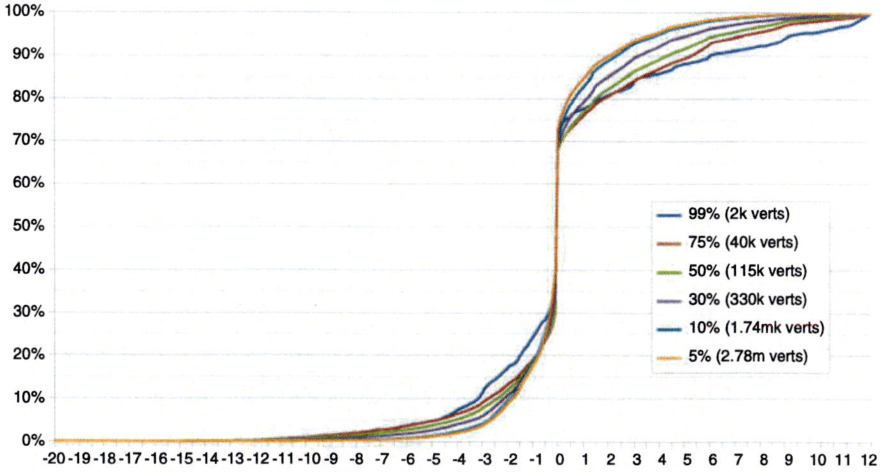

Fig. 9 Cumulative distribution functions of distortion values (horizontal axis) over increasingly fine meshes extracted from the CTA Brain dataset

Let us consider the distribution of the intensity and distortion values for the synthetic data set shown in Fig. 5. Figure 10 shows the stable and unstable Morse decompositions computed based on the intensity and on the distortion fields. It is clear how the distribution of the intensity and of the distortion values influences the corresponding segmentations. Both stable and unstable Morse decompositions obtained from the intensity field consists of 1,331 cells and have a regular structure. The stable decomposition obtained from the distortion field consists of 12,972 cells, while the unstable one consists of 3,738 cells. The decomposition pattern in the stable and unstable distortion-based complex varies in different portions of the mesh. This is due to the function sampling that is different from its period.

Figure 11 shows Morse decompositions built from the full-resolution tetrahedral mesh discretizing the Neghip dataset. We thresholded the visualization along an isovalue to better illustrate the structure of the molecules. The stable and unstable Morse decompositions obtained from the intensity field consist of 104 cells and of 41 cells, respectively. The stable and unstable Morse decompositions obtained from the distortion field consist of 3,654 stable cells and 23,334 cells, respectively. Some components of the unstable decomposition represent the location of atoms (i.e. maxima of the density) and the proper space in which electrons revolve around. Due to the interference of electron density of adjacent atoms, some components are created and correspond to some maxima of the density field. These components do not properly contain atoms.

Figure 12 illustrates the intensity field, the corresponding distortion values and the segmentations obtained from a uniform resolution mesh $\Sigma_{10\%}$ extracted from the CTA-Brain dataset (see also a view as set of slices in Fig. 7). The decomposition obtained from the intensity field consists of 37,631 stable cells and of 23,835 unstable cells, while the decomposition obtained from the distortion one consists

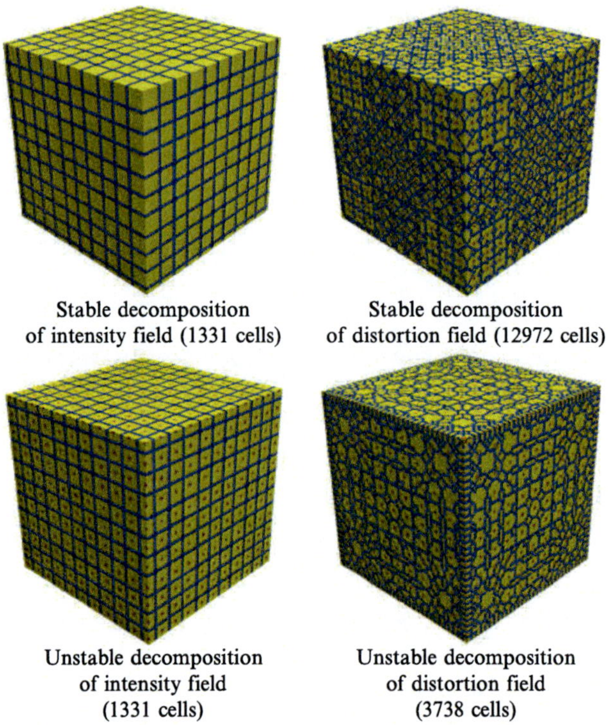

Fig. 10 Morse decompositions for the synthetic data set defined by intensity function $f(x, y, z) = sin(x) + sin(y) + sin(z)$. Minima (stable) or maxima (unstable) vertices are colored in red, vertices on the boundary of several regions in blue and vertices within a region in yellow

of 136,641 stable cells and 128,687 unstable cells. Figure 13 shows the largest segments from the segmentations. Observe that, while the unstable regions are more structured and follow the field values, the stable regions are much more influenced by the boundary and by the less relevant regions of the original scalar field. The former therefore seem to provide a more meaningful decomposition. The large number of cells in the unstable decomposition computed on the basis of distortion is due to the fact that there is a large number of small areas in which the concentration of the contrast agent changes abruptly (i.e., distortion has a maximum).

8 Concluding Remarks

We have presented an innovative approach to the analysis of 3D images based on the notion of discrete distortion, which generalizes discrete curvature to triangulated hypersurfaces in 4D space, and on Morse decomposition.

We have proposed the use of a multi-resolution model based on clusters of tetrahedra, called diamonds, which enables the analysis of a 3D image through

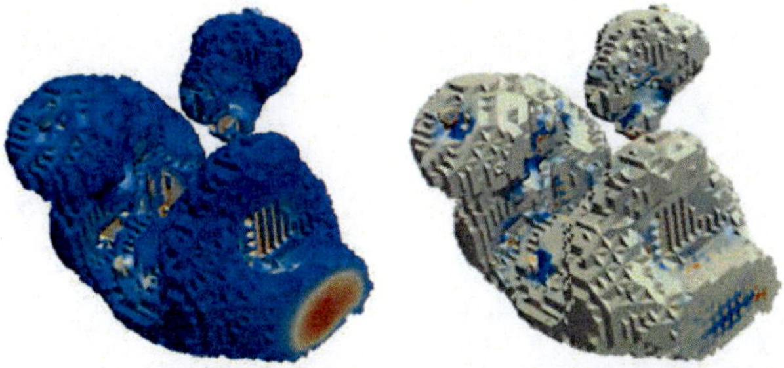

Neghip field | Neghip distortion

Stable decomposition of original field (104 cells) | Stable decomposition of distortion field (3,654 cells)

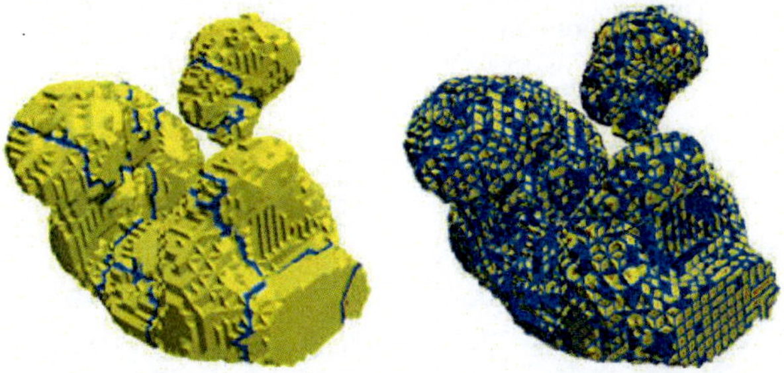

Unstable decomposition of original field (41 cells) | Unstable decomposition of distortion field (23,334 cells)

Fig. 11 Original field and distortion field, and segmentations, for variable resolution Neghip data set at 0% approximation error. Segmentations are shown with minima (stable) or maxima (unstable) vertices in red, vertices on the boundary of more than one region in blue and vertices within a region in yellow

crack-free approximations encoded as tetrahedral meshes. One important aspect of using mesh-based multi-resolution models is that the image can be analyzed by using much fewer samples than in the full image. This facilitates our analysis of large 3D volume datasets by using significantly fewer resources. The other aspect that we have shown through our experiments is the utility of discrete distortion in analyzing approximated images, thus giving good insights about the field behavior already at low resolutions.

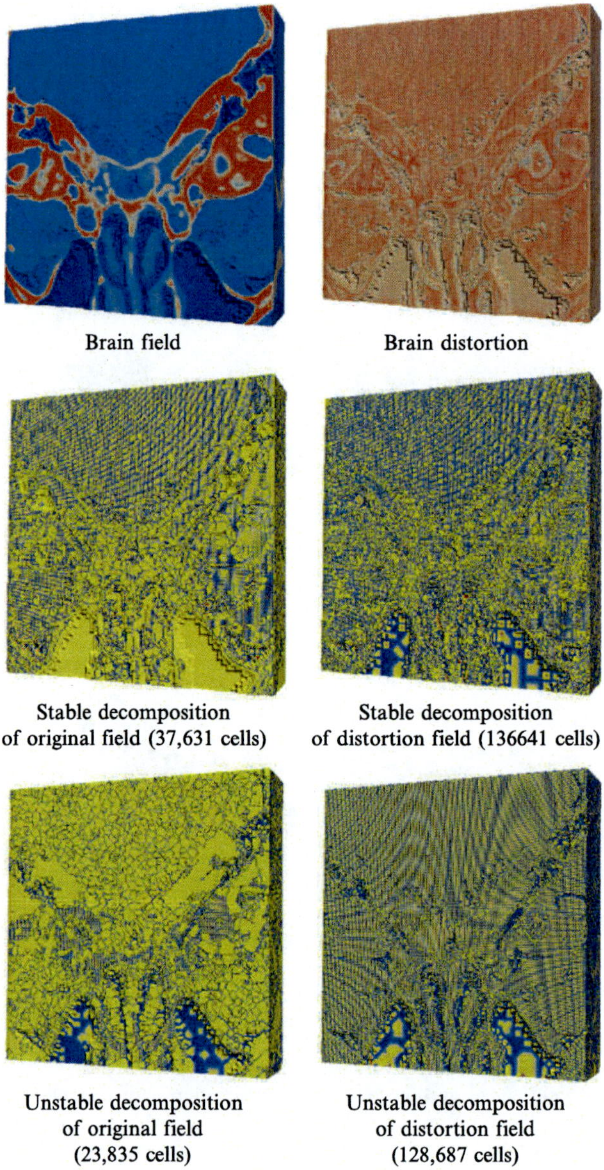

Fig. 12 Original field and distortion field, and segmentations, for CTA-brain data set at 10% resolution

We are currently investigating methods for incorporating distortion into the selective refinement query, i.e., the operation that extracts a variable-resolution mesh from the multi-resolution model, according to user-defined error criteria. This would enable the extraction of variable-resolution tetrahedral meshes with higher resolution in regions with greater distortion while reducing the resolution in regions with lower distortion.

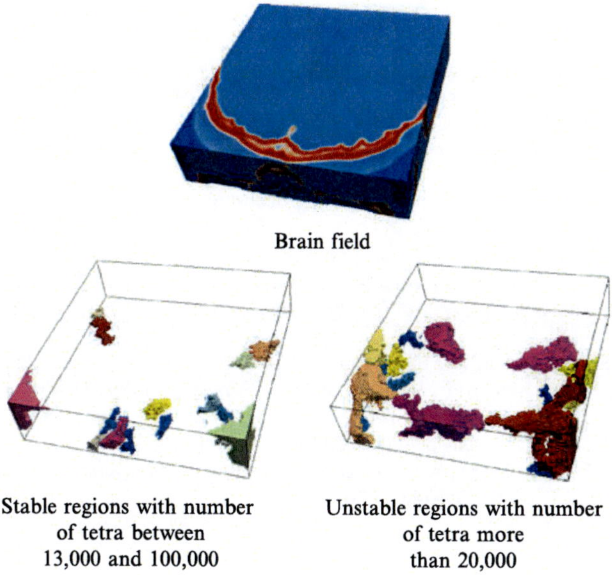

Fig. 13 Stable and unstable decompositions using a threshold to visualize distinct regions formed by a large number of tetrahedra

The definition of discrete distortion can also be extended to define a distortion tensor, when multiple scalar fields exist in the same domain (e.g., pressure, temperature, density). We plan to define a distortion tensor as well as concentrated curvature tensor in such a case, and use them to study the effects and the interactions among the various fields.

Discrete distortion can also be used in biochemistry to analyze the chemical structure of a molecule from the topological and geometrical points of view. Atoms and bonds in a molecule can be geometrically modeled as a tetrahedralized shape, where vertices correspond to atoms and edges to bonds. We believe that discrete distortion can provide a new approach to understand chemical reactions, the geometric structure of molecules, that are isomers, and properties of crystals.

Acknowledgments This work has been partially supported by the National Science Foundation under grant CCF-0541032 and by the MIUR-FIRB project SHALOM under contract number RBIN04HWR8.

References

1. P. Aleksandrov. *Combinatorial Topology*. Dover Publications Inc., 1998.
2. M. T. Anderson. Géométrisation des Variétés de Dimension 3 via le Flot de Ricci. *Société Mathématique de France, Gazette,*, 103:25–40, 2005.

3. E. J. Ansari, N. andE Delp. On detecting dominant points. *Pattern Recognition*, 24(5):441–451, 1991.
4. W. Beil, K. Rohr, and H. S. Stiehl. Investigation of approaches for the localization of anatomical landmarks in 3d medical images. *Computer Assisted Radiology and Surgery, CARS, Berlin, Germany*, 265–270, 1997.
5. J. Bey. Tetrahedral mesh refinement. *Computing*, 55:355–378, 1995.
6. S. Biasotti, L. De Floriani, B. Falcidieno, P. Frosini, D. Giorgi, C. Landi, L. Papaleo, and M. Spagnuolo. Describing shapes by geometrical-topological properties of real functions. *ACM Comput. Surv.*, 4(40), 2008.
7. L. Comic and L. De Floriani. Cancellation of critical points in 2d and 3d Morse and Morse-Smale complexes. In *14th IAPR International Conference on Discrete Geometry for Computer Imagery*, volume 4992 of *Lecture Notes in Computer Science*, 117–128, Lyon, France, 16-18 April 2008.
8. L. De Floriani and P. Magillo. Multiresolution mesh representation: models and data structures. In M. Floater, A. Iske, and E. Quak, editors, *Principles of Multi-resolution Geometric Modeling*, Lecture Notes in Mathematics, 364–418, Berlin, 2002. Springer Verlag.
9. N. Dyn, K. Hormann, K. Sun-Jeong, and D. Levin. Optimizing 3d triangulations using discrete curvature analysis. In T. Lyche and L. Schumaker, editors, *Mathematical Methods for Curves and Surfaces: Oslo 2000*, 135–146, 2000.
10. H. Edelsbrunner, J. Harer, V. Natarajan, and V. Pascucci. Morse-Smale complexes for piecewise linear 3-manifolds. In *Proceedings 19th ACM Symposium on Computational Geometry*, 361–370, 2003.
11. E. Fatemizadeh, C. Lucas, and H. Soltanian-Zadeh. Automatic landmark extraction from image data using modified growing neural gas network. *IEEE Transactions on Information Technology in Biomedicine*, 7(2):77–85, 2003.
12. T. Gatzke and C. Grimm. Estimating curvature on triangular meshes. *International Journal on shape Modeling*, 12:1–29, 2006.
13. T. Gerstner and M. Rumpf. Multiresolutional parallel isosurface extraction based on tetrahedral bisection. In *Proceedings Symposium on Volume Visualization*, 267–278. ACM Press, 1999.
14. B. Gregorski, M. Duchaineau, P. Lindstrom, V. Pascucci, and K. Joy. Interactive view-dependent rendering of large isosurfaces. In *Proceedings IEEE Visualization*, 475–484, 2002.
15. A. Gyulassy, P. Bremer, B. Hamann, and V. Pascucci. A practical approach to Morse-Smale complex computation: Scalability and generality. *IEEE Transactions on Visualization and Computer Graphics*, 14(6):1619–1626, 2008.
16. A. Gyulassy, V. Natarajan, V. Pascucci, P.-T. Bremer, and B. Hamann. Topology-based simplification for feature extraction from 3d scalar fields. In *Proceedings, IEEE Visualization 2005*, 275–280, October 2005.
17. A. Gyulassy, V. Natarajan, V. Pascucci, and B. Hamann. Efficient computation of Morse-Smale complexes for three-dimensional scalar functions. *IEEE Trans. Vis. Comput. Graph. (IEEE Visualization 2007)*, 6(13):1440–1447, 2007.
18. S. Hahmann, A. Belayev, L. Busé, G. Elber, B. Mourrain, and C. Rössl. *Shape Interrogation*. L. De Floriani, M. Spagnuolo (Eds.), Shape Analysis and Structuring (Mathematics+Visualization), 2009.
19. T. Hartkens, K. Rohr, and H. Stiehl. Evaluation of 3d operators for the detection of anatomical point landmarks in MR and CT images. *ComputerVision and Image Understanding*, 86(2):118–136, 2002.
20. D. Hebert. Symbolic local refinement of tetrahedral grids. *Journal of Symbolic Computation*, 17(5):457–472, May 1994.
21. A. Kimura, Y. Takama, Y. Yamazoe, S. Tanaka, and H. Tanaka. Parallel volume segmentation with tetrahedral adaptive grid. *International Conference on Pattern Recognition*, 2:281–286, 2004.
22. M. Lee, L. De Floriani, and H. Samet. Constant-time neighbor finding in hierarchical tetrahedral meshes. In *Proceedings International Conference on Shape Modeling*, 286–295, Genova, Italy, May 2001. IEEE Computer Society.

23. J. M. Maubach. Local bisection refinement for n-simplicial grids generated by reflection. *SIAM Journal on Scientific Computing*, 16(1):210–227, January 1995.
24. V. Mello, L. Velho, and G. Taubin. Estimating the in/out function of a surface represented by points. In *Symposium on Solid Modeling and Applications*, 108–114, 2003.
25. M. M. Mesmoudi, L. De Floriani, and P. Magillo. Morphological analysis of terrains based on discrete curvature and distortion. In *Proceedings of International Conference on Advances in Geographic Information Systems (ACMGIS 2008), Irvine, California, USA*, 2008.
26. M. M. Mesmoudi, L. De Floriani, and U. Port. Discrete distortion in triangulated 3-manifolds. *Computer Graphics Forum (Proceedings SGP 2008)*, 27(5):1333–1340, 2008.
27. J. Milnor. *Morse Theory*. Princeton University Press, 1963.
28. V. Natarajan, Y. Wang, P. Bremer, V. Pascucci, and B. Hamann. Segmenting molecular surfaces. *Computer Aided Geometric Design*, 23(6):495–509, 2006.
29. M. Ohlberger and M. Rumpf. Hierarchical and adaptive visualization on nested grids. *Computing*, 56(4):365–385, 1997.
30. V. Pascucci. Slow Growing Subdivisions (SGS) in any dimension: towards removing the curse of dimensionality. *Computer Graphics Forum*, 21(3):451–460, 2002.
31. S. C. Pei and C. Lin. The detection of dominant points on digital curves byscale-space filtering. *Pattern Recognition*, 25(11):1307–1314, 1992.
32. T. Regge. General relativity without coordinates. *Nuovo Cimento*, 19(3):558–571, 1961.
33. M. Reuter, F.-E. Wolter, and N. Peinecke. Laplace-Beltrami spectra as "Shape-DNA" of surfaces and solids. *Computer-Aided Design*, 38:342–366, 2006.
34. M. Rivara and C. Levin. A 3D refinement algorithm suitable for adaptive and multigrid techniques. *Communications in Applied Numerical Methods*, 8(5):281–290, 1992.
35. J. Roerdink and A. Meijster. The watershed transform: definitions, algorithms, and parallelization strategies. *Fundamenta Informaticae*, 41:187–228, 2000.
36. H. Samet. *Foundations of Multidimensional and Metric Data Structures*. The Morgan Kaufmann series in computer graphics and geometric modeling. Morgan Kaufmann, 2006.
37. A. Shamir. Segmentation and shape extraction of 3d boundary meshes. In *EG 2006 - State of the Art Reports*, Vienna, 2006.
38. T. Surazhsky, E. Magid, O. Soldea, G. Elber, and E. Rivlin. A comparison of gaussian and mean curvatures estimation methods on triangular meshes. In *Proceedings of Conference on Robotics and Automation, Proceedings. ICRA '03. IEEE International*, volume 1, 739–743, 2003.
39. C. Teh and R. T. Chin. On the detection of dominant points on digital curves. *IEEE Transactions on Pattern Analysis and Machine Intelligence*, 11(8):859–872, 1989.
40. M. Troyanov. Les surfaces Euclidiennes à singularités coniques. *L'enseignement Mathématique*, 32:79–94, 1986.
41. L. Vincent and P. Soille. Watershed in digital spaces: an efficient algorithm based on immersion simulation. *IEEE Transactions on Pattern Analysis and Machine Intelligence*, 13(6):583–598, 1991.
42. K. Weiss and L. De Floriani. Diamond hierarchies of arbitrary dimension. *Computer Graphics Forum (Proceedings SGP 2009)*, 28(5):1289–1300, 2009.
43. K. Weiss and L. De Floriani. Supercubes: A high-level primitive for diamond hierarchies. *IEEE Transactions on Visualization and Computer Graphics (Proceedings IEEE Visualization 2009)*, 15(6):1603–1610, 2009.
44. K. Weiss and L. De Floriani. Simplex and diamond hierarchies: Models and applications. In H. Hauser and E. Reinhard, editors, *EG 2010 - State of the Art Reports*, 113–136, Norrköping, Sweden, 2010. Eurographics Association.
45. K. Weiss, M. Mesmoudi, and L. De Floriani. Multiresolution analysis of 3D images based on discrete distortion. In *International Conference on Pattern Recognition (ICPR)*, 4093–4096, Istanbul, Turkey, 2010.
46. Y. Zhou, B. Chen, and A. Kaufman. Multi-resolution tetrahedral framework for visualizing regular volume data. In R. Yagel and H. Hagen, editors, *Proceedings IEEE Visualization*, 135–142, 1997.

Interactive Visualization–A Key Prerequisite for Reconstruction and Analysis of Anatomically Realistic Neural Networks

Vincent J. Dercksen, Marcel Oberlaender, Bert Sakmann, and Hans-Christian Hege

Abstract Recent progress in large-volume microscopy, tissue-staining, as well as in image processing methods and 3D anatomy reconstruction allow neuroscientists to extract previously inaccessible anatomical data with high precision. For instance, determination of neuron numbers, 3D distributions and 3D axonal and dendritic branching patterns support recently started efforts to reconstruct anatomically realistic network models of many thousand neurons. Such models aid in understanding neural network structure, and, by numerically simulating electro-physiological signaling, also to reveal their function.

We illustrate the impact of visual computing on neurobiology at the example of important steps that are required for the reconstruction of large neural networks. In our case, the network to be reconstructed represents a single cortical column in the rat brain, which processes sensory information from its associated facial whisker hair. We demonstrate how analysis and reconstruction tasks, such as neuron somata counting and tracing of neuronal branches, have been incrementally accelerated – finally leading to efficiency gains of orders of magnitude. We also show how steps that are difficult to automatize can now be solved interactively with visual support. Additionally, we illustrate how visualization techniques have aided computer scientists during algorithm development. Finally, we present visual analysis techniques allowing neuroscientists to explore morphology and function of 3D neural networks.

Altogether, we demonstrate that visual computing techniques make an essential difference in terms of scientific output, both qualitatively, i.e., whether particular

V.J. Dercksen and M. Oberlaender contributed equally to this work.

V.J. Dercksen (✉) · H.-C. Hege
Zuse Institute Berlin, Germany
e-mail: dercksen@zib.de; hege@zib.de

M. Oberlaender · B. Sakmann
Max Planck Florida Institute, Digital Neuroanatomy, Jupiter, USA
e-mail: marcel.oberlaender@mpfi.org; bert.sakmann@mpfi.org

L. Linsen et al. (eds.), *Visualization in Medicine and Life Sciences II*, Mathematics and Visualization, DOI 10.1007/978-3-642-21608-4_2,

goals can be achieved at all, and quantitatively in terms of higher accuracy, faster work-flow and larger scale processing. Such techniques have therefore become essential in the daily work of neuroscientists.

1 Introduction

One fundamental challenge in neuroscience is to understand how brains process sensory information about their environment and how this can be related to the animal's behavior. A widely used model system to study these relationships, from the molecular up to the behavioral level, is the somatosensory whisker system in rats. Like most nocturnal rodents, rats use their facial whiskers as complex tactile sensory organs to explore their environment. Such sensory input from single whiskers is processed by corresponding segregated neuronal networks in the primary somatosensory cortex (S1). Functional as well as anatomical evidence suggests that these networks resemble cylindric shapes [17, 22], hence being called *cortical columns* (Fig. 1(e)).

To shed light on how single whisker information is processed by a cortical column, recent attempts [12] aim to reconstruct its detailed anatomy and synaptic connectivity. Numerical simulations upon such high-resolution neural networks, with measured electrical input, caused by the deflection of a single whisker in the living animal (*in vivo*), will help to gain mechanistic understanding of sensory information processing in the mammalian brain.

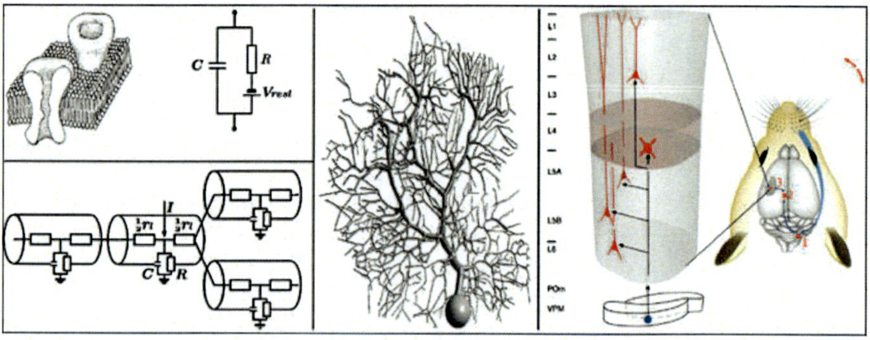

Fig. 1 (**a**) Ion channels provide a conductance through the membrane (Figure adopted from [7], courtesy of Sinauer Associates Inc., Publishers). (**b**) Associated passive electrical RC-circuit. (**c**) Spatial discretization of the cable equation results in multiple coupled compartments, each one locally equivalent to a uniform cable (Figure adapted from [3]). (**d**) Realistic full-compartmental model of a cerebral Purkinje neuron (Figure adopted from [19]). (**e**): Illustration of one-to-one correspondence between a whisker and a cortical column in the somatosensory cortex (3). A stimulus from a whisker is conveyed to its associated cortical column (schematically depicted on the left) via the brain stem (1) and the thalamus (2). Cells in the VPM nucleus of the thalamus project into the column and innervate cells in different layers (Figure adapted from [6])

As derived by [6, 8, 12], anatomical prerequisites to build realistic circuits comprise the number of neurons and neuron-types, their 3D distributions and characteristic dendrite and axon morphologies. Being composed of more than 17.000 neurons [10] of various anatomical as well as functional types, each type displaying complex axonal projection patterns (up to multiple centimeters per neuron), these prerequisites will hardly be satisfied by manual reconstruction approaches. Automated tools that extract the desired anatomical data from microscopic images are hence necessary and were recently described [14, 15].

Interactive visualization proved to be essential during the development of these methods, for validation of the results as well as for integrating the resulting data into neural networks. In this work, we illustrate how visual computing aids neuroscientists to extract and determine previously inaccessible anatomical data, how it facilitates the network reconstruction process and ultimately, how this might result in new scientific insights.

2 Biophysical Concepts of Neural Network Simulations

This section provides background for non-neuroscientists to better understand the anatomy and function of neurons and neuronal networks, as well as the modeling of electro-physiological signaling.

Functionally, neurons (or nerve cells) are the elementary signaling units of the nervous system, including the brain. These basic building blocks can mediate complex behavior, as potentially large numbers of them are interconnected to form neuronal networks. In general, neuronal networks are composed of chemically (synapses) or electrically (gap junctions) coupled local and long-range microcircuits of anatomically and physiologically classified neuron types. Specifically, each neuron is composed of a cell body (soma), multiple dendritic branches and one axonal tree, which receive electrical input from and transfer output towards other neurons, respectively. Hence, as found by the pioneer Ramón y Cajal about a century ago, neurons are usually grouped anatomically with respect to their soma shape, dendrite morphology and/or axonal branching pattern.

Neurons maintain a potential difference across their cell membrane, induced by deviating ion concentrations inside and outside the cell. This resultant resting potential (V_{rest}) is cell-type specific, but typically around -70 mV. Neuronal signaling involves changes in this potential.

The electro-physiological membrane properties can be described by electrical circuits containing a capacitance and resistance (RC-circuit, Fig. 1(b)). The membrane itself acts as a capacitance (C), which contains small pores (ion channels, Fig. 1(a)) with channel-type specific resistances (R_i). These pores mediate ionic currents across the membrane. Some channels are continuously open (passive or leaky) and others open and close actively with respect to voltage or chemical ligand concentration. Such active (e.g., voltage-dependent) conductances were first described in a neuron model by Hodgkin and Huxley (HH) in 1952:

$$C\frac{dV}{dt} = -\left(\sum_i I_i + I_{leak}\right), \tag{1}$$

where

$$I_i = \frac{V - E_i}{R_{i(V,t)}}, \tag{2}$$

represents voltage-dependent channel kinetics. E_i is the reversal potential of each ion channel type, i.e., the membrane potential at which there is no net flow of ions across the membrane. This model successfully explains a fast membrane depolarization, called action potential (AP) or spike, which is believed to be the fundamental electrophysiological unit in information processing.

Further development of electrical neuron models by Wilfred Rall, who believed that the complexity of the dendrites and axonal arborization would affect the neuronal processing, resulted in a cable theory for neurons. The cable equation

$$\frac{r_m}{r_l}\frac{\partial^2 V}{\partial x^2} = c_m r_m \frac{\partial V}{\partial t} + V \tag{3}$$

describes the flow of electric current (and change in potential) along passive one-dimensional neuronal fibers [18]. Here $r_m = \frac{R_m}{\pi d}$ and $c_m = C_m \pi d$ are the membrane resistance and capacitance per unit length, respectively, defined in terms of membrane resistance R_m and membrane capacitance C_m per unit area. The cable diameter is denoted by d; r_l is the (uniform) intracellular resistance per unit length along the cable.

Numerous bifurcations as well as variations in diameter and electrical properties along the neuronal branches, however, diminish the possibility to find analytical solutions for the cable equations. This leads to numerical solutions of spatially discrete, anatomy-based neuron models composed of multiple coupled HH-typed compartments (Fig. 1(d)), each locally resembling a uniform cable (Fig. 1(c)). Proper boundary conditions for the cable endings have to be specified. In particular at branching points, the voltage at the ends of the meeting cables is the same and the sum of all currents is 0. Input obtained from other cells at chemical synapses can be modeled electrically by adding appropriate currents to (3).

Initiated by Roger Traub and colleagues, many of these full-compartmental neuron models with active HH-type properties were synaptically interconnected, resembling realistic microcircuits and carrying out realistic neuronal operations. Following this tradition, attempts to reengineer the 3D anatomy and connectivity of functional networks of many thousand full-compartmental neurons were started [6,8,12]. Such goals became feasible with continuously increasing computing power and recently available large scale imaging, image processing and visualization techniques.

Hence, the reconstruction of a functionally well defined network, the 'cortical (barrel) column' in the rat's primary somatosensory cortex (S1) was started. Sensory information acquired by a single facial whisker is conveyed to dedicated region's in the brain stem, further to the ventral posterior medial nucleus of the thalamus (VPM) and finally to its corresponding cortical column (Fig. 1(e)). Excitatory thalamocortical input from this pathway into the column network, based on single whisker information, is essential to trigger simple behaviors, such as the decision to cross a gap. Hence, simulation of a realistically reengineered cortical column with single whisker input, measured *in vivo*, will potentially yield new insights and understanding of principle mechanisms that explain how the brain translates environmental input into behavioral responses.

Summarizing, the information processing of a neural network is simulated by modeling the propagation of electro-chemical signals through connected compartments representing the neuron geometry. The numerical solution of the resulting large system of ordinary differential equations consists of a potential value at each compartment at each time step. These time-dependent potentials defined on 3D graph structures are the input data for subsequent visual analysis.

A realistic and detailed anatomical model is a very important prerequisite to meaningfully simulate neural network activity. The 3D neuron morphology including neurite lengths, diameters and branching pattern is an important input parameter for the system of coupled cable equations. The number and distribution of nerve cells is important for defining the network and its connections. In the following, we focus on how this anatomical information can be obtained, and the role of image processing and visualization tools therein.

3 3D Reconstruction of Neuron Morphology from Microscopic Image Data

An important step in modeling neural networks is the reconstruction of three-dimensional morphology, particularly 3D axon and dendrite trajectories of an appropriate number of neurons. For modeling the thalamocortical part of the single whisker pathway, morphologies of all cell types within the cortical column as well as from its input region (VPM) are required.

As it is currently impossible to reconstruct all nerve cells from a single animal, a statistically representative number of cells of different anatomical types are reconstructed one by one from different animals and combined into a single network. Thus, for ~10 different cell types and a minimum of ~10 morphologies per type, at least ~100 cells have to be reconstructed. To make this feasible, an efficient reconstruction pipeline is required.

Typically, cells are filled *in vivo* with a tracer. A common choice is biocytin. As it is non-fluorescent, it does not bleach (making a re-scan possible) and causes no outshining of weakly stained axons by strongly stained dendrites. The part of the brain containing the stained cell is then cut into sections, as it is too thick too be imaged in its entirety at high resolution. Typically, there are 20 sections of 100 μm thickness per cell. Large areas have to be scanned to capture all axonal arbor. A transmitted light brightfield (TLB) microscope is used to image the sections at high resolution using a mosaic-scanning technique. This results in a stack of thin, but three-dimensional images of 20–90 gigabytes each.

3.1 *Original Workflow: Manual Tracing*

The original workflow [14] for tracing the neurons is based on the Camera Lucida technique using, for example, the Neurolucida system [11]. Special software displays a live camera image of a section captured by an attached computer-controlled microscope on the computer screen. The user then manually marks the neuronal structures. By moving the stage in the xy-plane and varying the focus in the z-direction, a progressively larger volume is inspected. The software combines the marked structures into a 3D graph representation.

The most common approach is to start with the section containing the soma and follow the dendritic and axonal branches from there. Whenever a branch reaches the section boundary, the section under the microscope is replaced with the next one. The user has to find structures in the new section that correspond to the end points of the current tracing and transform the current tracing, such that it is aligned with the new section. To find the correct alignment more easily, usually three or more branches are traced concurrently. After alignment, the tracing continues in the next section. This procedure is repeated until all branches have been entirely traced. This approach works well for dendrites, as they usually have localized branching patterns and relatively large diameters ($\sim$ 2–5 μm). Axonal branches, however, can be less than 1μm thick and extend further away. Tracing and finding correspondences between sections is, therefore, extremely tedious and time-consuming (up to 100 h for complex, wide spreading axons). It also requires experienced users to reach a reliable level of reconstruction quality. The number of skilled users is usually the limiting factor when large numbers of cells have to be reconstructed.

3.2 *First Improvement: Automatic Tracing*

In order to reduce the manual labor an automatic segmentation method for dendrites and axons is required. The segmentation is, however, non-trivial: the structures are very thin and contrast may be poor due to limited dye penetration, especially further away from the soma where the tracer is injected into the cell. Therefore, axons frequently appear fragmented. Also numerous background structures have a gray-value in the same intensity range as the axons and dendrites (see Fig. 2(a)), in particular glia cells that presumably contain biocytin contrast agent intrinsically.

Existing (semi-)automatic methods [9] focus mainly on tracing dendritic branches. Therefore, an algorithm was developed that automatically segments both dendritic and axonal structures from 3D TLB images and computes a graph

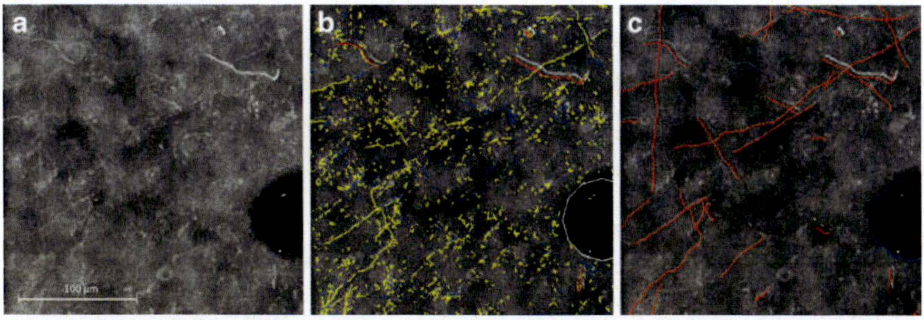

Fig. 2 (**a**) Small region of a maximum intensity projection of an inverted 3D TLB image of a single brain section. The bright filamentous structures are the axon fragments to be traced. However, the image contains a lot of background structures in the same intensity range. (**b**) Result of the automatic tracing algorithm, slightly shifted to show the correspondence with the image. To aid in the interactive post-processing, the fragments are colored according to their position within the section: touching top of the section (blue), bottom (green), both (red) or none (yellow). The round structure is a blood vessel cross-section. (**c**) Final tracing after post-processing

representation of the approximated centerlines [14]. The method is based on deconvolution [13], followed by a sequence of image filters and morphological operations, and finally skeletonization. The method finds virtually all neuron branches. It is, however, very conservative: oversegmentation is accepted in order to assure that no foreground structure is lost (see Fig. 2(b)).

To finish the reconstruction, a number of post-processing steps are performed interactively using the Serial Section Manager in the Neurolucida software. First, non-neuron artifacts are removed and fragmented branches are spliced (connected) within each section separately. The user is presented visual feedback in the form of three orthogonal projection images (xy-, xz- and yz-projections, see Fig. 3(a)). To support the user in deciding which fragments to delete and which to connect, the traced segments are overlaid onto a maximum intensity projection (MIP) image of the section.

Second, the sections are (rigidly) aligned. Alignment is done pair-wise, i.e., two neighboring sections at a time. One section serves as a reference and remains fixed, while the other is interactively rotated and translated. The user has to perform a pattern matching task: he needs to determine the correspondences between neuron fragments in the two slices. Again, having only 2D projection views, this can be extremely difficult, especially for dense, visually symmetric clouds of projected line fragments. In some cases, finding the correspondence is even impossible. To alleviate this problem, blood vessel contours are traced additionally. These contours are usually easier to match, as they are fewer and may have different diameters, giving a hint on which vessels may correspond. After a rough blood vessel match has been found, the neuron fragment correspondence can often be determined more easily.

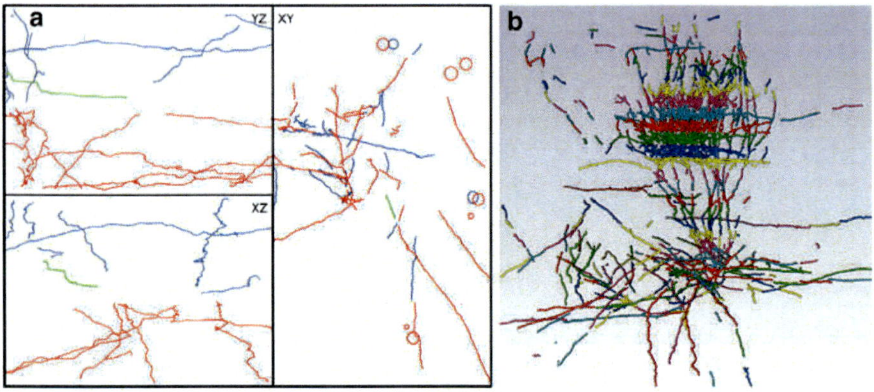

Fig. 3 (**a**) Two sections (red and blue) after alignment. The vessel contours (blue and red circles) are helpful during the manual alignment, as these features are easier to recognize and to match than the segment end points. Based on 2D projection views, it is often difficult to decide which axon fragments are to be spliced, e.g. in case of the green fragment. (**b**) Automatically aligned stack of reconstructed sections, colored by section

Finally, the corresponding fragments are spliced across the section boundaries. Again, the lack of a three-dimensional view on the data makes this step difficult and unnecessarily time-consuming.

3.3 Second Improvement: 3D Interactive Editing

Being based on 2D projection views, the interactive post-processing severely hampered the reconstruction process. Therefore, a 3D interactive environment, the *filament editor* [1], was developed as part of the Amira [20] visualization and data analysis software. Its main components are:

- A graph data structure, consisting of a set of nodes (the branching and end points) and segments connecting the nodes. The trajectory of each segment in 3D space is represented by a sequence of points. Attribute values can be associated with the nodes, segments and/or points. For example, a *radius* attribute can be associated with the points, defining a floating point value at each point along the segments. One could also define anatomical labels to all elements in the combined set of nodes and segments, resulting in the assignment of label values like *dendrite*, *axon* or *vessel* to subgraphs.
- A set of tools to select (parts of) the data structure, including clicking on or drawing a contour around target elements, selection of connected subgraphs, etc.
- A set of operations to manipulate the data structure, like deletion and connecting of nodes and segments. All operations can be undone/redone.

- A graphical user interface that – besides GUI elements for invoking the available selection and editing tools – contains a 3D viewer window for displaying the graph and other objects, like a MIP image. The graph can be colored according to its attribute values.

In the improved work flow, first, each slice is cleaned and spliced individually. The user loads the automatically traced neuron fragments and displays them colored by the attribute indicating whether a particular connected subgraph touches the top of the slice, the bottom, both or neither (see Fig. 2(b)). This information is exported by the tracing program, together with a MIP image of the section. The graph data is superimposed on the MIP image. In the MIP, most neuronal processes are easily distinguishable from falsely traced background structures. Hence, if not already traced perfectly, fragmented branches are spliced according to the MIP interactively. Afterwards, artifacts are removed by drawing a selection polygon around them. Only connected subgraphs that are entirely within the polygon are selected and removed. This feature makes it easy to remove many small artifacts at once, while avoiding the removal of parts of larger, already connected, neuronal branches. However, some processes are too faint to be recognized in the MIP. By rotating the tracing, human pattern recognition can compensate for that. Specifically the coloring of top and bottom structures proved to be helpful. Every branch has to touch either the top or bottom (in most cases both) borders of a slice. Therefore, in most cases, these weak processes resemble dashed 3D lines starting with blue (top), connecting to yellow (intermediate) and ending at green (bottom) fragments.

Second, all manually edited sections are combined into a single graph for interactive alignment in the *filament editor*. The nodes and segments of each section are labeled according to their section number for visualization and fast selection. A section is aligned with its fixed predecessor section by interactive translation and rotation in the 3D viewer using handles. All other sections can be hidden to not obscure the view. By repeated pair-wise alignment the entire stack is processed.

Third, the segments are spliced across the slice boundaries. The alignment and splicing process is regarded as the final quality control for each tracing. Missing or falsely connected branches as well as loops can be easily identified and corrected in the individual sections. As a final and optional step, subgraphs representing different anatomical structures can be labeled as such.

An important aspect of such interactive tools is usability. In the case of the *filament editor*, this not only includes fast access to the selection and editing tools by mouse and shortcut keys, it also requires stability and robustness against operating errors. Important is also responsiveness. All viewing, selection and editing operations have to be immediate, also for the relatively large data sets that may result from the automatic tracing (containing $\sim$1M points).

Altogether, the 3D environment enables efficient interactive post-processing and alignment of the automatically reconstructed sections. The additional depth cues, obtained primarily by interactive rotation of the scene, are the major advantage of the 3D editing, when compared to the 2D Neurolucida workflow. They are essential in order to quickly decide which fragments to discard and which to connect.

Based on the alignment by the interactive tool, all neuron morphologies could be reconstructed with satisfactory quality.

3.4 Third Improvement: Automatic Section Alignment

In order to make the alignment faster and more objective, an automatic alignment algorithm was developed [2]. Given two sections containing multiple segments (polylines), it computes an optimal transform, i.e., a rotation angle around the z-axis, a 2D translation in the xy-plane and an optional uniform scaling, using the following steps:

1. For a pair of sections to be aligned, find the set of fragment end points for each section. This results in two point sets P and Q, one for each section, that have to be matched. A matching M is a set of corresponding point pairs $\{(\mathbf{p}_i, \mathbf{q}_j)\}$.
2. Find a set of candidate matchings by searching for subsets P' and Q' for which the mutual distances between all points in P' are similar to the distances between the corresponding points in Q'. For each candidate matching, compute the optimal transformation. These are the starting transformations.
3. Starting from each starting transformation, further optimize the matching and the transformation T with respect to a scoring function that favors a large match size $|M|$, but penalizes large distances between corresponding points:

$$score(P, Q, M, T) = \frac{|M|}{min(|P|, |Q|)} \cdot e^{-\alpha \cdot \sqrt{\frac{\sum_{(\mathbf{p},\mathbf{q}) \in M} \|\mathbf{p} - T(\mathbf{q})\|^2}{|M|}}} \quad (4)$$

   The parameter α balances these contradicting goals. The final result is the matching and transformation leading to the highest score.

This method works very well for this type of data. It is also very fast, it takes a few seconds per slice pair for typical point set sizes of 30–100. As this automatic alignment method is integrated in the *filament editor*, the results are immediately displayed in the 3D viewer and can be visually validated. Whenever a slice pair cannot be successfully aligned, it can be transformed interactively, as before.

3.5 Reconstruction Time

The reconstruction time strongly depends on the neuron type, its spatial extent and axonal complexity. For instance, an axon of a pyramidal neuron in S1 can spread over millimeters, reaching a total length of several centimeters. Reconstructing L5 cells, approximately 20 sections (1.7×1.7×0.1 mm^3) at a resolution of 92×92×500 *nm* need to be imaged, resulting in a data set of 1.7 TB. For such neurons the reconstruction time can be divided as follows:

- Scanning time: $\sim$3 days ($\sim$4h per section).
- Automatic processing: $\sim$9 days ($\sim$10 h per section).
- Interactive post-processing and alignment: $\sim$2–3 days (5–90 minutes per section, 60 minutes on average).

Although this results in a total of two weeks per L5 cell, less than 20 hours of manual labor are required. The pipeline also greatly benefits from task and data parallelization. The time for automatic reconstruction decreases linearly with the number of microscopes, computers and human tracers. Finally, the interactive post-processing does not need to be done by a single neuroscience expert, but by multiple student assistants in parallel, after only a few hours of training. Altogether, this pipeline allows for a much higher throughput of individual cell models, making the modeling of large networks feasible.

4 Automatic Counting of Neurons in Large Brain Volumes

In order to model a realistic neuronal network in 3D, one needs to know how many nerve cells constitute the network and how they are distributed in space. The estimation of absolute numbers of neurons, densities or rates of density change in neuron populations has usually been based on random, sparse sampling methods, such as stereology [21]. These methods determine cell densities by inspecting a representative sub-volume of tissue and extrapolating the obtained density values to a reference volume. However, due to the lack of well-defined reference volumes and usually unjustified assumptions of homogeneous densities across them, it is preferable to detect and count all cells within the volume of interest.

Three-dimensional imaging techniques, e.g., confocal or widefield microscopy, and suitable neuronal stains, like NeuN, which stains all neuron somata, offer possibilities to achieve this. Using these techniques, neuron somata appear as roundish shapes, sometimes prolongated by the onset of the dendritic arbors that have taken up some stain.

4.1 Original Workflow

The simplest way to detect all somata in such 3D images is by manual marking. This is, for example, achieved by moving a 2D image slice through the data set and clicking on the center of each soma to place a landmark at its approximated centroid. The center is set on the slice where the soma has the largest diameter. This procedure, however, is very time-consuming, but can be regarded as rather accurate, with a reported inter-user variability of 2% [10].

An approach to make the workflow more three-dimensional is to display a rough binary segmentation by a semi-transparent isosurface and use this to place the landmarks. Alternatively, a direct volume rendering of the image can be used [16]. These approaches depend, however, on a suitable threshold value or transfer function and are, therefore, not very accurate when the brightness varies strongly across the image, which is typical for these kind of images.

The main problem for both of these approaches is the amount of required manual labor, making a quantification of some cubic millimeter large volumes impossible. Even a volume containing just one single cortical column and its direct surroundings ($500 \times 500 \times 2000\,\mu$m) embodies tens of thousands of cells, making manual marking of each neuron extremely time-consuming.

4.2 Automatic Detection and Counting Algorithm

To make the exhaustive counting of nerve cells in large tissue volumes feasible, we developed a method for automatic detection and counting of neuron somata in 3D images [15]. It consists of three steps (Fig. 5):

1. Segmentation. This step produces a binary image separating foreground from background. The main processing steps are contrast enhancement, thresholding and a number of morphological image filters for speckle removal and hole filling. The exact filters and parameters depend on the imaging and staining procedure. In [15] various combinations for different types of image data are presented. Due to locally high cell densities and limited resolution, clusters of touching somata may, however, remain.
2. Morphological splitting. Touching objects are separated at the connecting neck by marker-based watershed transform.
3. Model-based splitting. Any remaining touching objects are divided on the basis of the volume statistics of the object population. This is an optional step that assumes a dominant population of cells with a Gaussian-distributed soma volume. Any object that could not be split at a thin neck, but is statistically too large to consist of a single object, is divided into the most probable number of components.

The method results in cell counts and densities comparable with manual counting ($\sim$4% deviation). This makes it possible to refrain from manual counting and process large data sets completely automatically. Thus, also the exhaustive counting of entire column populations is now feasible.

4.3 Visualization for Algorithm Development and Validation

The development of a segmentation algorithm that consists of a sequence of image filters, each with its own parameters, is often a trial-and-error process that requires a frequent evaluation of intermediate results. As an objective quantitative measure is not always available, this evaluation is commonly done visually. For an efficient workflow during algorithm development, easy-to-use visualization tools are therefore indispensible.

A common way to judge the result of a binary segmentation algorithm is to display a gray-value image slice overlaid with the segmented foreground in a semi-transparent color, for example, using Amira [20] (see Fig. 4(a)). By moving the 2D slice, the algorithm developer can quickly verify the quality of the segmentation result. If a foreground voxel object needs to be separated into its constituent parts, as, for example, in step 2 and 3 of the automatic cell detection method, the 2D slicing approach is insufficient. The boundary between two touching objects should be located at a thin 'neck', which is very difficult to verify on 2D slices. In this case, the generation of object boundary surfaces using the Generalized Marching Cubes algorithm [5] has proven very useful. As in this surface representation each object has a distinct color, the correctness of the splitting result can be determined almost instantly.

Visualization is also very useful to compare differences between automatically computed soma positions and those of a reference data set, e.g., a 'gold standard', manually created by an expert. In order to discover whether a particular false positive or negative result is caused by the automatic algorithm, by expert error, data artifacts, or differences in treatment of only partially imaged boundary objects, etc., one has to look closely at each differing data point in combination with the image data. In Amira, one can for example display both landmark sets with different colors, and show the image data using 2D textured slices with contrast control, or 3D direct volume rendering. Together with a simple 3D viewer allowing for interactive zooming and rotating, this is a very effective environment for visual method validation.

Although the automatic method performs very well, it may occur that the results for a particular data set are not satisfactory. In such cases, it is useful to have an interactive landmark editor to make corrections.

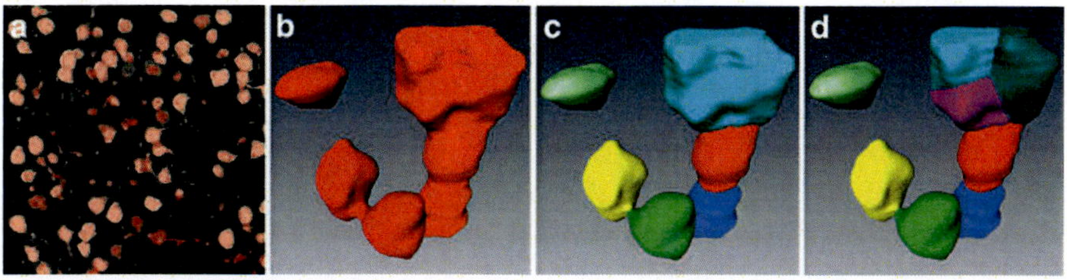

Fig. 4 (**a**) Segmentation validation by semi-transparent overlay. (**b**) Surface reconstruction of a cluster of touching cells after binary segmentation. (**c**) After marker-based watershed splitting. (**d**) After splitting based on soma volume statistics

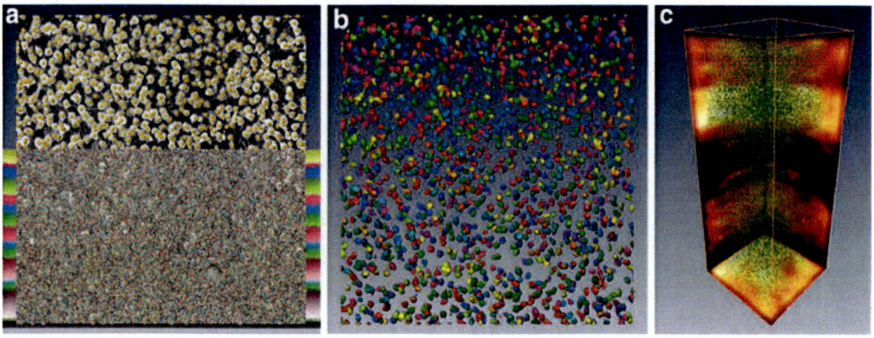

Fig. 5 (**a**) Maximum intensity image of 3D data set containing NeuN-stained somata and automatically detected centers. (**b**) Surface reconstruction of individual cell bodies after automatic segmentation and cluster splitting. (**c**) Automatically detected somata (green) and MIP of the cell density in a large tissue volume ($550\,\mu m \times 550\,\mu m \times 2\,mm$) containing a cortical column

4.4 Conclusion

Summarizing, we have seen how automatic counting of neuron somata allows the neuroscientist to reliably determine neuron numbers and 3D distributions in large tissue volumes, like a brain region containing cortical columns. Being free of assumptions, like homogeneity across a reference volume, the automated approach potentially leads to more valid results than sparse sampling methods. This example also illustrates that the use of visualization methods is indispensible during algorithm development, as it allows the computer scientist to perform the frequent evaluation of results quickly and effectively.

5 Visual Analysis of Neuron Morphology and Function

In this section we will present two examples of how even simple visualization tools for the analysis of 3D neuron morphology support neuroscientists in their daily work. We also discuss the requirements and challenges for analyzing functional properties of networks of neurons resulting from numerical simulations.

5.1 Evaluation of a Neuron Classification Algorithm

Given a reasonably large set of reconstructed neurons, one can manually and/or automatically group them in multiple morphological cell types. Such a classification is a key prerequisite for many anatomical investigations. For example, correlation of

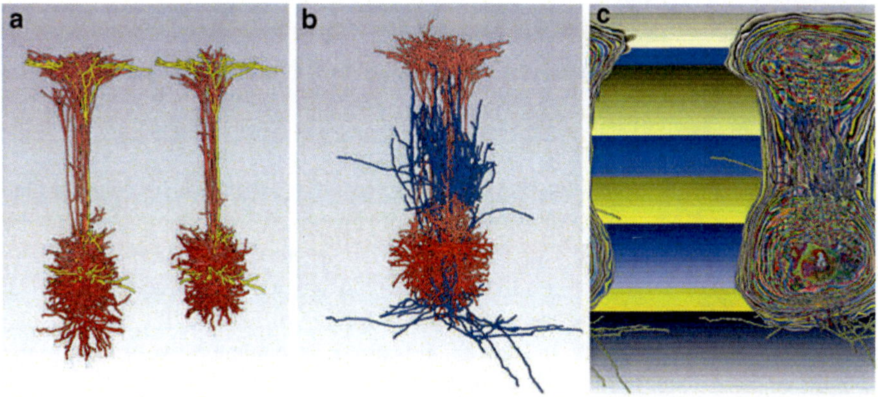

Fig. 6 (**a**) Visual validation of neuron morphology classification. The same highlighted neuron is displayed together with cells of the same type (left) and of a different type (right). In the latter case, the difference in shape is immediately visible. (**b**) Visualization of the spatial relation of two neuron types. The apical (pink) and basal (red) dendrites of a small number of L5B cells are displayed together with one VPM axon (blue). (**c**) Volume-rendered dendritic density field of a large population of L5B cells. Together with an axonal density field (not displayed) this allows for qualitative and quantitative analysis of the innervation domain of these two cell types

anatomical classes with physiological responses to stimuli or determination of the 3D spatial extent (with respect to anatomical position landmarks such as pia surface) of a certain cell type, help neuroscientists to gain insights into network structure and function.

For any given neuron clustering algorithm, for instance the one described in [12], a tool to visually validate and analyze its results is very useful. We developed a simple visualization tool to display sets of morphologies (see Fig. 6(a)). Neurons that have been assigned to the same cluster are overlaid. This gives insight into the typical properties of the particular class. The user can select and highlight one neuron from the list of all neurons. From this, the user can immediately see whether the selected neuron's morphology is similar to the morphology of the other neurons in the group. The user can also compare it to neurons in other clusters. The combined display of morphologies can also reveal errors in the reconstruction or in the registration into a common reference frame.

5.2 *Exploring 3D Innervation Volumes of Neuron Populations*

To understand the structure of a neural network, a tool that allows for 3D visual exploration of a population of neurons is very helpful. One property of interest is the 'innervation domain' of two neuron types, i.e., the region in space where axons of one type and dendrites of the other type overlap, indicating potential connections between the cell types.

Given a population of neurons that have been registered into a common coordinate frame, e.g., using the method described in [12], the innervation domain can be investigated by simply line-rendering some cells of both types with dendrites and axons colored differently (Fig. 6(b)). This allows one to see the relative positions of the two cell populations and their overlap regions.

For large neuron populations line renderings result in extreme visual clutter. One solution is to sample branch density on a uniform grid, i.e., to compute dendrite/axon length per volume unit. This density map can be interactively explored, using standard visualization techniques like volume rendering or iso-surfaces (Fig. 6(c)). Having such axonal and dendritic density maps for two different cell types, one can easily see where they overlap. In addition, it is possible to quantitatively analyze the individual density fields and their innervation domain.

5.3 Visual Analysis of Simulated Network Activity

Numerical simulation of the activity of single neurons (see Fig. 7) or large networks results in time-dependent potentials, defined on each compartment comprising the neuron morphology. Depending on the size of the network, the spatial resolution and the number of time steps, the simulation of a column-size network can easily lead to data sets that are hundreds of gigabytes in size. Tools for quantitative and visual analysis are required to gain knowledge from such large and complex data. Of interest are, for example, methods to correlate simulation results with measured quantities, like *in vivo* two-photon Ca^{2+} time series imaging [4] for

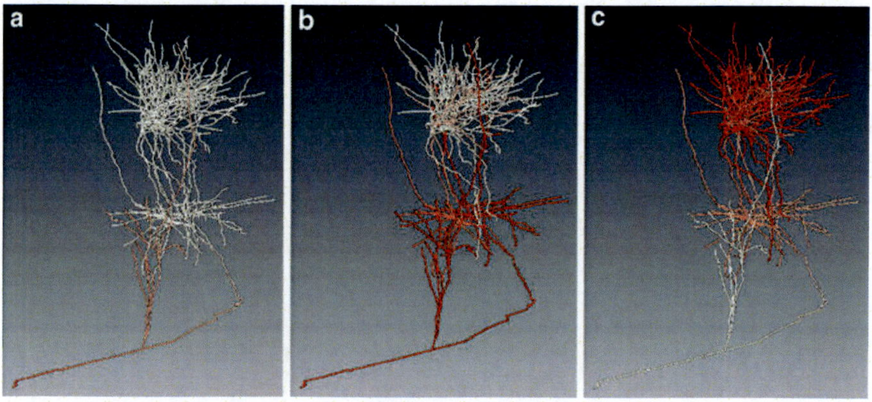

Fig. 7 Numerically simulated action potential propagating through a single VPM axon after 1 (**a**), 3 (**b**) and 5 (**c**) ms after artificial current injection at the axon onset (lower left)

validation purposes. Second, to find and understand yet unknown activity patterns, an interactive exploration tool is required that supports the formulation and verification of hypotheses, for example regarding correlations between anatomical and physiological data. Also the comparison of results of different simulation runs for different input conditions is important for network parameter tuning and sensitivity tests, requiring an even more challenging 5D analysis tool.

6 Discussion and Conclusion

We illustrated, by means of 'real-world' problems from the field of neuroscience, that visual computing tools, including interactive visualization, proved to be essential for the extraction and validation of previously inaccessible anatomical data from microscopic images. The resulting increase in throughput is important for obtaining required anatomical data for the reconstruction of large, high-resolution 3D neural networks, e.g., for numerical simulations. Interactive visualization tools also support the integration of anatomical data into such networks, and subsequent exploration and analysis.

The integration of all visualization tools into one software system (the visualization and data analysis software Amira [20]) has proven extremely beneficial as the user has access to many different modules to simultaneously display image, line-like and other data, allowing him to relate different objects in a single 3D interactive environment.

Although important anatomical data could be extracted for the reconstruction of a large neural network representing a cortical column, there are some important open issues that have to be addressed in order to be able to study the functioning of such large networks by numerical simulation of the electro-physiological processes. These problems include the determination of the wiring of the neuronal circuits, i.e., the number and distribution of synaptic connections between cells in the network ("connectomics"). Also other parameters that are important for the simulation need to be determined reliably for all cells in the network, like dendrite/axon radii, spine/bouton distributions, ion channel properties, etc. The current tendency of using higher resolution image modalities as an initial step to address these problems and the according increase in data set size, calls for adequate visual computing techniques to be developed in the near future. Also for the analysis of large-scale numerical simulations of neural networks, adequate visual computing methods will be indispensable.

Acknowledgements We would like to thank R.M. Bruno and C.P.J. de Kock for providing the sections for the neuron morphology reconstruction, H.S. Meyer for providing the stained somata images, P.J. Broser for initiating the collaboration, and finally S. Lang for his work on the NeuroDUNE simulation software.

References

1. Dercksen, V.J., Broser, P.J., Sakmann, B., Hege, H.C., Oberlaender, M.: Efficient 3D reconstruction of single neuron morphology from stacks of transmitted light brightfield microscopy images. In preparation
2. Dercksen, V.J., Weber, B., Günther, D., Oberlaender, M., Prohaska, S., Hege, H.C.: Automatic alignment of stacks of filament data. In: Proc. IEEE International Symposium on Biomedical Imaging, Boston, USA, 971–974 (2009)
3. Gerstner, W., Kistler, W.M.: Spiking Neuron Models: Single Neurons, Populations, Plasticity. Cambridge University Press (2002)
4. Göbel, W., Kampa, B., Helmchen, F.: Imaging cellular network dynamics in three dimensions using fast 3D laser scanning. Nature Methods **4**, 73–79 (2007)
5. Hege, H.C., Seebaß, M., Stalling, D., Zöckler, M.: A generalized marching cubes algorithm based on non-binary classifications. Tech. rep., ZIB Preprint SC-97-05 (1997)
6. Helmstaedter, M., de Kock, C.P.J., Feldmeyer, D., Bruno, R.M., Sakmann, B.: Reconstruction of an average cortical column in silico. Brain Research Reviews **55**(2), 193–203 (2007)
7. Hille, B.: Ionic channels of excitable membranes, 2nd edn. Sinauer Associates, Inc., Publishers (1992)
8. Markram, H.: The blue brain project. Nat Rev Neurosci **7**(2), 153–160 (2006)
9. Meijering, E.: Neuron tracing in perspective. Cytometry. Part A **77**(7), 693–704 (2010)
10. Meyer, H.S., Wimmer, V.C., Oberlaender, M., de Kock, C.P.J., Sakmann, B., Helmstaedter, M.: Number and Laminar Distribution of Neurons in a Thalamocortical Projection Column of Rat Vibrissal Cortex. Cerebral Cortex **20**(10), 2277–2286 (2010)
11. MicroBrightField Inc.: Neurolucida. www.mbfbioscience.com/neurolucida (2011)
12. Oberlaender, M.: Three-dimensional reengineering of neuronal microcircuits. the cortical column in silico. Ph.D. thesis, Univ. of Heidelberg, Germany (2009)
13. Oberlaender, M., Broser, P.J., Sakmann, B., Hippler, S.: Shack-Hartmann wave front measurements in cortical tissue for deconvolution of large three-dimensional mosaic transmitted light brightfield micrographs. J. Microsc. **233**(2), 275–289 (2009)
14. Oberlaender, M., Bruno, R.M., Sakmann, B., Broser, P.J.: Transmitted light brightfield mosaic microscopy for three-dimensional tracing of single neuron morphology. J. Biomed. Opt. **12**(6), 1–19 (2007)
15. Oberlaender, M., Dercksen, V.J., Egger, R., Gensel, M., Sakmann, B., Hege, H.C.: Automated three-dimensional detection and counting of neuron somata. J. Neurosci. Methods **180**(1), 147–160 (2009)
16. Peng, H., Ruan, Z., Long, F., Simpson, J.H., Myers, E.W.: V3D enables real-time 3D visualization and quantitative analysis of large-scale biological image data sets. Nature biotechnology **28**(4), 348–53 (2010)
17. Petersen, C.C.H.: The functional organization of the barrel cortex. Neuron **56**(2), 339–355 (2007)
18. Rall, W., Agmon-Snir, H.: Methods in neuronal modeling: from ions to networks, chap. Cable Theory for Dendritic Neurons, 27–92. C. Koch and I. Segev (eds.) (1998)
19. Schutter, E.D., Bower, J.M.: An active membrane model of the cerebellar Purkinje cell: II. Simulation of synaptic responses. J Neurophysiol **71**(1), 401–419 (1994)
20. Stalling, D., Westerhoff, M., Hege, H.C.: Amira: A highly interactive system for visual data analysis. In: C. Hansen, C. Johnson (eds.) The Visualization Handbook, chap. 38, 749–767. Elsevier (2005)
21. Sterio, D.C.: The unbiased estimation of number and sizes of arbitrary particles using the disector. J Microsc **134**(2), 127–136 (1984)
22. Wimmer, V.C., Bruno, R.M., De Kock, C.P.J., Kuner, T., Sakmann, B.: Dimensions of a Projection Column and Architecture of VPM and POm Axons in Rat Vibrissal Cortex. Cerebral Cortex **20**(10), 2265–2276 (2010)

MRI-Based Visualisation and Quantification of Rheumatoid and Psoriatic Arthritis of the Knee

Ben Donlon, Douglas Veale, Patrick Brennan, Robert Gibney, Hamish Carr, Louise Rainford, ChinTeck Ng, Eliza Pontifex, Jonathan McNulty, Oliver FitzGerald, and John Ryan

Abstract The overall goal of this project is to develop an application to quantify the level of synovial inflammation within patients suffering from Rheumatoid or Psoriatic Arthritis, based on automated synovium segmentation and 3D visualisation of MRI scan data. This paper discusses the direction we have taken during the development of the visualization components of the application, and an overview of the project in general. Current software methods of quantifying enhancement of inflamed synovial tissue by gadolinium contrast have several limitations. The clinician is required to classify individual regions of interest on a slice by slice basis which is a time-consuming process and suffers from user subjectivity. We propose a method of automating this process by reconstructing the slice information and performing quantification in three dimensions.

1 Background

1.1 Arthritis and Treatment

Rheumatoid Arthritis (RA) is a chronic disease, mainly characterized by inflammation of the lining, or synovium, of the joints. It can lead to long-term joint damage, resulting in chronic pain, loss of function and disability [1]. The condition affects about 1% of the worlds population, with women three times more prone than men. People of any age can be affected, but typically between the ages of

B. Donlon (✉)
UCD, Ireland
e-mail: ben.donlon@ucd.ie

J. Ryan
UCD, Ireland
e-mail: john.ryan@ucd.ie

L. Linsen et al. (eds.), *Visualization in Medicine and Life Sciences II*, Mathematics and Visualization, DOI 10.1007/978-3-642-21608-4_3,

30 and 50 [2]. It is a painful condition with symptoms including loss of mobility and joint functionality. Diagnosis is based on typical symptoms and signs, with blood tests, plain film radiography (X-Ray) images, and more recently Magnetic Resonance Imaging (MRI) also of diagnostic benefit.

Psoriatic Arthritis (PsA) is an inflammatory arthritis which affects approximately 10–30% of people suffering from chronic skin psoriasis [3]. While PsA can develop at any age, it frequently appears around 10 years after the first signs of psoriasis, with men and women equally affected. The treatment is similar to that of rheumatoid arthritis as the underlying process is inflammation, and treatments involve the reduction and control of the inflammation.

There are a number of disease modifying medications available for both RA and PsA including more novel treatments (biologics) which are designed to target particular proteins such as Tumour Necrosis Factor (TNF). Anti-TNF therapies such as etanercept are particularly effective and these agents are licensed for use in both RA and PsA.

A number of other biologics have been developed which target key pro-inflammatory proteins such as interleukin-1 (IL-1). Anakinra (Kineret) is a naturally occurring IL-1 receptor antagonist which blocks the biologic activity of IL-1, including its pro-inflammatory effect and its effect on cartilage degradation. Patients with rheumatoid arthritis have insufficient concentrations of the natural IL-1 receptor antagonist in synovium and synovial fluid to counteract their elevated IL-1 levels. Anakinra is licensed for use in the treatment of RA.

1.2 Imaging and Quantification

MRI is a medical imaging modality used in radiology to visualise the internal structure of the body in high detail, and has been chosen for use in this study for a number of reasons. Compared to Computed Tomography (CT), MRI provides greater contrast between different soft tissues such as our primary area of interest, synovial membrane tissue in joints. Additionally MRI does not use ionizing radiation, rather, images are generated using powerful magnetic fields acting upon hydrogen atoms in the water in the body. As patients require many imaging sessions during the course of their treatment, it is important that they are not exposed to large quantities of the harmful ionising radiation associated with CT and plain film radiography. While X-ray images have been used for arthritis diagnosis in previous studies, these images can lack the high contrast between soft tissues required to observe the fine details and features of synovium degradation associated with the disease [4]. MR generated images are therefore better suited for both image clarity and patient well being [5], with dynamic MR being used in clinical trials of biological agents in RA as early as 1999 [6].

In many cases, the use of ultrasound for imaging is an appropriate choice, as it can also detect early signs of bone erosion which X-ray cannot, and it does not rely on ionising radiation. There are however subtle differences which cause us to favour MR, as well as the fact that we access to an MR scanner, and a collaborative team of rheumatologists who favour this modality. The value of ultrasound depends on the experience of the user while MRI results are far less dependent on operator expertise. Ultrasound is particularly suited to smaller joints such as within a finger or toe, while with MRI you can get a complete overview of the hand, wrist or ankle area (the knee in our case) and you can visualise regions deeper in the body not readily accessible to ultrasound [7]. Our study aims to further promote MR as a viable diagnostic option.

Current methods of measuring disease activity is the use of quantitative and semi-quantitative analysis of synovial volume [8–10]. This technique generally requires each image to be graded with respect to scores from several criteria per image. Typically, this is a very time consuming process, and suffers from subjectivity. Our proposed application is designed to address these issues during use in a clinical environment.

2 Methodology

2.1 Image Acquisition

All of the MR images used in this study have been acquired under the guidelines of a protocol created specifically for this project. This document provides instructions for an entire set of imaging sequences, including various calibrations, patient preparation, image spacing and thickness, positioning of imaging planes etc. This has the aim of standardising all images used in the overall knee project at our clinical site [11].

The application we are developing currently supports common imaging formats such as JPEG, BMP etc, as well as more preferable DICOM standard images. DICOM images are preferred because of the extra header information available. From this header information many important details such as the exact spacing between images is available, which during visualisation is very important. Otherwise, this information has to be estimated. Typically, all data sets used during this study are DICOM standard.

2.2 Image Cleaning

Despite adhering to the image acquisition protocol above, our images still suffer from varying levels of noise (albeit reduced), so the first step upon loading these images into the application is a denoising process. Our aim is to remove all non-knee information from the image. This fundamentally involves identifying which pixels in an image are part of the knee, which are outside the knee, and setting these pixels to intensity zero. Our motivation for this denoising is twofold. Firstly, by isolating precisely which pixels are not part of the knee, we can more accurately calculate the volume of the knee. Secondly, processing the images like this enables us to improve the quality of our generated visualisations as discussed in Sect. 2.4.

When viewing a sequence of images, it is important that each image is contrast windowed based on the total pixel value range of the entire data set, and not just the pixel value range of an individual image. There needs to be continuity between images in a sequence, for changes in image brightness to appear as a gradient through the sequence. This often has the effect of images at the beginning or end of a sequence appearing very dark compared to images in the middle. If images are windowed individually, the sequence will appear to get brighter and darker erratically, and can suggest that objects appearing across several images change back and forth in tone/density quite dramatically.

To this end, two versions of the image are created during the cleaning process, one contrast windowed between the highest and lowest intensities in that image alone, and the second contrast windowed between *AverageMinValue* and *AverageMaxValue* for the entire sequence of images, where *AverageMinValue* is the sum of the minimum value for each image divided by the number of images, or

$$AverageMinValue = \frac{\sum_{i=1}^{d} i_{\min}}{d}$$

and *AverageMaxValue* is the sum of the maximum value in each image divided by the number of images, or

$$AverageMaxValue = \frac{\sum_{i=1}^{d} i_{\max}}{d}$$

where d is the number of images, i_{min} is the lowest pixel intensity in image i, and i_{max} is the highest pixel intensity in image i.

Our method for removing the extra noise works on the high contrast image, and the pixels identified as non-knee pixels are then removed from the lower contrast image.

The main step taken to determine a value to threshold is analysis of the image histogram. In the majority of images, there is a clear value which can easily be observed manually, a saddle point between two peaks on the left of the histogram. The histogram starts with an extremely high peak, on account of the mostly black area around the brighter knee object. The histogram falls and then rises again representing the object in the image. The saddle point between these two peaks is a good value to work with, as the darker noise values are becoming less frequent, and the brighter knee values are picking up.

While it is relatively easy to identify this point manually, choosing it automatically is made more difficult because of the fact that the histogram is never a smooth

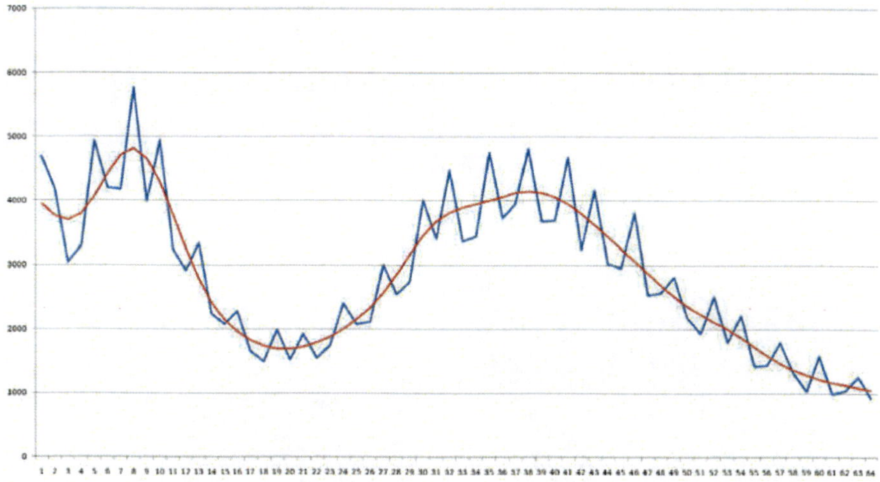

Fig. 1 Plot of erratic image histogram versus LOWPASS smoothed histogram

line, rather, it is a line that trends up and down but from datum to datum, increases and decreases erratically. To analyse the histogram automatically, it is necessary to first smooth it, as in Fig. 1.

We have used an application called LOWPASS to filter and smooth image histograms. LOWPASS is a zero lag low bandpass filter similar to the Jurik indicator (signal processing in the financial market). It uses a Fast Fourier Transform to translate the input to the frequency domain, applies a low bandpass filter and returns the filtered signal in the real domain [12].

Simply thresholding the blurred image and removing everything below the intensity picked by analysing the image histogram leaves a black and white mask, comprising the main knee object and usually some smaller outlier objects as in Fig. 2b. Parts of the knee will have disappeared also such as the darker bone material, but as long as the perimeter boundary of the knee is intact, this is not a problem. The next step is to choose the largest white object, which can be accomplished by iterating through the image, and flood filling each white object, recording the volume of each. The largest object is then identified and retained, as in Fig. 2c.

The next step is to identify which black pixels are outside the knee, versus which are inside, as simply removing all black pixels will also target dark pixels within the knee. Another flood fill operation seeded with the four corners of the image will highlight the black pixels outside the knee as in Fig. 2d.

These are the pixels which are finally set to zero and removed from the original image. During this step, we note the number of pixels set to zero for the purpose of calculating the volume of the knee within the dataset later. Figure 2e illustrates the final version of the image, and can be compared to the original version in Fig. 2f.

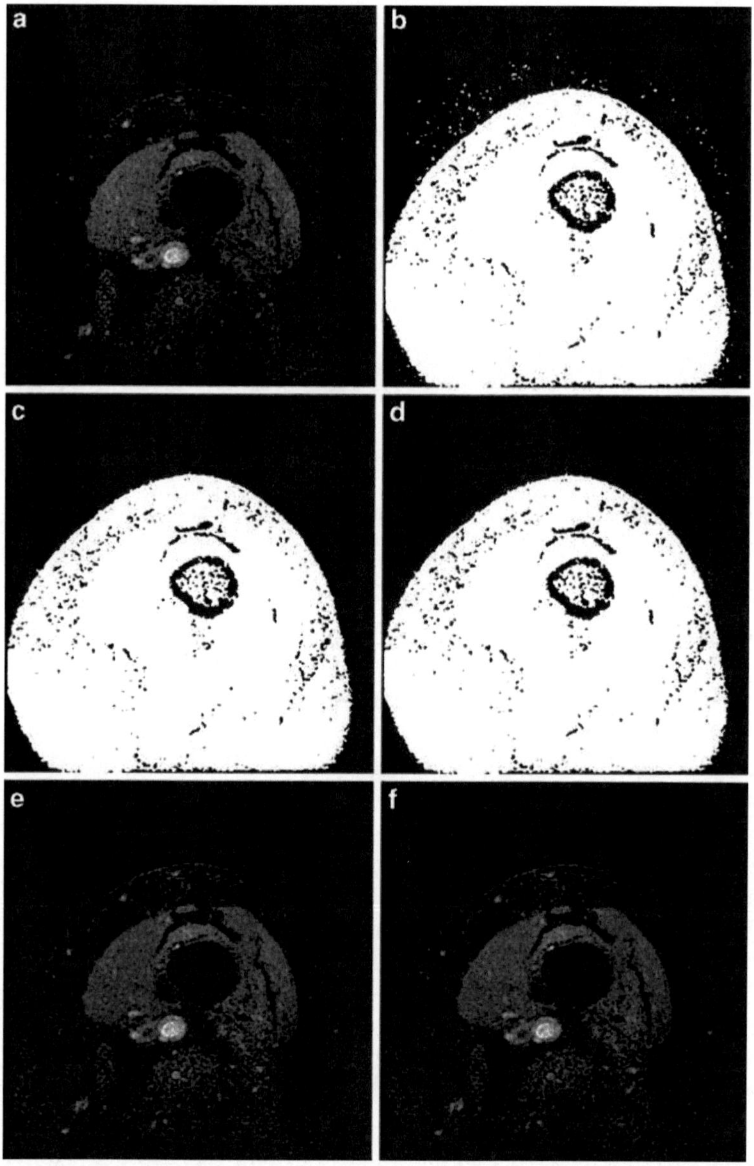

Fig. 2 Stages in the image denoising/cleaning process. A and F represent the original image, E represents the final image

In the event that our algorithm chooses a poor value to threshold, such that too much or too little of the image is removed, the user has the ability to re threshold any image to their satisfaction, by manually changing the threshold value previously calculated by histogram analysis.

2.3 Region of Interest Selection

Our application's Graphical User Interface (GUI) has been designed and created using Qt [13]. The application supports region of interest selection, with the intention of this being the method of segmenting out the inflamed synovium. Accurate segmentation and measurement of the inflammation of the synovium is critical for effective treatment of RA and PsA over time. Region of interest selection is implemented using a seed-filling algorithm in this application. The user chooses an initial seed point from any image, and an average intensity among the neighbouring pixels (5x5 grid) is calculated. The algorithm floods out from the seed point, and segments any neighbouring pixels that have an intensity greater than the intensity specified, until the boundary of the region is reached, where the boundary pixels will be less than or equal to the specified intensity.

This flooding effect works in all four cardinal directions on a given image, but can also penetrate the previous and following images in the sequence, producing the effect of segmenting regions in three dimensions. For a given pixel in an image, or voxel within a volume, if the pixel at the same (x,y) location in the next image has an intensity higher that the specified intensity, the algorithm will segment it, and mark each of it's neighbouring pixel for examination.

The user uses a slider to increase or decrease the isovalue, which has the effect of growing or shrinking the region of interest displayed on screen in real time. The user can cycle through the images in the sequence while continuing to adjust the isovalue slider, to see the effects of the 3D flood fill elsewhere in the dataset.

Once the user is satisfied with their selection for the initial seed, they may then choose an additional seed point and continue filling from there, using the isovalue slider as before. The user has the ability to segment several unconnected regions using this method, for example to segment inflamed synovial tissue at the front and back of the knee.

A problem sometimes encountered is where extremely bright regions exist at the perimeter of the knee, due to coil interference during image acquisition. The algorithm can easily flood into these regions accidently, especially in cases where the algorithm has been initially seeded on a different image with less interference, and from there it can feed back along the stack of images flooding into areas of higher intensity than it should. To reduce this from happening, the user is able to view thumbnails of all images in the sequence at once, and can identify at a glance when the algorithm is flooding into areas it shouldn't, and act accordingly (usually increasing the segmentation isovalue).

As a preventative measure however, the application generates special image masks at runtime which are designed to prevent the algorithm flooding towards the perimeter of the knee at all, as our region of interest should never appear there. During segmentation, the algorithm can only flood pixels marked within the white region which is smaller than the actual knee region in the image, as in Fig. 3 (outer white contour included for illustrative purposes). We generate this mask by taking

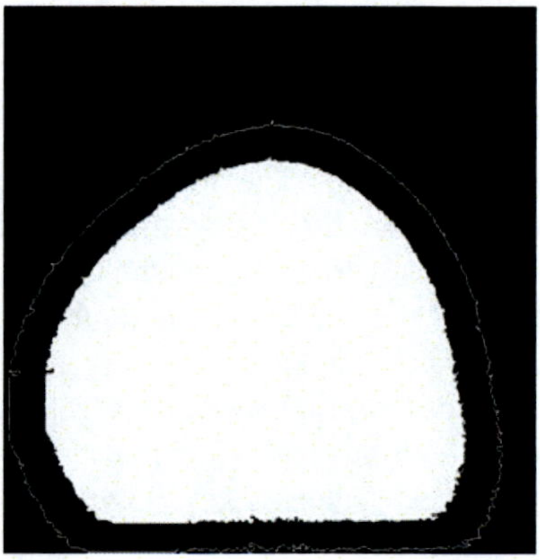

Fig. 3 Image mask generated by scaling the knee contour by 0.85 towards the knee centroid

each point along the outer knee contour, and scaling it towards the knee centroid by a factor of usually 10% or 15%. This value can be changed as the user desires.

Each knee boundary point is shifted to form a new boundary by the following, where the border pixel $p_1 = (x_1, y_1)$, the centroid $p_2 = (x_2, y_2)$ and the new shifted border pixel $p_3 = (x_3, y_3)$

$$x_3 = x_1 + s \times (x_2 - x_1)$$

$$y_3 = y_1 + s \times (y_2 - y_1)$$

Where s is the scaling factor, so using a value of 0.5 would make p_3 be half way between p_1 and p_2.

Once the user is satisfied with the entire finalised selection, clicking the quantify button causes the boundary of the segmented regions to be highlighted in red on each appropriate image in the 2D viewing window. This segmented region is also loaded into the visualisation widget, and the segmented data can be displayed there in 3D. Quantification of the segmented region begins, as discussed in Sect. 2.5.

2.4 Visualisation

One of the primary goals of this project is to visualise the knee data in a manner other than individual MR image slices, to allow the user (clinician) to manipulate and view all of the data at once. By reconstructing the image data in 3D, the clinician can get a better feel for the volumetric space involved, and the spatial confines of the disease process. We have implemented two visualisation techniques, both of which are currently achieved using OpenGL [14].

Our first visualisation technique used the marching cubes algorithm. First published in 1987 SIGGRAPH [15], the algorithm is used to extract an isosurface (a surface that represents points of constant value) as a polygonal mesh from within a three dimensional volume.

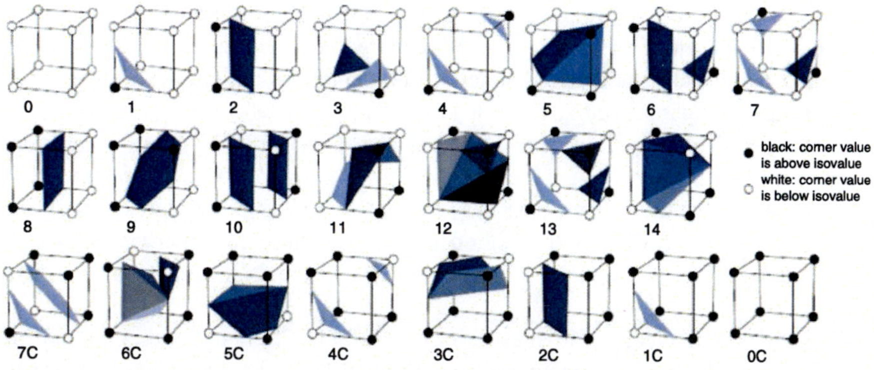

Fig. 4 Marching Cubes Cases as described in [16]. Image courtesy of G.Johansson and H. Carr [17]

To achieve this, the algorithm sweeps through the volume taking 8 neighbouring points at a time (forming an imaginary cube), determining what polygons are needed to represent the part of the isosurface that passes through this cube for a specific isovalue, rendering them and then proceeding to the next cube of eight points, marching through the entire volume. For each cube there are 256 possible configurations of the values at each corner of the cube, with values being either above or below the specified isovalue. If we describe the values above the isovalue as 1, and the values below as 0, we can construct an 8-bit integer which acts as an index to an array of 256 possible polygon configurations within the cube. From the 256 polygon configurations, there are 15 unique configurations (Fig. 4) which can be reflected and rotated to produce one of the 256 total polygon configurations. An issue with holes appearing in surfaces under certain conditions was resolved by Montani et al '94 [16], by adding complementary cases to the original 15 basic cases.

Lighting models can then be used to shade the resulting mesh. The application currently supports flat shading and smooth shading via central differencing. In this application, the MR images are converted to a three dimensional array of vectors, and it is this array of vectors that the marching cubes algorithm works upon. The user can specify a desired isovalue and the corresponding shaded isosurface is displayed in the OpenGL viewing window, and can be rotated via an arc ball implementation. If the user has segmented out a region of interest with the 2D image manipulation tools, this segmented volume may be viewed alone, or as part of the overall structure highlighted as a different colour.

Our second visualisation technique is volume rendering. Volume rendering is a technique used to display a 2D projection of a 3D discretely sampled data set. Compared to algorithms such as marching cubes, traditional volume rendering is a computationally intensive task. There are several methods available when volume rendering, and in this application, a variation of texture mapping is used to provide fast, although imperfect results. For this technique, each image in the dataset is textured onto a quad (a simple square in OpenGL). Each quad is layered on top of the previous one, and each texture is rendered semi-transparent. The user can manually manipulate the alpha channel of the images which will make the images more or less transparent, until a suitable level is found that allows the structure of the dataset to be viewed, without it being too bright or too dull. This semi transparent stack of images appears in the OpenGL viewing window, and can be rotated via an arc ball implementation. If the user has segmented out a region of interest with the 2D image manipulation tools, this region appears red within the otherwise greyscale 3D volume (Fig. 6). Regions segmented in the 2D manipulation window appear here in real time.

The process of cleaning images described in Sect. 2.2 benefits this method of visualisation. While viewing each image individually, the noise which was removed would be relatively difficult to see, but the effect of stacking all the images together makes the noise appear quite apparently. Cleaning the images of this noise (non knee data present in the image) makes the volume appear much more clearly.

2.5 *Quantification*

The second major goal of this project is the quantification of the disease throughout treatment. This will manifest as tracking of RA and PsA progression in a patient, allowing us to compare a patient in a before and after sense, consider if the drug treatment is proving effective, or help make decisions to adjust treatment dosage. A quantification metric will also easily allow the comparison of one patient with another, indicating which has more severe development by comparing either individual sets of scan data or several taken over a period of time.

When the user segments out a region of interest, the application calculates what percentage of the overall knee within the dataset has been segmented. This is achieved by counting each segmented voxel, and calculating this number as a percentage of the number of voxels within the dataset that make up the knee object. The number of knee voxels is calculated during the cleaning process when the images are first loaded. Each pixel that is cleaned (set to zero) is counted, and this number is subtracted from the total number of voxels in the dataset, giving us the number of knee voxels. Note that pixels within the boundary of the knee which have an intensity of zero are not counted as cleaned, only pixels which have been specifically flagged as outside (as in Fig. 2d) are counted as cleaned.

Thus, the number of segmented voxels as a percentage of the knee is simply

$$\frac{\sum_{i=1}^{d} s[i]}{\left(\frac{(h\times w\times d)-\left(\sum_{i=1}^{d} c[i]\right)}{100}\right)}$$

where $s[i]$ is the number of segmented pixels in image i, h is the height in pixels of each image, w is the width in pixels of each image, d is the number of images in the sequence, and $c[i]$ is the number of cleaned pixels in image i.

This alone is a good way of measuring inflammation from one examination to the next. As long as the user picks a suitable intensity when segmenting the region of interest, the percentage of inflamed synovium can be measured, and improvements or degradation can be noted.

3 Preliminary Results

3.1 *Visualisation*

The application has been developed to a stage which currently supports MR image sequence data sets as DICOM standard images or common image formats such as JPEG, BMP etc. These images are processed to remove background noise while extracting the anatomical features within, can be viewed in sequence and have a 3D region of interest segmented using seeded points. The dataset can be rendered in the visualisation pane using the marching cubes algorithm as in Fig. 5, or using the textured quads method of rendering the volume as in Fig. 6. If a region of interest has been selected, the highlighted region is reflected in the 3D render.

3.2 *Quantification*

Before planning a protocol for clinical tests, initial tests were performed on twelve MRI datasets, including five cases where images were acquired twice during treatment with 3-5 months between imaging. In each of these datasets, synovial tissue has been segmented using the process described in Sect. 2.3, generating a value of inflamed synovium as a percentage of knee tissue. These preliminary results are displayed in Table 1.

For patients 1–5 for which we have examined two datasets, it is encouraging to note that the percentage of inflammation has decreased in each case between examinations as expected, due to the effects of disease treatment. In each of the

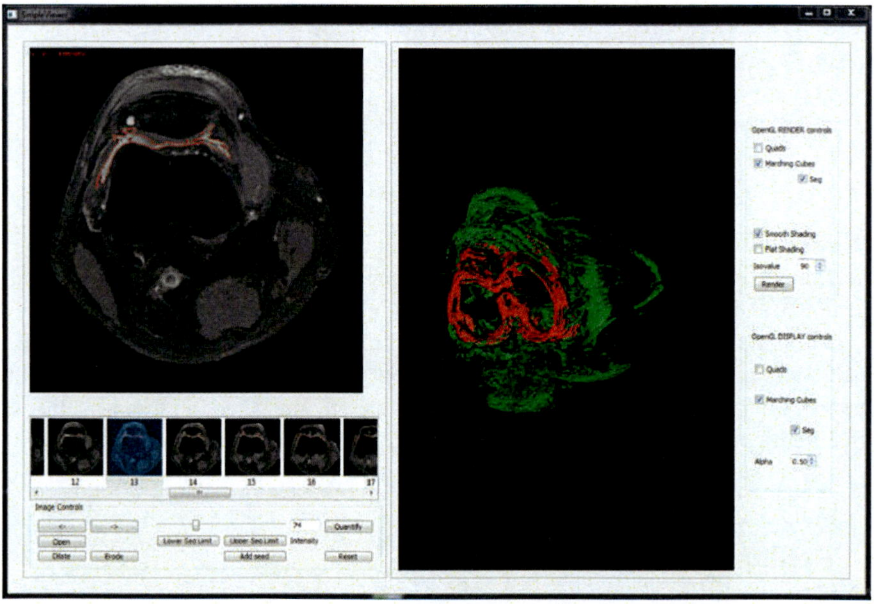

Fig. 5 Application rendering a data set using marching cubes. Region of interest visible in red

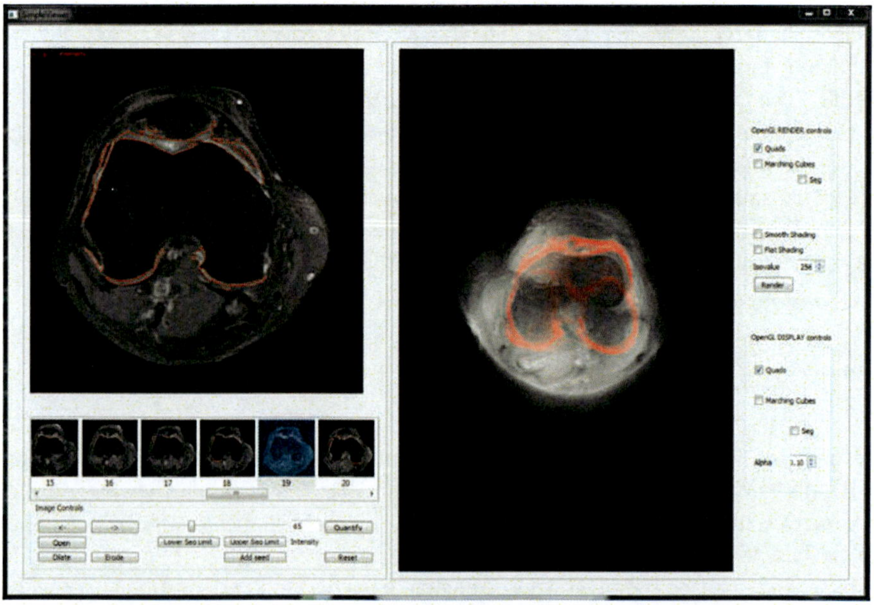

Fig. 6 Application rendering a data set as textured quads. Region of interest visible in red

Table 1 Initial Quantitative Results for MRI Scans Tested

Dataset	Days Between Imaging	Slices	Knee Voxels	Segmented Synovial Voxels	Inflammation
P1 Before	132	33	2435202	50124	2.06%
P1 After		32	4562338	90548	1.98%
P2 Before	159	40	5752372	265480	4.62%
P2 After		33	4428521	111769	2.52%
P3 Before	155	35	5666527	203475	3.59%
P3 After		30	4167073	108896	2.61%
P4 Before	140	36	5185089	111917	2.16%
P4 After		33	4303240	61286	1.42%
P5 Before	101	34	3750738	115430	3.08%
P5 After		33	3402642	56234	1.65%
P6		33	3442896	179993	5.23%
P7		36	3733268	130726	3.50%

twelve datasets, we were satisfied with our ability to precisely segment a region of interest using the seeded segmentation method we have implemented.

4 Future Work

As this project is a work in progress, there are several avenues for further development to consider. We would like to investigate using the GPU to render geometry as there are very obvious advantages afforded by this. The use of the GPU to accelerate visualisation (such as isosurface extraction with the marching cubes algorithm) can feasibly provide increases in speed in excess of tenfold in medical data sets [17].

The current implementation of marching cubes is somewhat rough and suffers from staircase artefacts. We intend to implement "smooth" marching cubes in the near future, possibly using a solution such as Gibson's surface net technique to smooth binary segmented data [18]. Our current implementation is suitable for the time being, during the initial testing phase.

The current volume rendering approach is not entirely suitable for fully viewing data sets from certain angles, and we aim to implement a true volume rendered solution utilising ray casting [19]. There are a number of algorithms available for implementation including optimised versions of the original technique such as Levoy's shear-wrap optimization [20].

Our method of seed filling based segmentation has been an experimental investigation, which so far has proved satisfactory. Despite this, we are actively researching other methods of segmentation which may prove more suitable, with a contour tree based topological approach as another area of ongoing research [21].

The quantification algorithm is currently being developed beyond the current iteration, as the current method is not yet sophisticated enough for deployment in a medical environment. Factors under consideration include the actual intensity of voxels identified as inflammation within the synovium when calculation a measure of the disease progression.

During the clinical validation phase of the project, scores generated with the use of this application will be compared to results generated using a semi-quantitative scoring technique developed in house, and histological information. The semi-quantitative technique has been developed by our radiology and rheumatology collaborators [11] based on multiple clinical and general radiographic criteria. Each image is scored manually against a set of 10-20 criteria making it time consuming and subjective. The performance of the application will also be evaluated against that of another application developed as part of our overall arthritis project, considering usability, functionality, stability etc. [22].

Ultimately, there is scope to develop a framework application capable of analysing a variety of conditions and illnesses. The Visualisation modules will effectively render a dataset of images for any part of the body, and specific quantification algorithm plugins could be specifically developed for use in a wide range of medical studies, beyond the observation of the arthritic condition in different human joints, but possibly including a number of completely different medical disorders which manifest as physical deterioration within different tissue types.

5 Conclusions

With continued work, this application can develop further, becoming a tool that can hopefully provide further insight to the mechanics and treatment of rheumatoid and psoriatic arthritis. Current progress has been met with encouraging feedback from our clinical collaborators [11], and a clear set of ideas and goals for future progress will develop this into a powerful tool for the assistance of diagnosis and treatment of rheumatoid and psoriatic arthritis.

While it is true that there are many commercial and open source software packages available which allow high quality interactive visualisations of volumetric datasets such as Voreen, SCIRun, AMIRA [23–25], the key benefit of developing a custom designed application is that it can be tailored to suit a specifically defined problem, with a high level of input from the medical professionals that will use it, as to their exact requirements.

References

1. Arthritis Foundation. *Arthritis Disease Centre.* http://www.arthritis.org/disease-center.php?disease_id=31 Cited 01 July 2010
2. Arthritis Foundation.*Disease Centre: Rheumatoid Arthritis Who Gets It?* http://www.arthritis.org/disease-center.php?disease_id=31&df=whos_at_risk Cited 01 July 2010
3. National Psoriasis Foundation. *About psoriatic arthritis.* http://www.psoriasis.org/about/psa/ Cited 01 July 2010
4. A. Giovagnoni, W. Grassi, F. Terilli, P. Blasetti, E. Paci, P. Ercolani, C. Cervini (1995). *MRI of the hand in psoriatic and rheumatical arthritis.*

5. Vital E, Emery P (Sep 15 2005). *Advances in the treatment of early rheumatoid arthritis.* Am Fam Physician 72 (6): 1002, 1004
6. Veale DJ, Reece RJ, Parsons W, Radjenovic A, O'Connor PJ, Orgles CS, Berry E, Ridgway JP, Mason U, Boylston AW, Gibbon W, Emery P. *Intra-articular primatised anti-CD4: efficacy in resistant rheumatoid knees. A study of combined arthroscopy, magnetic resonance imaging, and histology.* Ann Rheum Dis. 1999 Jun;58(6):342–9.
7. P. Gould (March 10 2009) Musculoskeletal Network *MRI and ultrasound reveal early signs of rheumatoid arthritis* http://www.musculoskeletalnetwork.com/rheumatoid-arthritis/content/article/113619/1387429 Cited 01 July 2010
8. Gaffney K, Cookson J, Blades S, Coumbe A, Blake D. *Quantitative assessment of the rheumatoid synovial microvascular bed by gadolinium-DTPA enhanced magnetic resonance imaging.* Ann Rheum Dis 1998; **57**: 152157.
9. Klarlund M, Ostergaard M, Lorenzen I. *Finger joint synovitis in rheumatoid arthritis: quantitative assessment by magnetic resonance imaging.* Rheumatology 1999; **38**: 6672.
10. RomingerMB, Bernreuter WK, Kenney PJ, Morgan SL, Blackburn WD, Alarcon GS. *MR imaging of the hands in early rheumatoid arthritis: preliminary results.* RadioGraphics 1993; **13**: 3746.
11. St Vincent's University Hospital *Department of Rheumatology* http://www.stvincents.ie/Departments/Department_of_Rheumatology.htm Cited 01 July 2010
12. Qtstalker - Commodity and stock market charting and technical analysis (2008). *Low Band Pass Filter - LOWPASS* http://qtstalker.sourceforge.net/lowpass.html Cited 01 July 2010
13. Nokia Corporation. *Qt A cross-platform application and UI framework.* http://qt.nokia.com/ Cited 01 July 2010
14. Khronos Group. OpenGL *The Industry Standard for High Performance Graphics.* http://www.opengl.org/ Cited 01 July 2010
15. William E. Lorensen, Harvey E. Cline: *Marching Cubes: A high resolution 3D surface construction algorithm.* In: *Computer Graphics*, Vol. 21, Nr. 4, July 1987
16. Claudio Montani, Riccardo Scateni, and Roberto Scopigno. *A modified look-up table for implicit disambiguation of marching cubes.* The Visual Computer, 10(6):353–355, December 1994.
17. Gunnar Johansson, Hamish Carr: *Accelerating Marching Cubes with Graphics Hardware* CASCON 2006
18. Gibson, S. *Constrained Elastic SurfaceNets: Generating Smooth Surfaces from Binary Segmented Data.* Proceedings MICCAI, 888–898, MIT, 1998
19. Marc Levoy *Display of Surfaces from Volume Data* In: *IEEE Computer Graphics and Applications*, Vol. 8, No. 3, May, 1988, 29–37
20. Philippe Lacroute, Marc Levoy: *Fast volume rendering using a shear-warp factorization of the viewing transformation* Proc. SIGGRAPH 1994
21. Carr H, Ryan J, Joyce M, Fitzgerald O, Veale D, Gibney R, Brennan P. *A Topological Approach to Quantitation of Rheumatoid Arthritis* Visualization in Medicine and Life Sciences 27–37, 2008
22. Garraud C, Ryan J, Carr H, Gibney R, Veale D, Fitzgerald O, Brennan PC. *A volumetric reconstruction method for semi-automatic quantification of rheumatoid arthritis within the knee.* Virtual Reality Interactions and Physical Simulations 2007
23. Westfälische Wilhelms - Universität Münster. *Voreen - Volume Rendering Engine* http://www.voreen.org/ Cited 01 July 2010
24. The University of Utah. *SCIRun* http://www.sci.utah.edu/cibc/software/106-scirun.html Cited 01 July 2010
25. Visage Imaging GmbH. *Amira - Visualize Analyze Present* http://www.amira.com/ Cited 01 July 2010

An Application for the Visualization and Quantification of HIV-Associated Lipodystrophy from Magnetic Resonance Imaging Datasets

Tadhg O'Sullivan, Patrick Brennan, Peter Doran, Paddy Mallon, Stephen J. Eustace, Eoin Kavannagh, Allison Mcgee, Louise Rainford, and John Ryan

Abstract HIV-associated lipodystrophy is a syndrome characterized by the abnormal distribution or degeneration of the body's adipose tissue. Since it was first described in 1999, the complications associated with the condition have become a very real problem for many HIV patients using antiretroviral drug treatments. Accurate visualization and quantification of the body's adipose tissue can aid in both the treatment and monitoring of the condition. Manual segmentation of adipose tissue from MRI data is a challenging and time consuming problem. It is for this reason that we have developed an application which allows users to automatically segment a sequence of MRI images and render the result in 3D.

1 Introduction

Human immunodeficiency virus (HIV) is a major health problem which affects millions of individuals worldwide. In 2007 the Joint United Nations Programme on HIV/AIDS (UNAIDS) reported that an estimated 33 million people were living with HIV [1]. Since the introduction of highly active antiretroviral treatments (HAART) in 1996 there has been a substantial decrease in the morbidity and mortality rates of HIV-infected patients. HAART, which is now the standard treatment for HIV infection, involves the use of a combination of antiretroviral drugs with the aim

T. O'Sullivan (✉) · P. Doran · P. Mallon · A. Mcgee · L. Rainford
University College Dublin, Dublin, Ireland
e-mail: t.osullivan@ucd.ie

P. Brennan · J. Ryan
University of Sydney, Sydney, Australia
e-mail: john.ryan@sydney.edu.au

S.J. Eustace · E. Kavannagh
Mater Misericordiae University Hospital, Dublin, Ireland

L. Linsen et al. (eds.), *Visualization in Medicine and Life Sciences II*,
Mathematics and Visualization, DOI 10.1007/978-3-642-21608-4_4,

of suppressing the replication of a virus within host cells. Antiretroviral drugs, such as nucleoside reverse transcriptase inhibitors (NRTI) or Protease inhibitors (PIs), attempt to interrupt specific stages of the life-cycle of a retrovirus. In most cases, antiretroviral drugs will inhibit the activity of certain enzymes which are need by a retrovirus in order for it to successfully replicate [2]. While it is without question that HIV-infected patients have benefitted from the use of HAART, there are also a number of adverse effects associated with this form of treatment. In 1997 one such adverse effect was reported by the US Food and Drug Administration (FDA). The FDA stated that a small number of HIV-infected patients, who where currently prescribed Protease Inhibitors, were presenting with hyperglycemia and type 2 diabetes [3]. Over the next few years there were increased reports of HIV-infected patients presenting with maldistribution of adipose tissue. The condition, known as lipodystrophy, is characterized by central adiposity, dorsocervical fat accumulation (also known as the "Buffalo Hump"), and peripheral fat wasting. It has also been found that the condition is associated with an increased risk of cardiovascular disease due to elevated cholesterol levels and insulin resistance or type 2 diabetes [4]. The complications associated with lipodystrophy have become a very real problem for many HIV-infected patients using antiretroviral drug treatments. Accurate visualization and quantification of the body's adipose tissue can aid in both the treatment and monitoring of the condition. When dealing with large amounts of data, such as with MRI datasets, the task of interpreting said data becomes increasingly complex. It is for this reason that techniques developed in the field of scientific visualization are employed. Through the use of computer graphics we can accurately and realistically render large datasets in 3D (as volumes, surfaces, etc...) [5, 6]. Segmentation is the process of separating image data into structures which have a relevance to the task at hand. There are a number of different methods used to segment data and it is generally the task at hand and the type of data which determines which method to employ. Manual and semiautomatic segmentation methods can be challenging and time consuming with datasets of a significant size. It is for this reason that statistical analysis methods are commonly used in order to automatically delineate structures of interest [7]. Through the implementation of scientific visualization algorithms, statistical analysis based segmentation and a number of image processing techniques we are developing an application which will allow users to automatically segment a sequence of MRI images, render the results in 3D and quantify said results.

2 Methods and Materials

The modality employed for this research is MRI. There are two main reasons for choosing MRI datasets over more conventional methods such as Dual energy X-ray absorptiometry (DEXA) or Computed tomography (CT). Firstly, the lack of radiation exposure with MRI is a particular advantage, especially if multiple datasets over an extended period of time. Secondly, MRI has shown to produce datasets with superior contrast between tissues in comparison with other imaging modalities. This increased contrast is especially noticeable with visceral adipose tissue (VAT) and subcutaneous adipose tissue (SAT) [8, 9]. A graphical user interface (GUI) has been developed using the QT Software Development Kit [10] which allows for the deployment of cross-platform applications. Through the implementation of the Insight Toolkit (ITK) [11] the GUI allows users to load and view MRI DICOMs. Users of the application have the option to load either a single DICOM image or an entire series of images.

3 Adipose Tissue Segmentation

The segmentation process, which can be applied once a series of DICOM images have been loaded into the application, involves several steps. Firstly, an intensity histogram is computed for each image. With T1-weighted MRI images adipose tissue will have a high signal intensity in relation to other tissues present [15]. For this reason we can extract the initial seeds required for the segmentation algorithm by iterating back to front along each intensity histogram and stopping at the first non-empty bin. Using the intensities found it is then possible to iterate through the images and extract seed points within the tissue in question to be used at a later point in the process. The Canny edge detection algorithm is then applied to each image in the dataset in order to determine the boundaries of different tissues [16]. This algorithm is chosen over other common edge detection operators (Roberts, Prewitt, Sobel for example) for two reasons. Firstly, the primary step of the algorithm is to apply a convolving Gaussian kernel in order to reduce noise and smooth the images. Secondly, the Canny algorithm detects horizontal, vertical and diagonal edges (most techniques detect horizontal and vertical edges). Both of these features ensure more accurate edge detection, especially in medical imaging where noise is invariably present and features of interest will most likely have rounded edges. Once the edge images have been computed a morphological closing operation is applied to the images. It is imperative for the next stage of the segmentation process that all edges within the images are closed. The first two stages of the algorithm can be seen in Fig. 1

A flood fill algorithm is then applied using the previously gathered seed points and the morphologically closed edge images as its primary inputs. If a seed point occurs within a black area of an image the flood fill begins. Masks of the images are created using Otsu's method [17] and are used to ensure seed points are not outside the feature of interest in the images. With most flood fill operations the filled region grows based on whether the north, south, east and west pixel intensities are within a certain range of the current pixel intensity. For the purposes of our algorithm the region continues to grow until it reaches an edge and it is for this reason that the morphological close operation is important. Choosing the highest intensity within each image to determine seed points will lead to a small amount of initial seeds (between 3 and 10 per image) and this, in turn, causes only small areas of the images

Fig. 1 Segmentation Process. On the left the original DICOM image from which an intensity histogram is calculated and the seed points are extracted. In the middle the resulting image after the Canny edge detection algorithm is applied. On the right the edge image after a morphological closing operation is performed

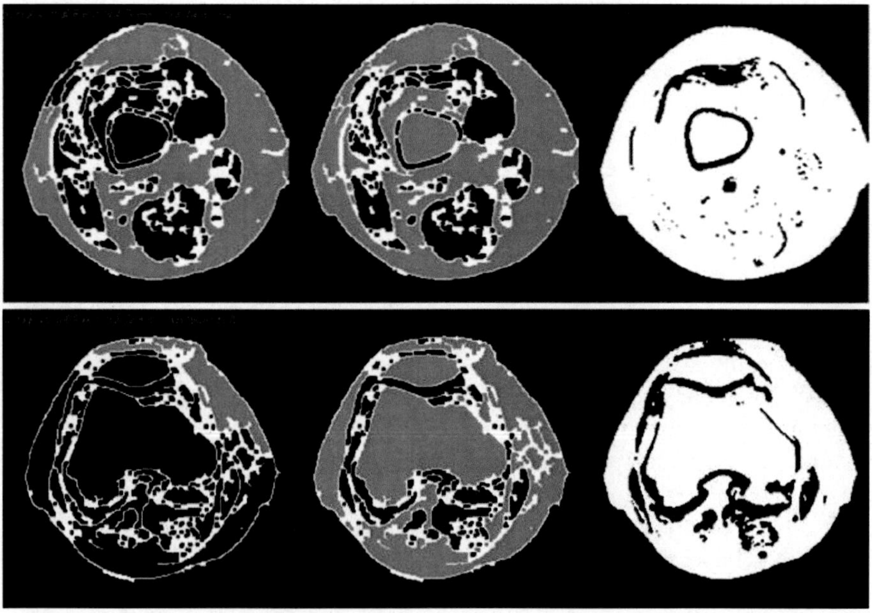

Fig. 2 Segmentation Process. On the left the initial results of the flood fill algorithm. In the middle the results of the flood fill after more seed points are calculated. On the right the mask images produced using Otsu's method

to be filled. From the resulting filled areas a mean value representing the tissue of interest is calculated. The flood fill operation is then performed one more time using new seed points based on said mean value. See Fig. 2.

The last stage of the algorithm produces binary images representing the segmented tissues (See Fig. 3). These images are later used for both visualization and quantification of adipose tissue within the whole volume.

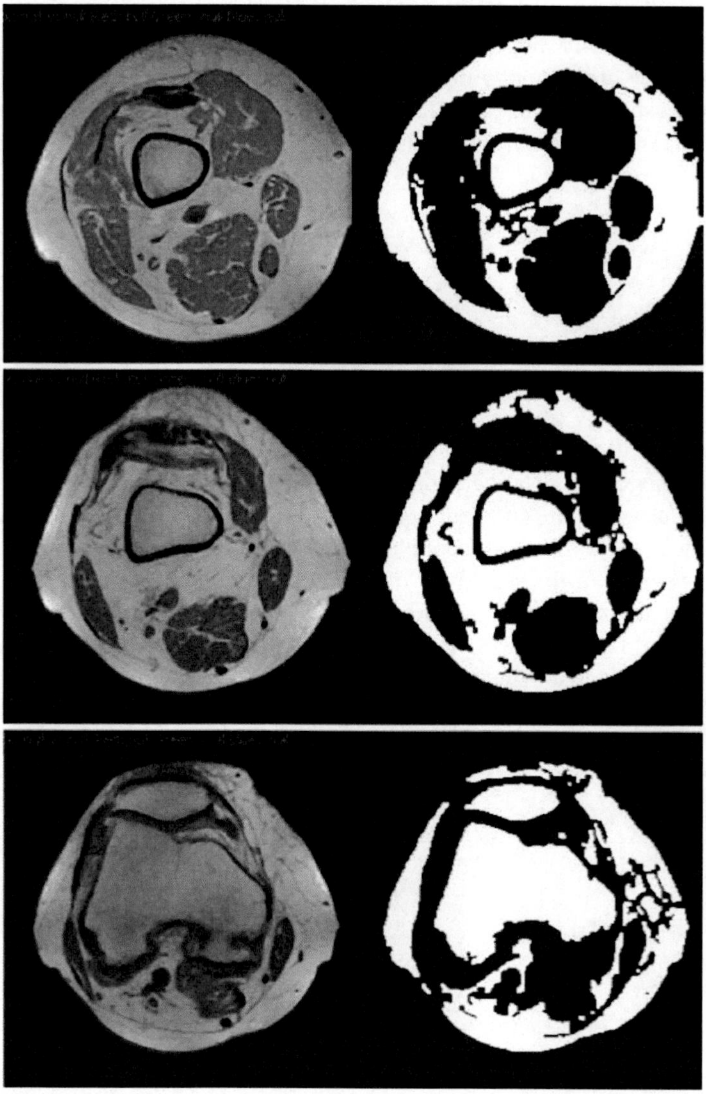

Fig. 3 Segmentation Process. Original images and the final results of the segmentation algorithm

4 Visualization

Once the segmentation process has been completed, the resulting images are used to produce an isosurface for the purposes of 3D visualization. The algorithm used to represent the dataset in 3D is called Marching Cubes. The Marching Cubes algorithm is known to produce accurate surfaces from datasets and is used widely in the field of visualization [6]. Other visualization techniques, which have been implemented, would also suit the needs of this application (such as volume rendering using ray casting). Due to time taken to produce 3D images using these techniques they are currently not included in the application. This 3D isosurface is rendered using OpenGL [12]. The percentage of the volume which consists of adipose tissue is also calculated by the application. The application and many of its features can be seen in Fig. 4.

5 Future Work

5.1 Bias Field Correction

In order to accurately segment and quantify MRI data one must deal with an inherent artefact related to the acquisition process. This artefact is known as the bias field, image inhomogeneity or image nonuniformity. The bias field can be caused by a number of factors such as patient position within the MRI scanner, certain tissues which may be more or less susceptible to the magnetic frequency of the scanner or an inhomogeneity of the sensitivity of the radio frequency receiving and emitting coils. The resulting artefact is a lowering of pixel intensity/brightness levels in certain areas of the data [13]. Through the implementation of a 3D height map and an RGB transfer function, the inhomogeneity of intensities within a given tissue can be clearly seen (see Fig. 5). There are a number of methods available for correcting

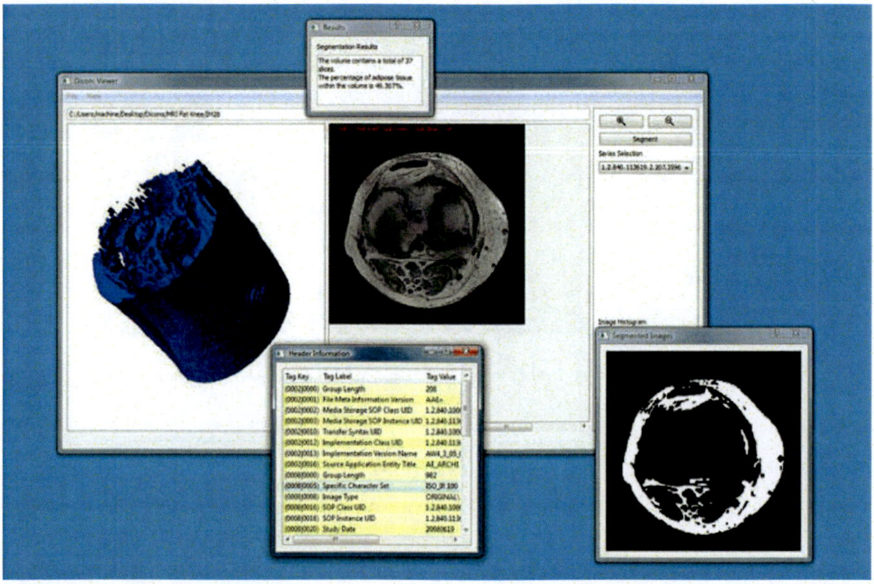

Fig. 4 A screenshot of the application running. The primary window displays the 2D images which the user has loaded and the rendered isosurface of the segmented tissue. A number of additional features such as the retrieval of DICOM header information are also available

Fig. 5 The effect of image inhomogeneity; the original greyscale image, a 3D heightmap of the data and the same data with a RGB transfer function applied to it

this artefact but many of them are slow and computationally expensive. A method for correcting intensity inhomogeneities within an image has been found and is currently being implemented and tested. The method is called adaptive contrast enhancement [14]. Once fully implemented, we will be able to test its ability to correct bias within MRI data.

5.2 *Increased Accuracy in Segmentation Algorithm*

One of the central aims of this piece of research is to have a highly accurate segmentation process. It is for this reason that an analysis of a number of statistical methods for tissue identification is currently being undertaken. Many techniques, from histogram equalization to bivariate scatter plots are being tested using a number of MRI datasets. Once a superior tissue classification method has been identified it will be integrated into the current segmentation algorithm. The resulting algorithm will be both robust and highly accurate.

5.3 *Lipodystrophy Quantification*

Work has begun on developing a method for quantifying lipodystrophy from MRI datasets. Currently the application gives an overall percentage of adipose tissue present in an entire dataset. The future goal is to have a curve representing the distribution of adipose tissue throughout the entire body. Subdividing the human body (head, trunk and limbs) will allow users of the application to easily identify any abnormalities in the distribution of adipose tissue. Once complete, this method will also allow users to compare the current MRI scan of a patient against older ones. This will be done through the use of a database system which will store the results from older datasets. Datasets of individuals with normal distributions of adipose tissue will also be used to create a curve representing an expected normal distribution (based on gender and age group). This will also aid in identifying abnormal adipose tissue distribution.

5.4 Evaluation of Application

Once the application has reached the final stage of development qualitative and quantitative assessments will be carried out. These will involve a comparison of the segmentation results and visualization of data gathered using MRI versus DEXA and CT. The finalised application will also allow users to grade their overall experience of using the application (which features are useful, ease of use, response time of application, etc...) and professionals to assess the accuracy of the segmentation results.

References

1. (UNAIDS), J.U.N.P.o.H.A., Report on the global HIV/AIDS epidemic. 2008.
2. Palmer, C., HIV treatments and highly active antiretroviral therapy. Australian Prescriber, 2003. 26(3): 59-61.
3. Administration, F.a.D., Protease inhibitors may increase blood glucose in HIV patients. FDA, 1997. 27(2).
4. Shevitz, A., et al., Clinical perspectives on HIV-associated lipodystrophy syndrome: an update. AIDS, 2001. 15(15): 1917-1930.
5. Levoy, M., Efficient ray tracing of volume data. ACM Trans. Graph., 1990. 9(3): 245-261.
6. Lorensen, W.E. and H.E. Cline, Marching Cubes: A high resolution 3D Surface construction algorithm. 1987. 21(4): 163-169.
7. Preim, B. and D. Bartz, Visualization in Medicine: Theory, Algorithms, and Applications. 2007: Morgan Kaufmann. 736.
8. Abate, N., et al., Estimation of adipose tissue mass by magnetic resonance imaging: validation against dissection in human cadavers. J. Lipid Res., 1994. 35(8): 1490-1496.
9. Jin, Y., et al., Segmentation and Evaluation of Adipose Tissue from Whole Body MRI Scans Lecture Notes in Computer Science, 2003. 287: 635-642.
10. QT. QT. [cited 2009 24/09/09]; Available from: http://qt.nokia.com/products.
11. ITK. Insight Toolkit. [cited 2009 24/09/09]; Available from: http://www.itk.org/.
12. OpenGL. OpenGL. [cited 2009 24/09/09]; Available from: http://www.opengl.org/.
13. Gispert, J.D., et al., Method for Bias Field Correction of Brain T1-Weighted Magnetic Resonance Images Minimizing Segmentation Error. Hum. Brain Mapp., 2004. 22: 133–144.
14. Srinivasan, S. and N. Balram, Adaptive Contrast Enhancement Using Local Region Stretching. ASID, 2006: 8-12.
15. Nicola Abate, Dennis Burns, t Ronald M. Peshock, Abhimanyu Garg, and Scott M. Grundy., Estimation of adipose tissue mass by magnetic resonance imaging: validation against dissection in human cadavers, Journal of Lipid Research, 1994. 35: 1490-1496.
16. John Canny, A Computational Approach to Edge Detection. IEEE Transactions on Pattern Analysis and Machine Intelligence, Vol. 1986. 6: 679-698.
17. Nobuyuki Otsu, A Threshold Selection Method from Gray-Level Histograms. IEEE Transaction on Systems, Man, and Cybernetics, 1979. 9(1): 62-66.

Part II
Classification

Semi-Automatic Rough Classification of Multichannel Medical Imaging Data

Ahmed Elmoasry, Mohamed Sadek Maswadah, and Lars Linsen

Abstract Rough set theory is an approach to handle vagueness or uncertainty. We propose methods that apply rough set theory in the context of segmentation (or partitioning) of multichannel medical imaging data. We put this approach into a semi-automatic framework, where the user specifies the classes in the data by selecting respective regions in 2D slices. Rough set theory provides means to compute lower and upper approximations of the classes. The boundary region between the lower and the upper approximations represents the uncertainty of the classification. We present an approach to automatically compute segmentation rules from the rough set classification using a k-means approach. The rule generation removes redundancies, which allows us to enhance the original feature space attributes with a number of further feature and object space attributes. The rules can be transferred from one 2D slice to the entire 3D data set to produce a 3D segmentation result. The result can be refined by the user by interactively adding more samples (from the same or other 2D slices) to the respective classes. Our system allows for a visualization of both the segmentation result and the uncertainty of the individual class representations. The methods can be applied to single- as well as multichannel (or multimodal) imaging data. As a proof of concept, we applied it to medical imaging data with RGB color channels.

A. Elmoasry · L. Linsen (✉)
Jacobs University, Bremen, Germany
e-mail: a.elmoasry@jacobs-university.de; l.linsen@jacobs-university.de

M.S. Maswadah
South Valley University, Egypt
e-mail: Maswadah@hotmail.com

L. Linsen et al. (eds.), *Visualization in Medicine and Life Sciences II*,
Mathematics and Visualization, DOI 10.1007/978-3-642-21608-4_5,

1 Introduction

Segmentation of imaging data is an important and crucial step in the medical visualization pipeline. Many segmentation algorithms for grayscale images exist in the image processing community. Some of them have been extended to multichannel (or multimodal) data. In the context of medical imaging, the segmentation approaches face the problem of uncertainty. This uncertainty is introduced at various levels in the processing pipeline including image acquisition, image reconstruction, registration, and image processing. One prominent example is the partial volume effect.

In this paper, we propose a segmentation method that allows us to quantify the uncertainty in the data. The segmentation method is based on a classification using rough sets. Such a classification allows us to derive lower and upper approximations as well as probabilities. With that information, we can visualize the segmented images as well as the uncertainty for each segment.

Rough set theory introduced by Zdzislaw Pawlak in the early 1980s [16, 17, 20] offers an approach to granular computing, which is part of the computational intelligence area [23]. Computational intelligence techniques approximately solve real-world problems in decision making, pattern classification, and machine learning. Prominent examples are fuzzy sets, neural networks, genetic algorithms, decision trees, and rough sets. The basic idea underlying the rough set approach to information granulation is to discover to what extent a set of given objects (in our context: groups of pixels) approximates another set of objects of interest. Objects are compared by considering their descriptions. An object description is modeled as a vector of function values representing the object's features or attributes [22]. We explicitly make use of this property when applying rough set theory to multichannel imaging data.

One advantage of the rough set theory is the generation of readable if-then decision rules as the output of the classification process when applied to a training set. Such rules allow us to reveal new patterns in the data material and to use them as classifiers for unseen data sets. Unlike many other computational intelligence techniques, rough set analysis requires no external parameters and uses only the information presented in the given data.

Another nice features of rough set theory is that it can be determined whether the data is complete, i.e., whether there are enough samples in the training set, and, if so, whether there is redundant information in the data. In the case of redundant information, only the minimum amount of data needed for the classification model is used. This property of rough sets is extremely helpful for applications where domain knowledge is limited [17–19, 21]. We make use of this property by enhancing the vector of function values with a number of derived properties. In addition to feature space properties, we also apply several statistical properties, which insert spatial (or object space) information. Using rough sets, we do not need to know, which statistical properties are necessary. We can just add all of them and the relevant ones are selected during the generation of the classification rules.

We have built a semi-automatic interactive system, where rough set classification rules are automatically computed from a set of samples provided by the user using a simple and intuitive interaction mechanism. Segmentations are obtained using a k-means approach. The system visualizes the derived segmentation and its uncertainty.

The main contributions of this paper are:

- Rough classification (modeling uncertainty).
- Combined feature-/object-space classification of multichannel data.
- Automatic rule generation from rough classification (using k-means).
- Uncertainty visualization.

We structure the paper as follows: Sect. 2 provides an explanation of the basic framework of rough set theory along with some of the key definitions. Section 3 provides an overview of related work in the area of rough image processing and clustering. Section 4 describes our approach to rough set classification including methods for uncertainty visualization and how we obtain the final segmentation using k-means clustering. In Sect. 5, we explain how we apply this framework to combined feature-/object-space segmentation of multichannel imaging data. The interactive and visual components of our semi-automatic system are described in Sect. 6. Finally, Sect. 7 provides some results and discusses them. Although our methods are applicable to any type of single- or multichannel imaging data, for a proof of concept we restrict ourselves to applications to RGB color imaging data.

2 Rough Set Theory

2.1 Rough Set Approximations

Rough set theory goes back to the ideas of Pawlak [16], who introduced the idea of describing a set with a lower and upper approximation. The description of these sets are derived from an information system (sometimes referred to as an attribute-value system). An information system is a pair (U, A), where U is the universe, i.e., a non-empty, finite set of objects, and A is the set of attributes, i.e., a non-empty, finite set of attributes (or features) $a \in A$ such that $a : U \rightarrow V_a$, where V_a denotes the set of values that feature a may have. Thus, the information table assigns a value $a(u) \in V_a$ to each attribute $a \in A$ and object $u \in U$. Information systems are often given in form of a decision table that lists these value, i.e., the rows of a decision table correspond to the objects and the columns correspond to the features. In the context of image segmentation, the objects represent pixels and the features are the attributes of the pixels such as RGB color. While the decision table contains all pixels, we will in our interactive approach use selected and manually classified pixels as a training set to determine a classification for all pixels. As such, we introduce another column for the training set, which adds the classification label. This additional row is referred to as the decision feature, while the others are referred to as the conditional features.

Given an information system (U, A) and a subset $B \subset A$ of conditional features, we define an equivalence relation Ind_B by

$$Ind_B = \{(x, y) \in U \times U : \forall a \in B \; a(x) = a(y)\}.$$

This relation is typically referred to as indiscernibility relation, i.e., if $(x, y) \in Ind_B$, then x and y are indiscernible (or indistinguishable) by the attributes of B. The equivalence relation imposes a partition (family of disjoint equivalence classes) onto the universe U that is denoted by U/Ind_B.

The motivation for using rough sets for segmentation purposes comes from the possibility to express uncertainty in terms of an upper and lower approximation of a segmentation. Assume that the user selects a certain set of pixels $X \subset U$ and marks them as belonging to one class. Then, we would like this selection to induce a classification onto the entire universe U based on the attribute values $a \in B$ of X. However, this is not always possible, as the class may include and exclude objects, which are indistinguishable when just looking at the attributes of B. In rough set theory, one defines a lower approximation of X with respect to B by

$$L_B(X) = \bigcup \{Y \in U/Ind_B : Y \subseteq X\}$$

and an upper approximation of X with respect to B by

$$U_B(X) = \bigcup \{Y \in U/Ind_B : Y \cap X \neq \emptyset\}.$$

Hence, the lower approximation $L_B(X)$ of a set X contains all equivalence classes $[x]_B$ that are subsets of X and the upper approximation $U_B(X)$ contains all equivalence classes that have objects in common with X. The difference

$$Bnd_B(X) = U_B(X) - L_B(X)$$

of upper and lower approximation is called the boundary region of X with respect to B. Any set X with a non-empty boundary region $Bnd_B(X)$ is called a rough set. Figure 1 shows an illustration of these sets. Assuming that the red curve depicts the a crisp boundary of a set X, the green set depicts the lower approximation, the blue set the boundary region, and the green and blue cells together the upper approximation. The lower approximation $L_B(X)$ of X with respect to B is also referred to as the positive region, while the complement of the upper boundary $U - U_B(X)$ is referred to as the negative region. Consequently, the boundary region can also be described as what remains from the universe when subtracting both the positive and the negative region. In Fig. 1, the green set depicts the positive region and the white set the negative region.

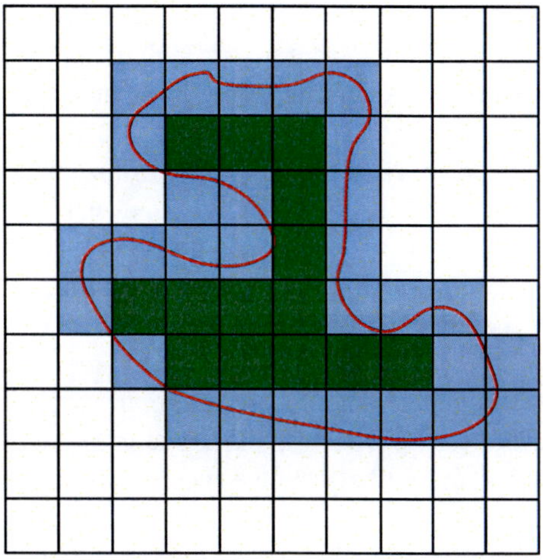

Fig. 1 Illustrated example of approximations for set X: Let the red curve depict the crisp boundary of X. Then, the green set depicts the lower approximation, the blue set the boundary region, and the green and blue cells together the upper approximation. Also, the green set depicts the positive region and the white set the negative region

2.2 *Decision Rules*

Our goal is to use the concept of rough sets to express uncertainty in segmentation tasks. The framework we are using rough sets in is that of supervised learning. Thus, we are interested in learning from a given training set how we can classify the entire universe. To do so, we derive decision rules that describe a classification.

In the context of supervised learning, the task is to estimate sets of the entire universe that are conform with the attribute values of the training set. These sets can often be described by a few rules to a subset of the attributes. Hence, the goal of formulating decision rules is to identify a minimum number of rules that separate the classes. Those attributes that are necessary for defining the rules are called the reduct. When the reducts are found, the largest part of the generation of the rules is done. To transform the reduct into rules, one first has to bind the condition feature values of the object class from which the reduct originates to the corresponding features of the reduct. Then, to complete the rule, a decision part comprising the resulting part of the rule is added. This is done in the same way as for the condition features. These rules are then applied to the entire universe and define a classification. How we generate the classifiers for the segmentation of multichannel medical imaging data, is described in Sect. 4.

3 Related Work

Many different approaches to the segmentation of multichannel imaging data exist. A big group among those approaches are those that operate on the multidimensional feature space and try to establish a separation of classes. To be mentioned are the approaches based on median cut [8] and k-means clustering [13]. These approaches typically impose a partition of the universe into disjoint classes. When modeling uncertainty, we want to "soften" the boundary of that classes in the sense of rough set theory. A famous approach that imposes a similarly "rough" segmentation is that of fuzzy c-means clustering [2, 5]. The main disadvantage of fuzzy c-means clustering is that the imposed clusters are convex. Hence, it is impossible to separate interlocked non-convex clusters, even if they can easily be separated by non-convex clusters. Using rough set theory we are able to define a minimum number of rules that separate any clusters without inconsistencies. Even inconsistencies can be handled though, as the respective objects lie in the boundary regions.

In this paper, we will use the k-means clustering approach in two ways: When the boundary regions are non-empty, we either depict them as uncertain or, if the user desires, we enforce a decision by mapping the objects in the boundary regions to the more likely choice (positive or negative region). This decision can be based on the c-means clustering idea, see Sect. 4.4. Also, we compare our segmentation results to those obtained by the standard k-means approach. in addition, we compare our segmentation results to those obtained by fuzzy c-means clustering. Therefore, we would like to explain those two approaches in more detail.

The main idea of the k-means algorithm [13] is to define k centroids in the multidimensional feature space, one for each cluster. These centroids should be placed as far from each other as possible. Then, each object of the universe is associated with the nearest centroid, where the distances are measured in the multidimensional feature space. When all the points have be assigned centroids, the k centroids are recalculated as the average centers of each cluster. Then a new binding has to be done between the same data set points and the nearest new centroid. This process is iterated until no more changes are happening or until a given maximum number of iterations has been reached. There are many variants to this general idea including different distance metrics and different strategies for finding the initial cluster assignment. Lingras [12] proposed an approach to incorporate the rough set concept into a k-means clustering approach by using lower and upper approximations. The centroid computations, then, take into consideration whether an object lies in the positive or boundary region. If the boundary region is empty, the approach reduces to the standard k-means approach. A similar extension was proposed by Sushmita [24] in form of an evolutionary rough c-means clustering algorithm. Genetic algorithms are employed to tune the threshold and relative importance of upper and lower approximations of the rough sets modeling the clusters. The approaches by Lingras [12] and Sushmita [24] have not been applied to image segmentation tasks yet. Although they are introducing the concept of rough sets, they still only extract convex clusters. Note that this is different from what we propose in Sect. 4.4, where we only use k-means within boundary regions of possible non-convex clusters (and only when enforcing a decision is desired). Another major difference between our approach and the approaches by Lingras [12] and Sushmita [24] is that we explicitly derive the boundary regions to use them for uncertainty visualization.

The idea of the fuzzy c-means method [2, 5] is similar to the k-means approach, but it allows one data point to belong to two or more clusters. Fuzzy partitioning is carried out through an iterative optimization of the data points membership in the clusters and the corresponding update of the cluster centers. Again, the iteration terminates when there is no difference between the membership results with respect to some given precision. It is based on the minimization of the objective function

$$J_m = \sum_{i=1}^{N} \sum_{j=1}^{C} u_{ij} ||x_i - c_j||^2, \qquad 1 \leq m \leq \infty ,$$

where m is a real number larger than 1, u_{ij} is the degree of membership of x_i to the jth cluster, x_i is the ith object of the d-dimensional measured data, c_j is the d-dimensional centroid of the jth cluster, and $||.||$ denotes any norm expressing the similarity between any measured data objects and the centroid. In each iteration the membership degree u_{ij} and the cluster centroids c_j are updated by

$$u_{ij} = \frac{1}{\sum_{j=1}^{C} \left(\frac{||x_i - c_j||}{||x_i - c_k||} \right)^{\frac{2}{m-1}}}, \qquad c_j = \frac{\sum_{i=1}^{N} u_{ij}^m x_i}{\sum_{i=1}^{N} u_{ij}^m}.$$

More recently, extensions to the standard clustering approaches like k-means or fuzzy c-means have been developed that incorporate spatial information [1, 3, 4, 11]. We incorporate information about the spatial neighborhoods of a pixel into our approach by including statistical measures that are derived in object space and combined with the feature space attributes, see Sect. 5.

Rough set theory has been applied to image processing before, cf. [6]. In this context, the objects represent pixels and the equivalence classes define a partition. Often, the segmentation task is reduced to a certain region of interest [9]. The idea is that boundaries between object regions within the image are often ill-defined because of uncertainties (such as partial volume effect) that arise during acquisition and processing. This uncertainty can be handled by describing the different objects as rough sets with upper and lower approximations [14, 15]. Hirano and Tsumoto [9] and Shoji and Shusaku [7] use rough representation of regions of interest in medical images to handle inconsistencies between the knowledge-driven shape and the image-driven shape using the respective approximations. Some approaches combine rough set representations with other classifiers. Swiniarski and Hargis [25] use rough sets as a front-end of neural-network-based texture image recognition, and Jiang et al. [26] proposed a rough neural network approach that integrates neural networks with reduction of rough set theory to classify digital mammography. The approaches by Hirano and Tsumoto [9] and by Shoji and Shusaku [7,9] are the ones that are closest to out work. While they focus on detecting a single region of interest, i.e., one positive region, our approach also allows for the segmentation of the entire medical imaging data in the sense of partition. Moreover, our approach combines feature- with object-space considerations and deals with multichannel data. Finally, we derive uncertainty measures that we use for uncertainty visualization.

4 Rough Classification

Given an information system according to Sect. 2, we can apply rough set theory to compute lower and upper approximations as well as negative and positive regions and boundaries. The computational step that needs to be taken for a rough classification is the formulation of rules that serve as a classifier. Such rules are based on the concept of best cuts.

4.1 Best Cuts

Let $A = (U, A \cup \{d\})$ be an information system with conditional features A and decision feature d. For the decision feature d we can assume that it distinguishes between m different classes and we assign to those classes the labels $1, \ldots, m$. Let $a \in A$ be a conditional attribute. It induces a sequence $v_1^a < v_2^a < \cdots < v_{n_a}^a$, where $\{v_1^a, v_2^a, \cdots, v_{n_a}^a\} = \{a(x) : x \in U\}$ are values of the attribute a with n_a denoting the number of different values that appear for attribute a. Then, the set of all possible cuts on attribute a is given by

$$C_a = \left\{\left(a, \xi_1^a\right), \left(a, \xi_2^a\right), \ldots, \left(a, \xi_{n_a-1}^a\right)\right\}$$

with $\xi_i^a \in \left(v_i^a, v_{i+1}^a\right)$ for $i = 1, \ldots, n_a - 1$. The set of possible cuts on all attributes is denoted by $C_A = \bigcup_{a \in A} C_a$.

The idea for finding the best cut is to search for a cut $(a, c) \in C_A$ that discerns the largest number of pairs of objects. Let $D = \{(x, y) \in U \times U : d(x) \neq d(y)\}$. Then, we compute for all $(a, c) \in C_A$ and $X \subseteq D$ the discernibility number

$$W^X(a,c) = L^X(a,c) \cdot R^X(a,c) - \sum_{i=1}^{m} l_i^X(a,c) \cdot r_i^X(a,c),$$

where

$$L^X(a,c) = card\{x \in X : a(x) < c\}$$

and

$$R^X(a,c) = card\{x \in X : a(x) > c\}$$

refer to the number of objects from X below or above the cut (a, c), respectively, and

$$l_j^X(a,c) = card\left\{x \in X : [a(x) < c] \bigwedge [d(x) = j]\right\}$$

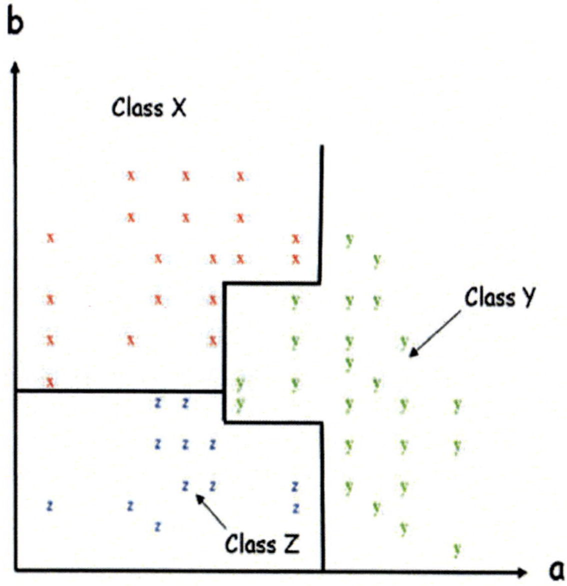

Fig. 2 Best cuts for a two-dimensional attribute space with three classes

and

$$r_j^X(a,c) = card\left\{x \in X : [a(x) > c] \bigwedge [d(x) = j]\right\}$$

refer to the number of objects from X belonging to the jth decision class and being below or above the cut (a,c), respectively. We choose the cut (a,c) with the largest discernibility number $W^X(a,c)$ and remove all the discerned pairs from D. Moreover, the chosen cut is removed from the candidate list of possible cuts C_A. This process is iterated until all pairs of objects in D are discerned, i.e., $D = \emptyset$. In case $D \neq \emptyset$ and no more cuts are left, the data is incomplete and more training data samples need to be added.

Figure 2 shows an example for cuts in a two-dimensional attribute space. The cuts separate all the pairs of object that are distinguishable within this two-dimensional attribute space. The best cuts are optimal with respect to the discernibility number and try to keep the number of cuts as small as possible. Note that the classes that are specified by a conjunction of the cuts can be non-convex. The conjunctions are defined by the classification rules, see Sect. 4.2.

Example. Consider the simple example with 10 objects, 3 attributes, and 3 decision classes in Table 1. Moreover, let us use the values $\xi_i^a = \dfrac{v_i^a + v_{i+1}^a}{2}$ for cuts in attribute $a \in A$ $(i = 1, \ldots, n_a - 1)$. We find that the cut $(a_3, 138)$ divides the set of objects into two subsets $U_1 = \{u_1, u_3, u_4, u_5, u_9, u_{10}\}$ and $U_2 = \{u_2, u_6, u_7, u_8\}$. Then, if the algorithm chose the best cuts locally on U_1 and U_2, the cuts $(a_1, 134)$, $(a_2, 134)$, and $(a_2, 142)$ would be best cuts for U_1 (discerning 6 pairs of objects from U_1) and the cuts $(a_1, 142)$ and $(a_3, 146)$ would be best cuts for U_2 (discerning 4 pairs of objects from U_2). However, we search for the best cut globally computing

$W^X(a, c)$ for each cut for the two subsets U_1 and U_2. We determine $(a_2, 134)$ to be the best one, as it discerns 10 pairs of objects. It divides U_1 into $X_1 = \{u_1, u_9, u_{10}\}$ and $X_2 = \{u_3, u_4, u_5\}$ and U_2 into $X_3 = \{u_7, u_8\}$ and $X_4 = \{u_2, u_6\}$. X_1, X_3, and X_4 each are in one decision class, i.e., no more cuts necessary. X_2 is further split by cut $(a_1, 134)$ into $\{u_4, u_5\}$ and $\{u_3\}$. For the former we find the last cut $(a_2, 146)$. So, the best cuts are the presented 4 from the 15 available cuts.

4.2 *Classification Rules*

Having derived the best cuts, we know the reducts. Let R be the set of reducts. Now, we look into all objects and set up causalities for the reducts as shown in Table 2. The union of all such rules can be expressed as a few if-then statements. Such if-then statements can, then, be efficiently used to classify a large set of objects.

Table 1 Decision table

		A		
U	a_1	a_2	a_3	d
u_1	124	128	132	0
u_2	128	140	140	1
u_3	132	148	124	2
u_4	132	144	124	1
u_5	136	144	132	0
u_6	140	144	140	1
u_7	144	124	152	2
u_8	148	152	152	2
u_9	148	124	124	0
u_{10}	152	124	124	0

Table 2 Rule generation

Algorithm 1: Rule generation

Input: Information system $(U, C \cup D)$ and reducts $R = \{a_1, ...a_m\} \subseteq C$
Output: Set of decision rules $RULES(R)$
1: for $u \in U$ do
2: for $a_i \in R$ do
3: $v_i = a_i(u)$;
4: end for
5: $v_d = d(u)$;
6: $RULES(R) = RULES(R) \cup a_1 = v_1 ... a_m = v_m \Longrightarrow d = v_d$;
7: end for
8: Return $RULES(R)$;

Table 3 Classification rules

If:	$\Longrightarrow$ *Class*
$(-Inf < a_2 < 134.0) \wedge (-Inf < a_3 < 136.0)$	$\Longrightarrow 0$
$(134.0 < a_1 < Inf) \wedge (-Inf < a_3 < 136.0)$	$\Longrightarrow 0$
$(134.0 < a_2 < 146.0) \wedge (136.0 < a_3 < Inf)$	$\Longrightarrow 1$
$(-Inf < a_1 < 134.0) \wedge (134.0 < a_2 < 146.0)$	$\Longrightarrow 1$
$(146.0 < a_2 < Inf)$	$\Longrightarrow 2$
$(-Inf < a_2 < 134.0) \wedge (136.0 < a_3 < Inf)$	$\Longrightarrow 2$

Example. For the example from above, the reducts are all conditional features and the union of all rules can be expressed in form of the if-then statements shown in Table 3.

4.3 Uncertainty Visualization

The generated rules define rules that are based on the choice of suitable cuts (a, ξ_i^a) with a being an attribute of the reduct and $\xi_i^a \in (v_i^a, v_{i+1}^a)$, see Sect. 4.1. The choice of ξ_i^a leaves some flexibility within the interval (v_i^a, v_{i+1}^a). Assuming that we want to define the set $X \subset U$ of objects that lies left to cut (a, ξ_i^a), choosing $\xi_i^a = v_i^a$ gives a lower approximation of X (positive region), while choosing $\xi_i^a = v_{i+1}^a$ gives an upper approximation of X (its complement is the negative region). The interval (v_i^a, v_{i+1}^a) defines the boundary region. The boundary region represents the area of uncertainty.

If we perform this consideration for each segment X of the image and for all cuts that are used for the classification rules, we derive positive regions for each segment X surrounded by boundary regions that express the uncertainty at the transition between two regions. We visualize the uncertainty for each segment by color coding positive and boundary regions.

4.4 K-Means Segmentation

When applying the rough classification to segmentation of imaging data, we get a partitioning that includes areas of uncertainty. If we want to induce a decision for pixels in the uncertain region or if we want to compute a probability for those pixels belonging to a certain class (or segment), we propose to couple our rough classification with k-means clustering.

To do so, we compute the means for each attribute in each rule. For those objects x that are in the uncertain regions, we compute the distances

$$d(x, k_l) = \sqrt{\sum_{j=1}^{m}(x_j - k_{lj})},$$

where x_j denotes the jth feature value of object x and k_{lj} denotes the mean value of the jth feature for the lth class. Then, object x can be assigned to the nearest matching rule with respect to the defined distance.

The same approach can be used in case of inconsistencies, i.e., when there are more than one matching rule for an object. The defined distances represent votes. Every matching rule contributes votes to its decision value. The votes are added and the class with the largest number of votes is chosen. Quality measures such as support, strength, accuracy, and coverage (cf. [10]) associated with the decision rules can be used to eliminate some of the decision rules.

5 Combined Feature-/Object-Space Segmentation

When applying the rough classification approach to multichannel medical images, we can rely on the fact that only the reducts are used to compute the rules. Hence, we can add as many derived properties as we want without affecting significantly the number of rules. As a proof of concept, we apply our method to RGB color data. We further enhance the feature space by converting the RGB colors to the CIE L*a*b* color space and adding those 3 channels to the RGB channels. As a pure feature space segmentation of medical images is not desirable, we use statistical measures to correlate the RGB colors to some spatial properties. The statistical measures compute values in the neighborhood of each pixel. We use the 8 surrounding pixels of each pixel to compute mean, variance, standard deviation, mode, and median for the RGB values. These statistical measures are added to the feature vector. Together with the RGB and the CIE L*a*b* channels, the statistical measures formulate a 21-dimensional feature space. This are the conditional features we use for rough classification.

One additional nice feature of using this 21-dimensional feature space is that we barely observed any inconsistencies, even when assigning to one and the same RGB color different classes. If there really happens to appear inconsistent data, our system replaces these inconsistent data with the mode of the respective class.

6 Interactive System

We use a user-based interactive framework to apply the rough classification approach to multichannel medical imaging data. This semi-automatic approach provides the user with a view of a 2D slice of the data set within a graphical user interface. Within that slice, the user can interactively select certain regions and assign them to a certain class. These regions form the training set, which are used to automatically build the rough classification. Then, the segmentation decisions are automatically made for all pixels of the 2D slice. The segmentation result can be further refined by interactively selecting further regions and assigning them to the respective classes. Our system provides a visualization of the segmentation results as well as uncertainty visualizations for all classes. Hence, if a certain class represents a region of interest, the uncertainty in the segmentation result of that region becomes apparent. The segmentation decisions can also be transferred to other slices or to the entire 3D data set. Because of the simplicity of the rules, our system remains interactive. Of course, one can also refine the result by interacting with other slides.

7 Results and Discussion

7.1 *Quality Measures*

In order to judge the quality of our segmentations, we have to define quality measures. When using synthetic data, the ground truth is known and we can compute the segmentation accuracy

$$S_A = \frac{M_P}{T_{NP}},$$

where M_P is the number of misclassified pixels and T_{NP} is total number of pixels. For real data, one may use other measures such as compactness (low within-class variance) and isolation (high distance between class centroids).

7.2 *Comparison to K-Means and Fuzzy C-Means*

We have applied our approach to the synthetic images shown in Figs. 3 and 4. The images are simple examples of size 300×100 and 400×400 pixels, to which we added 60% noise using the GIMP noise generator. We compare our approach to the results obtained by k-means and fuzzy c-means, where all three approaches have been applied to the 21-dimensional feature vector. All three approaches are initialized with the user-defined classes. In particular, the initial centroids that are used for the k-means and fuzzy c-means approach are computed by the user-given classification. The results are, again, shown in Figs. 3 and 4, respectively. The first observation is that the combined feature-/object-space approach supported the reconstruction of the original regions. Moreover, we can observe that our rough classification obtains the best result. This is also documented in Table 4, where the segmentation accuracies for the 3 approaches and the 2 images are listed. Rough classification achieves about 99% accuracy, k-means around 93%, and fuzzy c-means about 95%. In terms of speed, all three approaches were fast enough to allow for an application in an interactive system.

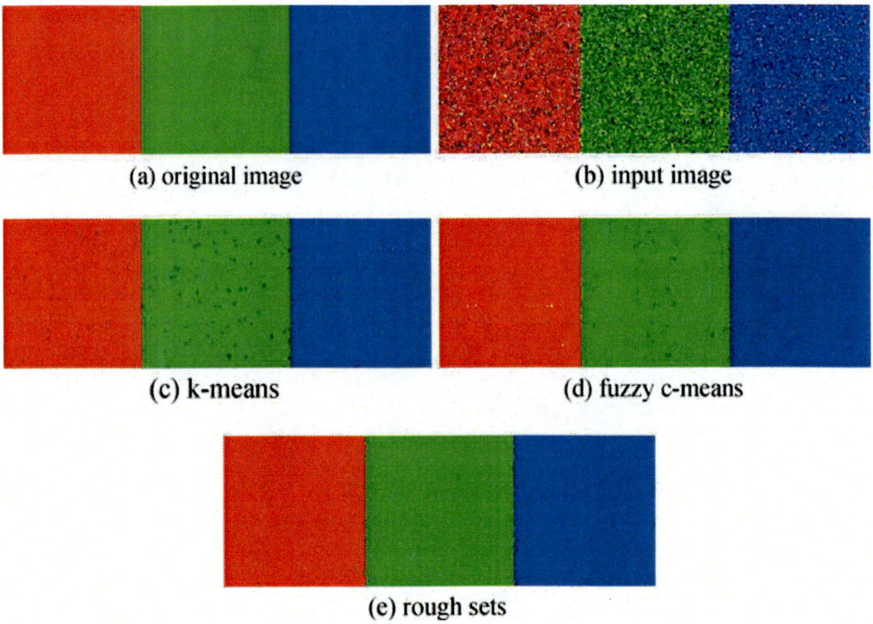

Fig. 3 Comparison of segmentation result of our rough classification approach to k-means and fuzzy c-means when applied to a synthetic image with added noise

7.3 Multichannel Medical Image Segmentation

We applied our semi-automatic system to the segmentation of multichannel medical image segmentation. The data set we used are cryosections of a human brain of resolution $336 \times 411 \times 100$[1]. We have chosen the 2D slice depicted in Fig. 5 for assembling the training set. A few pixels from that slice have been selected to interactively choose six different classes. The chosen regions are shown in Fig. 5. We established the 21-dimensional feature space and applied our segmentation based on rough classification. Figure 6a shows the result. The colors for each segment can be chosen interactively. Given the few samples that have been provided, the results are very reasonable. Next, we took the rules we obtained from the first slice and assigned them to other slices. Figures 6(b–d) show the slices and the respective segmentations.

We also compute the class accuracy (not to be confused with the segmentation accuracy) as the quotient of the number of pixels in the lower approximation and the number of pixels in the upper approximation. In Table 5, we give the accuracy numbers for all 6 classes and 6 different slices. We also derive a quality term $Q(I)$ for the segmentation by combining the accuracy numbers for the 6 classes.

[1]Data courtesy of Art Toga, UCLA.

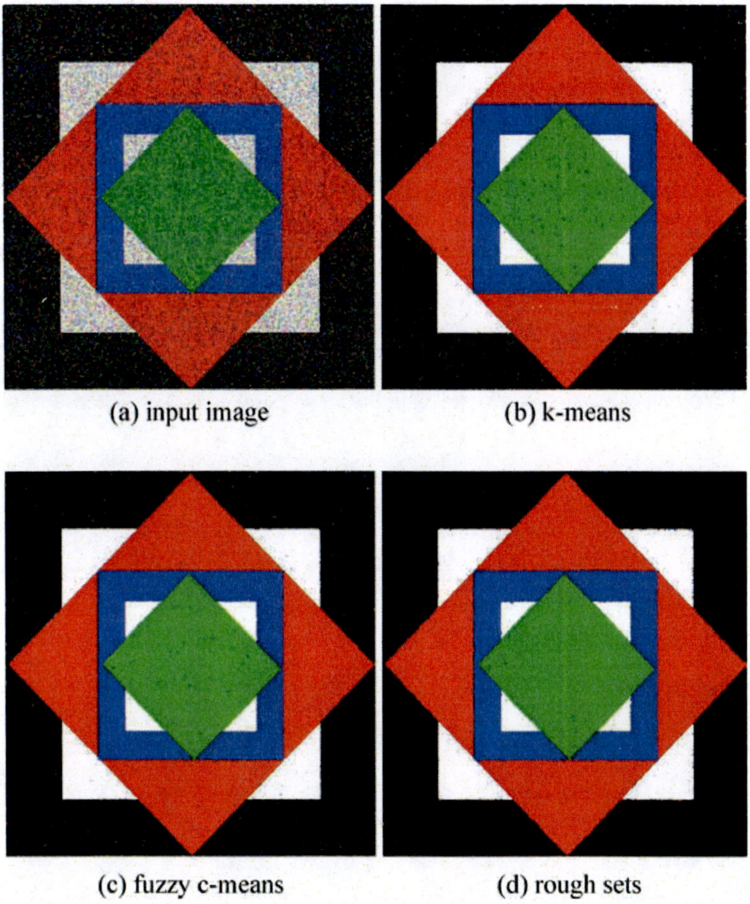

Fig. 4 Comparison of segmentation result of our rough classification approach to k-means and fuzzy c-means when applied to a synthetic image with added noise

Table 4 Comparison of segmentation accuracy of our rough classification approach to k-means and fuzzy c-means

	Rough sets	k-means	Fuzzy c-means
Figure 3	.998	.924	.951
Figure 4	.988	.933	.948

7.4 Uncertainty Visualization

The class accuracy in Table 5 is a measurement that documents the uncertainty in the segmentation for each of the classes. In order to visualize the uncertainty, we compute for each pixel the probabilities that it belongs to each class. In Table 6, we give respective numbers for all the samples of a 2D slice and all 6 classes. The

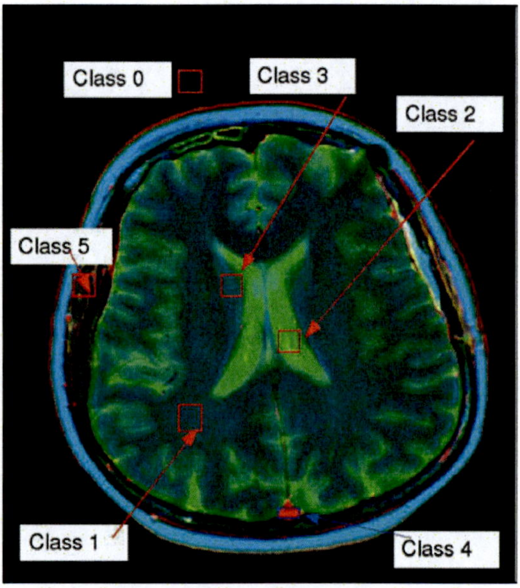

Fig. 5 Assigning the training set by interactively selecting image regions and respective classes

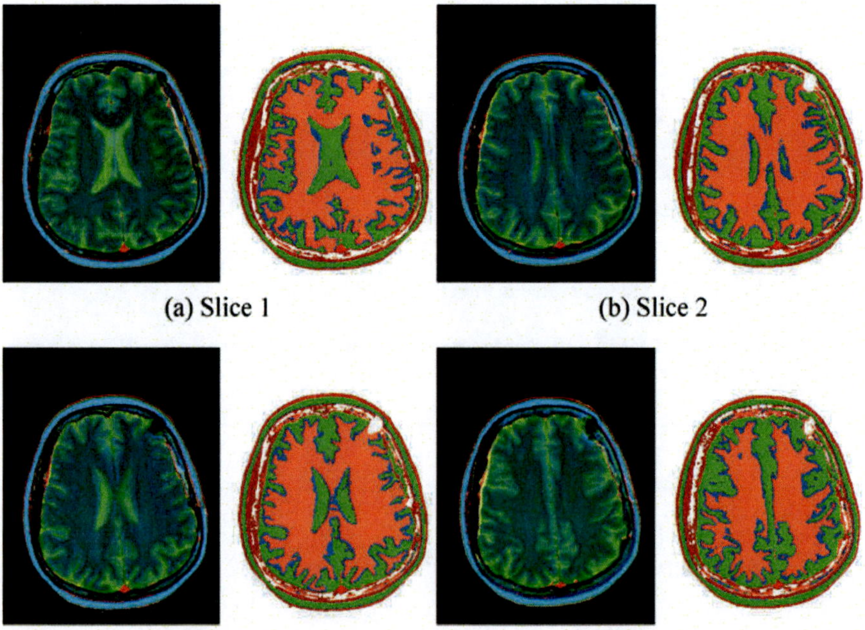

(a) Slice 1 (b) Slice 2

(c) Slice 3 (d) Slice 4

Fig. 6 Multichannel medical imaging data segmentation based on rough classification

Table 5 Class accuracy and respective segmentation quality

	Class						
Slice	0	1	2	3	4	5	Q(I)
1	0.998	0.934	0.963	0.799	0.969	0.812	0.955
2	0.988	0.943	0.961	0.799	0.918	0.847	0.957
3	0.99	0.938	0.962	0.801	0.906	0.854	0.956
4	0.985	0.931	0.958	0.834	0.913	0.913	0.956
5	0.989	0.921	0.96	0.851	0.908	0.850	0.958
6	0.978	0.90	0.964	0.847	0.917	0.848	0.956
Average	0.988	0.928	0.961	0.822	0.922	0.854	0.956

Table 6 Percentage of pixels assigned to the respective class which have probabilities 100%, 75–100%, 25–75%, and 0–25%. The last column gives the least probability of a sample assigned to the respective class

	Positive		Boundary		
	100%	75–100%	25–75%	0–25%	Least probability
Class 0	0.999983	0.000016574	0	0	0.854422
Class 1	0.99098	0.000415694	0.00860167	0	0.498528
Class 2	0.970603	0.000171409	0.0292252	0	0.36323
Class 3	0.932032	0.000289226	0.067679	0	0.399448
Class 4	0.997379	0	0.00262054	0	0.512198
Class 5	0.891966	0.00461506	0.103419	0	0.391854

numbers state, what percentage of the pixels assigned to the respective class have probabilities 100%, 75-100%, 25-75%, and 0-25%. It also gives the least probability of a pixel assigned to the respective class. We use these probabilities to partition the samples into three regions with respect to each class: The positive region is that with probabilities >75% and the boundary that with probabilities 0–25%. All other pixels belong to the negative region. We display the uncertainty for each class individually by showing color-coded 2D slices, where green depicts the positive region, yellow the boundary, and red the negative region. Hence, the yellow areas depict uncertain regions. Figure 7 shows the uncertainty visualization of one slice and three different classes.

8 Conclusions and Future Work

We have presented an approach for semi-automatic rough segmentation of multichannel imaging data. Our approach uses rough set theory to compute lower and upper approximation and to define classification rules based on best cuts. The segmentation is obtained using a k-means approach on the rough classification. The segmentation results have higher accuracy than the ones obtained by the standard k-means approach or the fuzzy c-means approach. The underlying feature vector integrates a combined feature-/object-space approach using statistical measures,

(a) Class 1 (b) Class 2 (c) Class 3

Fig. 7 Uncertainty visualization for 3 different classes: green represents positive region, yellow boundary, and red negative region

which allowed us to segment connected areas and distinguish separated areas. We computed uncertainty and visualized the uncertainty for the different classes using color-encoded 2D slices. A possible extension for future work is to develop 3D uncertainty visualizations that are as intuitive and precise as our 2D examples.

References

1. Mohamed N. Ahmed, Sameh M. Yamany, Nevin Mohamed, Aly A. Farag, and Thomas Moriarty. A modified fuzzy c-means algorithm for bias field estimation and segmentation of mri data. *IEEE Transactions on Medical Imaging*, 2002.
2. J. C. Bezdek. *Pattern Recognition with Fuzzy Objective Function Algorithms*. Plenum Press - New York, 1981.
3. Songcan Chen and Daoqiang Zhang. Robust image segmentation using fcm with spatial constraints based on new kernel-induced distance measure. *IEEE Transactions on Systems, Man, and Cybernetics*, 2004.
4. Keh-Shih Chuang, Hong-Long Tzeng, Sharon Chen, Jay Wu, and Tzong-Jer Chen. Fuzzy c-means clustering with spatial information for image segmentation. *Computerized Medical Imaging and Graphics*, 2006.
5. Dunn, J.C.. A Fuzzy Relative of the ISODATA Process and its Use in Detecting Compact, Well Separated Clusters, *J. Cyber.*, 2004, 3, 32–57.
6. Hassanien, Aboul Ella, Abraham, Ajith, Peters, James F., and Kacprzyk, Janusz. Rough Sets in Medical Imaging: Foundations and Trends. *Computational Intelligence in Medical Imaging: Techniques and Applications*, G. Schaefer, A. Hassanien, J. Jiang, Eds. 2009.
7. Hirano S., and Tsumoto S., Segmentation of Medical Images Based on Approximations in Rough Set Theory, Rough Sets and Current Trends in Computing., *Third International Conference, RSCTC 2002, Malvern, PA, USA*, 2002, 950–951.
8. P. Heckbert, Color image quantization for frame buffer display, *Computer Graphics (Proceedings of ACM SIGGRAPH 82)*, 297–307.
9. Hirano S. and Tsumoto S., Rough representation of a region of interest in medical images., *International Journal of Approximate Reasoning*, (2005), vol. 40, 2334.

10. Komorowski J., Pawlak Z., Polkowski L. and Skowron A. Rough sets: A tutorial. In S. K. Pal and A. Skowron, editors, Rough FuzzyHybridization. A New Trend in Decision-Making. *Springer-Verlag*. 1999, 398.
11. A.W.C.Liew, S.H.Leung, and W.H.Lau. Fuzzy image clustering incorporating spatial continuity. *IEE Proc.-Vis. Image Signal Process*, 2000.
12. Lingras P. Applications of rough set based k-means,kohonen, ga clustering. *Transactions on Rough Sets*, 7:120–139., 2007.
13. J. B. MacQueen. Some methods for classification and analysis of multivariate observations. *Proceedings of 5-th Berkeley Symposium on Mathematical Statistics and Probability*, 1967, volume 1, 281–297.
14. Pal, Sankar K. (2001) Fuzzy image processing and recognition: *Uncertainties handling and applications, Internat. J. Image Graphics* 1 (2001) (2), 169–195.
15. Pal, Sankar K., B. Uma Shankar, Pabitra Mitra: Granular computing, rough entropy and object extraction. *Pattern Recognition Letters* 26(16): 2509–2517, 2005.
16. Pawlak Z. Classification of Objects by Means of Attributes. *Institute for Computer Science, Polish Academy of Sciences*. 1981 Report 429.
17. Pawlak Z. Rough sets. *International J. Comp. Inform. Science*. 1982 vol. (11), (341–356).
18. Pawlak Z. Rough sets: Theoretical aspects of reasoning about Data Dordrecht, *Kluwer Academic Publishers.*, 1991.
19. Pawlak Z. Grzymala-Busse J. Slowinski R. and Ziarko W. Rough Sets. *Communications of the ACM*. 1995, vol. 38(11), (88–95).
20. Pawlak Z. and Skowron A. Rudiments of rough sets. *Information Sciences*. 2007 vol. 177, 3–27.
21. Peters, J.F., Pedrycz, W. Rough Sets: *Mathematical Foundations. Physica-Verlag*. 2002.
22. Peters, J.F. Classification of Perceptual Objects by Means of Features. *International Journal of Information Technology and Intelligent Computing*. 2007.
23. Peters, J.F., Pedrycz, W. Computational intelligence, in Electrical and Electronics Engineering Encyclopedia. *NY: John Wiley and Sons, Ltd.*, 2008.
24. Sushmita M. An evolutionary rough partitive clustering., *Pattern Recognition Letters.*, 2004, volume 25, 1439–1449.
25. Swiniarski, Roman W., and Larry Hargis., Rough sets as a front end of neural-networks texture classifiers., *Neurocomputing.*, 2001, volume 36(-4), 85–102.
26. Yun J., Zhanhuai L., Yong W. and Longbo Z., A Better Classifier Based on Rough Set and Neural Network for Medical Images., *Proceedings of the Sixth IEEE International Conference on Data Mining*, 2006, 853–857.

An Evaluation of Peak Finding for DVR Classification of Biological Data

Aaron Knoll, Rolf Westerteiger, and Hans Hagen

Abstract In medicine and the life sciences, volume data are frequently entropic, containing numerous features at different scales as well as significant noise from the scan source. Conventional transfer function approaches for direct volume rendering have difficulty handling such data, resulting in poor classification or undersampled rendering. Peak finding addresses issues in classifying noisy data by explicitly solving for isosurfaces at desired peaks in a transfer function. As a result, one can achieve better classification and visualization with fewer samples and correspondingly higher performance. This paper applies peak finding to several medical and biological data sets, particularly examining its potential in directly rendering unfiltered and unsegmented data.

1 Introduction

Direct volume rendering (DVR) is a ubiquitous method for visualizing 3D raster data, including medical and biological scan data. Volume data in the life sciences are often noisy and contain features at numerous different scales. This poses difficulties for many classification schemes and automatic transfer function generators. While gradient-based 2D transfer functions [8] deliver clear improvements, they can be difficult to create and manipulate compared to 1D approaches. As a result, high-quality DVR classification methods are often abandoned in favor of more approximate methods such as unshaded grayscale maximum-intensity projection (MIP), or simple isosurface rendering.

Similarly, when rendering discrete isosurfaces within a volume, the preferred method is often to specify the isosurfaces explicitly, or even extract them separately

A. Knoll (✉) · R. Westerteiger · H. Hagen
University of Kaiserslautern, Germany,
e-mail: knoll@rhrk.uni-kl.de; rwester@informatik.uni-kl.de;hagen@informatik.uni-kl.de

L. Linsen et al. (eds.), *Visualization in Medicine and Life Sciences II*,
Mathematics and Visualization, DOI 10.1007/978-3-642-21608-4_6,

as geometry, rather than to embed them directly in a transfer function. While it handles the most glaring artifacts resulted from point-sampled postclassification, preintegrated classification [4] must still sample adequately with respect to the Nyquist frequency of the data, and omits features when it does not.

Peak finding [10] is a recently-proposed classification scheme that bridges DVR classification and discrete isosurfacing. Rather than sampling uniformly along the ray, peak finding explicitly solves for peaks in the transfer function at their corresponding isovalues in the data field. By sampling and shading directly at these peaks, this method achieves better results than postclassification or preintegration when a transfer function is sufficiently sharp, particularly in the case of Dirac peaks. More significantly, in the case of entropic data, peak finding can successfully identify features at peaks that are omitted by standard methods, even when the chosen transfer function is modest. While the original paper detailing the peak finding algorithm revealed the general appeal of this method in handling noisy data, it only examined one biological data set.

This paper investigates the merits of peak-finding in noisy medical and biological volume data. In particular, we are interested in the application of peak finding as an exploratory classification early on in the data analysis pipeline, especially in visualizing unsegmented and unfiltered data directly from scan source. While peak finding is clearly useful in some scenarios, we are careful not to oversell its merits. The goal of this work is to reveal the strengths as well as limitations of peak-finding as an exploratory classification technique.

2 Related Work

Levoy [13] employed ray casting in the first implementation of direct volume rendering. The advent of z-buffer hardware and built-in texture interpolation units allowed for interactive performance with slice-based rasterization approaches [2, 3]. While slower per-sample than slicing, volume ray casting methods are feasible on programmable GPU hardware [11, 22] and currently represent the state-of-the-art [21].

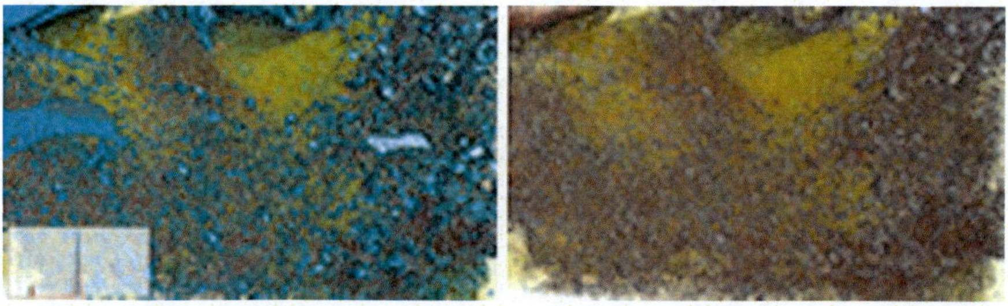

Fig. 1 Zoomed-out zebrafish optic tract rendered with peak finding (left) and preintegration (right), at 3.0 and 3.4 fps, respectively with a 1440×720 frame buffer

Isosurface rendering has conventionally been accomplished via mesh extraction using marching cubes [14] or more sophisticated offline methods [24]. Volume-isosurface ray casting methods were first conceived by Sramek [26]. Parker et al. [20] implemented a tile-based parallel ray tracer and achieved interactive rendering of isosurfaces from large structured volumes, employing a hierarchical grid as a min-max acceleration structure and an analytical cubic root solving technique for trilinear patches. Hadwiger et al. [6] combined rasterization of min-max blocks with adaptive sampling and a secant method solver to ray cast discrete isosurfaces on the GPU. Peak finding [10] is inspired by this approach in that it uses discrete sampling to isolate roots.

A large body of volume rendering literature deals with transfer functions, both in how to construct them and employ them in classification. To limit artifacts when sampling high-frequency features of a transfer function, the best existing approaches are preintegration [4, 15, 23] and analytical integration of specially constructed transfer functions [9]. Hadwiger et al. [5] analyze the transfer function for discontinuities to generate a pre-compressed visibility function employed in volumetric shadow mapping. Our approach is similar except that we search for local maxima, and use these directly in enhancing classification.

Higher-order filters have previously been the subject of study for structured [17, 19] and unstructured [12] volumes. In our higher-order reconstructions, we employ a method similar to [25], in which a 4-point stencil filter is efficiently reconstructed on the GPU using two linearly interpolated fetches. We also experiment with more rigorous filters which are processed offline, namely the anisotropic diffusion filter proposed by Weikert [27].

3 Background

In direct volume rendering, irradiance I is computed as a discrete approximation of the radiative light transport equation through a continuous scalar field. On a segment of a ray, this is given by

$$I(a,b) = \int_a^b \rho_E(f(s))\rho_\alpha(f(s))e^{-\int_a^s \rho_\alpha(f(t))dt}ds \tag{1}$$

where ρ_E is the emissive (color) term and ρ_α is the opacity term of the transfer function; a, b are the segment endpoints, and $f(t) = f(\mathbf{O} + t\mathbf{D}) = f(\mathbf{R}(t))$ is the scalar field function evaluated at a distance t along the ray. Computing this integral entails approximating it with discrete samples. With uniformly spaced sampling, we break up the ray into equally spaced segments and approximate the opacity integral as a Riemann Sum,

$$e^{-\int_a^s \rho_\alpha(f(t))dt} = \prod_{i=0}^{n} e^{-\Delta t\ \rho_\alpha(f(i\ \Delta t))} = \prod_{i=0}^{n}(1-\alpha_i) \tag{2}$$

where Δt is the uniform sampling step, $n = (s - a)/\Delta t$, and

$$\alpha_i \approx 1 - e^{-\Delta t \ \rho_\alpha(f(i \ \Delta t))} \tag{3}$$

By discretizing the integral on $[a, b]$ in (1) as a summation, we have the following discrete approximation for I,

$$I \approx \sum_{i=0}^{n} \rho_E(i) \prod_{j=0}^{i-1} (1 - \alpha_j) \tag{4}$$

where $\rho_E(i) = \rho_E(f(i\Delta t))$ is given by the transfer function. Evaluating the transfer function after reconstruction is known as postclassification. While it is equally possible to classify before filtering, preclassification results in worse reconstruction of the convolved scalar field and is unnecessary on current hardware.

3.1 Classification Methods

Postclassification (Fig. 2a) usually delivers adequate sampling when the transfer function and data field are both sufficiently smooth. To accomplish this, one generally requires a sampling rate beneath the Nyquist limit of this convolved signal. In rendering high-frequency features, and particularly discrete isosurfaces, postclassification with a uniform step size will invariably fail. Theoretically, an isosurface is a subset of a DVR transfer function, consisting of a discrete Dirac impulse at that isovalue. To visualize this surface in a volume renderer, one must sample that impulse with an infinite (continuous) sampling rate. In practice, it is generally sufficient to specify fuzzy isosurfaces, sacrificing some classification precision in the interest of visual aesthetics.

Engel et al. [4] note that we can effectively sample at the minimum of either the transfer function or the data frequency. Preintegrated transfer functions build upon this notion by integrating separately over transfer function and scalar field domains (Fig. 2b). This allows isosurface-like features to be rendered more accurately using a lower sampling rate. However, preintegration still assumes the scalar field is sampled adequately. When the data itself is noisy, it is often impractical to sample at the Nyquist limit of the data for reasons of performance. As a result, both postclassification and preintegration can fail to reconstruct high-frequency features of interest when undersampling the scalar field. Moreover, preintegration assumes a piecewise integration over the transfer function domain between two samples, which does not correspond to the actual scalar field value along the ray, particularly when the source data is entropic. Although interpolating between endpoints can remove obvious shading artifacts [15], preintegration can still fail to reproduce features when the data domain is undersampled.

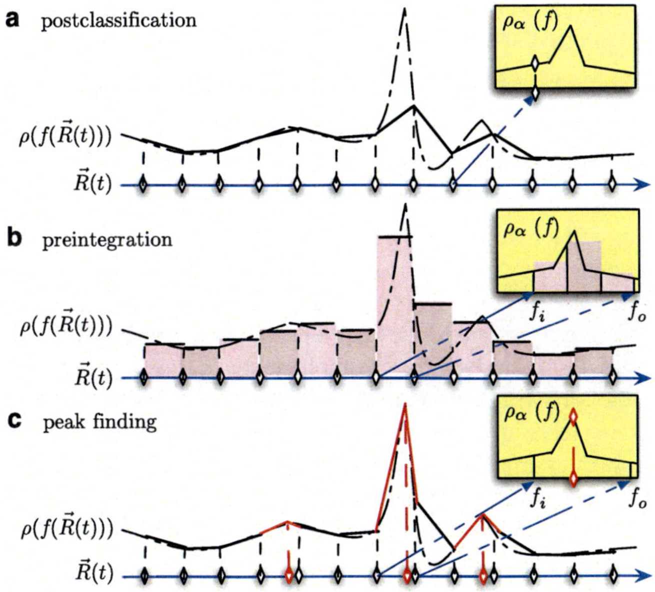

Fig. 2 Postclassification, preintegration and peak finding illustrated

Peak finding [10] performs a similar function as preintegration, but employs standard point-sampled postclassification when the transfer function frequency is low, and an explicit isosurface root-solving technique when the transfer function contains a peak on a given range. This consists of a precomputation step to identify peaks in the transfer function, and a root-solving method to solve for them during ray casting. As a result, sharp features in the transfer function are always approximated by discrete isosurfaces, which a ray will either hit or miss. In this sense, peak finding can be seen as a modification to postclassification, but using an adaptive sampling mechanism to identify sharp features as necessary (Fig. 2c).

Unlike preintegration, peak finding relies on Descartes' rule of signs to determine whether a (root) feature lies between two endpoints of a segment. With preintegration, the integral of ρ on $[a,b]$ is a piecewise summation of all ρ values between $f(a)$ and $f(b)$. For this to be accurate, f must vary smoothly between a and b; beyond general continuity it should be Lipschitz, e.g.

$$L(a,b) = \frac{|\rho(f(b)) - \rho(f(a))|}{b-a} < k \tag{5}$$

where $b - a = \Delta t$, and this sample rate is chosen according to k. Conversely, peak finding only requires that the scalar field be monotonic on $[a,b]$ to find and accurately reproduce the high-frequency sample. Clearly, when the data is noisy and sampled below the Nyquist limit, both conditions can fail. However, monotonicity is a far weaker requirement than Lipschitz continuity, and the success of peak finding is due in great part to this difference.

3.2 Peak Finding vs. Preintegration

Peak finding is attractive in that its algorithm is not significantly different from either volume rendering or isosurface ray casting. Both algorithms employ regular sampling, in the case of DVR to compute the volume rendering integral and in the case of isosurfacing to isolate roots. Peak finding takes advantage of this and does both. As a result, this technique can be implemented quickly by extending existing renderers. Although we propose peak finding in conjunction with differential sampling, the two techniques are orthogonal. It is equally possible to employ peak finding in a uniform sampling ray caster, a slice-based volume renderer, or a shear-warp system. Moreover, as described in Sect. 4.1, it is simple to extend an existing preintegration scheme to handle peak finding, with a small modification to the lookup table construction and classification algorithm.

Overall, peak finding and preintegration are similar, but make different assumptions about the integral over a given segment. Preintegration assumes this integral can be accurately approximated by piecewise summation. This works well when the transfer function and convolved field are smooth, but encounters difficulties when they are not. Peak finding assumes this integral can be approximated by one or several discrete impulses. This introduces bias, but is better suited for noisy data and sharp C_0 transfer functions for which standard techniques fail.

4 Peak Finding

This section discusses peak finding, which was originally proposed in [10]. We again note this is not a new contribution, but rather a reiteration of the technique that will be evaluated in more detail in the remainder of this paper.

Peak finding combines standard postclassification with an explicit ray-isosurface intersection in high-frequency points of interest at transfer function peaks. Ray-isosurface intersection consists of solving the reconstruction filter function $f(x, y, z) = \upsilon$ at an isovalue υ. Substituting the ray equation $\mathbf{R}(t) = \mathbf{o} + \mathbf{d}t$, we have a one-dimensional equation in terms of the ray parameter t:

$$f(x, y, z) - \upsilon = f(\mathbf{R}(t)) - \upsilon = 0 \tag{6}$$

In peak finding, we simply need to know the isovalue υ if and where a peak exists between any two sample values. Then, we can employ a common root-finding technique to find the t at which to sample.

4.1 Building the Peak Finding Table

The peak finding table is a 2D lookup table as with preintegration. For each discretized segment, we stores an isovalue υ or (optionally) a set of isovalues υ_i that possibly exist within this segment. These values are sorted from the first to last peak value encountered on a given segment defined by the entry and exit values of the scalar field function, $[f_i, f_o]$. In cases where ρ is not monotonic on $[f_i, f_o]$ and multiple peaks are encountered, we reverse the order when necessary.

We first build a sorted 1D array of peaks from the transfer function ρ_α. A peak is simply defined at a local maximum. The set of peaks consists of at most half the number of actual data points in our piecewise-linear transfer function, but typically it is far less. Smooth 1D functions such as splines would have relatively fewer peaks, existing at the critical points of these functions. We, we use piecewise-linear transfer functions, as we are primarily interested in specifying and handling sharp features.

The lookup table construction then proceeds as follows: for a range of values $[i, j]$ corresponding to lookup entries from our volume $f(\underline{t}), f(\overline{t})$. If $i < j$, we search our transfer function for the next peak point (or in the case of multiple peaks, next 4 points) such that the opacity $\rho_\alpha(\upsilon) > i$ and $\rho_\alpha(\upsilon) \leq j$. If $i > j$, we search in descending order for peaks with $\rho_\alpha(\upsilon) \leq i$ and $\rho_\alpha(\upsilon) > j$. When necessary, a segment spanning multiple peaks will reverse the sorting order to register all possible peaks within that segment. This process is again similar to preintegration, except that separate discrete peak values are stored instead of a single integral approximation. In each table entry, we store the domain isovalue(s) υ corresponding to each peak. When no peak exists, we use a flag outside of the range of scalar values in the volume. Building the lookup table is relatively undemanding, and proceeds in $O(N^2)$ time, similarly to the improved algorithm of [15] for preintegration. In practice, building a peak-finding table is roughly twice as fast as building a preintegrated table at the same resolution due to the lack of floating point division. More importantly, in most cases a coarser discretization (128 or 256 bins) is sufficient for peak finding, whereas preintegration would require a larger table (512-1024 bins) for comparable quality when rendering near-discrete isosurfaces, limiting interactivity when changing the transfer function.

4.2 Classification

In the main ray casting loop (for example in a fragment shader), we can perform peak finding between samples in the place of a preintegrated lookup. We first query the peak finding table from a 2D texture to determine whether or not there is a peak. If the peak exists, we subtract that isovalue from the entry and exit values, and employ Descartes' rule of signs. If this test succeeds, we assume the segment contains a root. Bracketed by $\underline{t}, \overline{t}$, we use three iterations of a secant method (also employed by [6, 16]) to solve the root:

$$t_1 = t_0 - f(t_0)\frac{t_1 - t_0}{f(t_1) - f(t_0)} \tag{7}$$

When the secant method completes, we have an estimate for the root t along the ray segment. We now sample at this position and perform postclassification. However, sampling at the peak requires two subtle choices. First, we do not evaluate our field $f(\mathbf{R}(t))$, but rather assume that the value at this point is our desired isovalue. This works because we are solving for the root position, not its value; moreover for sharp transfer functions it is crucial in avoiding Moire patterns. Second, we do not scale ρ_α by the segment distance Δt (in (3)) but instead use a constant $\Delta t = 1$. Although this may seem counterintuitive, the scaled extinction coefficient is itself a correction mechanism for the inherently discrete approximation of the volume rendering integral. Lastly, one can choose either to solve for a single peak or up to 4 multiple peaks, depending on their spacing within the volume. In practice, multiple peaks are seldom necessary, and we do not use them in the examples in this evaluation.

4.3 Implementation

As in the original peak finding paper [10], our implementation consists of a straightforward GLSL shader in OpenGL. The 1D transfer function is given as a set of points $\{v, \{r, g, b, a\}\}$, then processed into a fairly wide (8K elements) 1D texture, allowing for rapid access on the GPU and generally sufficient transfer function precision $\Delta f > 1e - 4$.

For space skipping, we employ a uniform macrocell grid. This is generated directly from the transfer function on the CPU, stored in a 3D texture on the GPU, and then traversed directly within the shader via a 3DDDA algorithm [1]. We use a simple measure of local gradient to adapt the step size within each macrocell. In this scheme, each macrocell computes a metric based on the ratio of the maximum standard deviation of its voxels to that of the entire volume, and uses this as a rough multiplier for the frequency:

$$m = \sqrt{\lceil \mathrm{Var}(f_{cell}) \rceil / \lceil \mathrm{Var}(f_{vol}) \rceil} \tag{8}$$

As this corresponds to frequency, its inverse can be used to vary the sampling step size Δt. In practice we wish this to be a positive integer, and a multiplier $M = 2m^{-1} + 1$ delivers good results.

For dynamic higher-order filtering, we implemented a tricubic B-spline filter using the method of [25], with the BC smoothing ($B = 2, C = 1$) kernel of [18]. We optionally employ this for both sampling and root solving processes.

5 Results

Results were gathered on an NVIDIA 285 GTX at resolutions around 1 MP (1024×1024). Unless otherwise stated, we disabled differential sampling [10]. While this results in roughly 1.5x-3x worse performance at equivalent quality, we wished to evaluate peak finding alone, without this additional control variable. As a result of this, and of most images requiring over 300 samples per ray for adequate sampling, rendering these data sets is noninteractive (though generally explorable, over 1 fps) for most images in this paper, even on the NVIDIA 285 GTX. We stress that differential sampling is a practical technique to improve performance at little cost in visual accuracy. Peak finding can also be implemented in a slicer framework, which could resulting in improved performance when differential sampling is not employed. Unfortunately, aggressive rasterization-based culling mechanisms would be of little use for these data sets, as there is little exploitable empty space. Moreover, the goal of this paper (and peak finding in general) is to show better exploratory classification rather than ensure interactivity.

We consider three data sets in this paper that we believe are fairly representative of biological and medical volume data. The first is a moderately large (910x512x910) segmented data set of a zebrafish optic tract acquired through scanning electron microscopy [7]. The next is a highly noisy volume containing a zebrafish embryo, also acquired through microscopy. Finally, we consider a brain MRI before and after segmentation. Most of these datasets benefit from moderately sharp (though not Dirac) transfer functions when performing 1D classification; in most cases peak finding aids in highlighting sharp surface features, though not necessarily in correcting other artifacts stemming from scan, segmentation or reconstruction deficiencies.

5.1 Zebrafish Optic Tract

In Fig. 3, we consider a close-up of the zebrafish optic tract with postclassification, preintegration, isosurfacing only, and peak finding. This scene is similar to that in the original peak finding paper [10] but with a more revealing transfer function, and with results considered in greater detail. The transfer function consists of a moderately sharp peak, as shown in the teaser (Fig. 1). Both postclassification and preintegration significantly undersample the sharp feature corresponding to the axons. While increasing the sampling rate can ultimately alleviate this problem, it is computationally prohibitive; with preintegration we needed roughly 12 times as high as sampling rate to achieve comparable feature reconstruction, resulting in a frame rate of 0.1 fps for these images.

Not all surface features are omitted by standard DVR when sampling below the data Nyquist limit. Indeed, large membranes such as the feature in the upper-left corner are successfully reproduced by standard methods. However, features defined beneath the sampling frequency are generally omitted. For moderately large

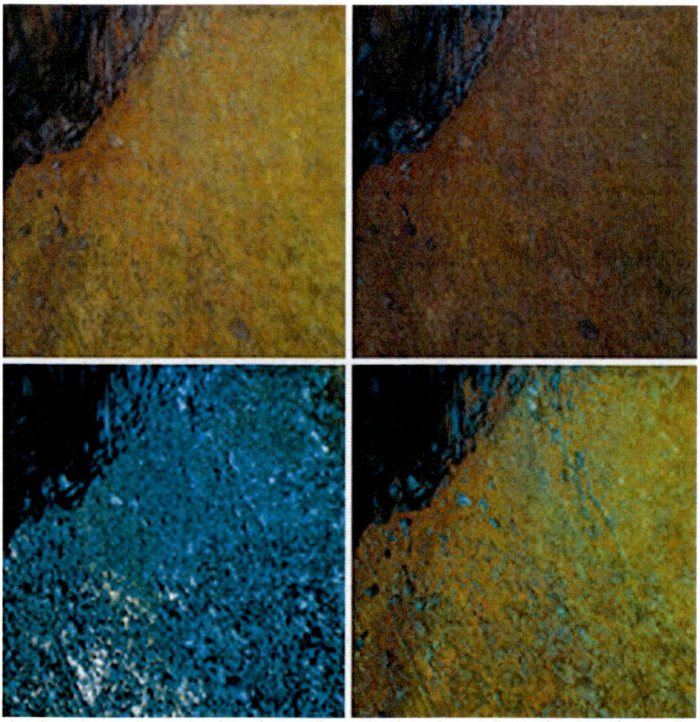

Fig. 3 Axons and glial cells in a zebrafish optic tract acquired through confocal microscopy. Top left to bottom right: postclassification, preintegration, semi-transparent peak isosurfaces, and peak finding. These images rendered at 1.9, 1.7, 2.6 and 2.0 fps, respectively at 1024×1024

data such as the zebrafish, this in fact encompasses most features the user would be interested in, i.e. axons and glial cells. Finally, while rendering of discrete isosurfaces at the transfer function peaks is useful, it tells us nothing about the density of the interstitial media. Rendering with peak finding has the advantage of improved depth perception, and more flexible classification.

5.2 *Filtered vs. Unfiltered*

To evaluate peak finding on filtered data, we consider zebrafish embryo data also acquired through electron microscopy. Though smaller, it is noisier than the optic tract and more difficult to classify. One goal in analyzing this data is to identify distinct cells and their boundaries as they undergo mitosis. Even with a smoothing filter, it is difficult to analyze this data without more sophisticated segmentation or analysis. With peak finding, our goal is simply to better visualize cell boundaries where they might exist.

5.2.1 Dynamic Filtering

We use a strong smoothing filter from the BC family ($B = 2, C = 1$), which consists of an blend between tri-cubic Catmull-Rom (interpolation) and B-spline (smoothing) bases. More details are discussed by Mitchell [18].

Figure 4 illustrates the difference between standard trilinear interpolation and tricubic BC filtering. In all cases, peak finding helps us find a particularly sharp boundary near the cell membrane, which helps visually separate cell interiors from the outside glial fluid. The original data is sufficiently noisy that it is difficult to discern cells with both standard DVR and semi-transparent isosurfacing. With peak finding, it is still difficult to interpret, but specifying a sharp feature at the cell membrane provides better intuition. Combining this with clipping planes yields reasonable first-glance classification, without needing to classify data offline or employ more complicated 2D transfer functions.

One advantage of peak finding is that when dynamic higher-order filters are used to reconstruct data, the root-solving technique can use those as well, effectively allowing for direct raycasting of higher-order isosurfaces within the DVR framework. Conversely, employing a smoothing filter can improve the entropy of the original scalar field function, effectively making $\rho(f)$ more Lipschitz and lessening the need for peak finding. In most cases, peak finding remains useful nonetheless,

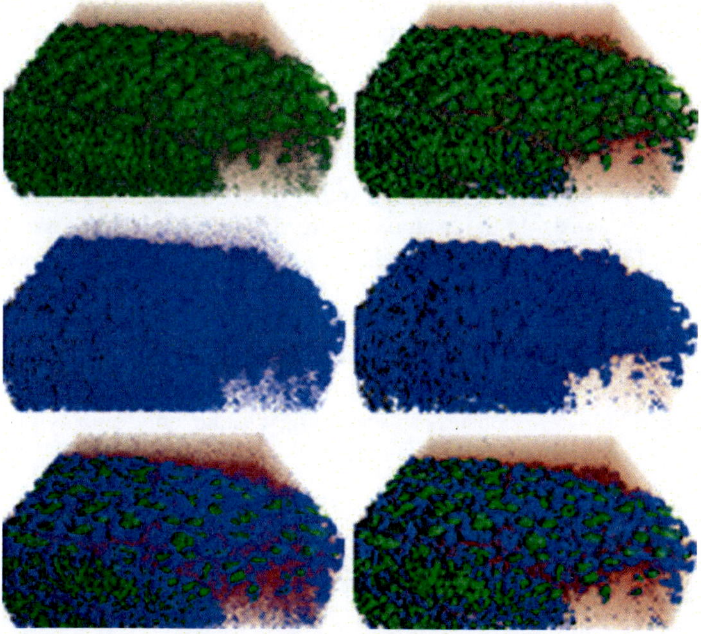

Fig. 4 Zebrafish embryo with trilinear (left), dynamic B-spline (right) filtering, using preintegration (top), isosurfacing (middle) and peak finding (bottom)

Fig. 5 Embryo dataset preprocessed offline with an anisotropic diffusion filter, with peak finding (top), postclassification (bottom left) and semitransparent isosurfacing (bottom right)

as strong smoothing reconstructions can remove features from the data before they become renderable with standard techniques at lower frequencies.

5.2.2 Static (Preprocessed) Filtering

An wider assortment of reconstructions, such as Kalman filters [7] or anisotropic diffusion filtering [27] can be afforded via offline filtering of the volume data. These are often employed as a first step for highly noisy data such as the zebrafish embryo. Unfortunately, we cannot use this filter kernel in DVR sampling or peak finding between discrete voxels. In the example in Fig. 5, where data is processed with an anisotropic diffusion filter, the scalar field is smoothed to the point that conventional classification techniques begin to work. Unlike in the previous example (Fig. 4), it is possible to distinguish cell boundaries with preintegration, and peak finding helps subtly, if at all. It is interesting to note that with high sampling rates and turbulent (though not excessively noisy) data such as the example in Fig. 5, peak finding optical effects similar to subsurface scattering, despite only employing straight primary rays. This arguably helps accentuate boundaries.

5.3 Limitations of Peak Finding

When beginning this evaluation, we were hopeful that peak finding would provide useful insights towards two difficult problems in visualization: directly rendering unsegmented volumes, and handling anisotropic gaps in data scanned slice-by-slice.

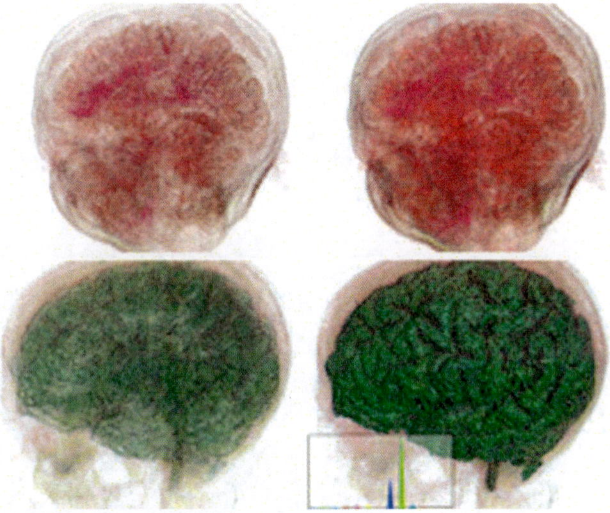

Fig. 6 Top: unsegmented MRI scan. Neither preintegration (left) nor peak finding (right) result in useful visualization of boundaries. Bottom: in some cases, preintegration fails to reconstruct sharp boundaries that peak finding detects

Unfortunately, these pose issues beyond the realm of surface reconstruction from a given filter, and are not addressed by peak finding.

5.3.1 Unsegmented Volume Data

We applied preintegration and peak finding to an MRI scan of a brain before and after segmentation (Fig. 6). Unfortunately, peak finding if anything exhibits worse results than standard DVR, due to the worse occlusion of boundary features. Considering this problem more carefully, it is clear why peak finding is not helpful: segmentation is really an art of identifying connected 3-manifolds and labelling them correctly, rather than simply identifying fine 2-manifold boundaries (which peak-finding successfully does).

5.3.2 Anisotropic Gaps

Individual slices can often be omitted in CT or microscopy acquisition, leaving conspicuous gaps in the volume. Ideally, it would be useful to reconstruct features between these gaps. For small gaps, we considered using peak finding and a higher-order smoothing filter to simply interpolate values across neighboring voxels. However, peak finding only accentuates these omissions, as shown in Fig. 7. This is

Fig. 7 Anisotropic gaps in the volume data, with peak finding (top) and preintegration (bottom)

not necessarily undesirable; and in a sense is to be expected. It is again worth noting that peak finding does not repair artifacts inherent to the original scalar field.

6 Discussion

Peak finding is a simple classification technique that delivers concrete advantages to biological and medical visualization, as well as any visualization of noisy data with sharp 2-manifold boundaries. When rendering sharp transfer functions and reconstructing surfaces, it poses significant advantages over preintegration, which has been conventionally used for this purpose. While not a replacement for advanced surface reconstruction or poor segmentation, it is a useful exploratory tool for scientific visualization.

The classifications used in this paper were all 1D transfer functions. It is important to note that many of the abilities of peak finding can be achieved with 2D gradient-magnitude classification at even lower sampling rates. However, it is theoretically possible to perform multidimensional peak finding, therefore a comparison of multidimensional classification methods should consider both methods. We are greatly interested in pursuing such an evaluation as future work.

Acknowledgements This work was supported by the German Research Foundation (DFG) through the University of Kaiserslautern International Research Training Group (IRTG 1131); as well as the National Science Foundation under grants CNS-0615194, CNS-0551724, CCF-0541113, IIS-0513212, and DOE VACET SciDAC, KAUST GRP KUS-C1-016-04. Additional thanks to Liz Jurrus and Tolga Tasdizen for the zebrafish data, to Rolf Westerteiger, Mathias Schott and Chuck Hansen for their assistance, and to the anonymous reviewers for their comments.

References

1. J. Amanatides and A. Woo. A Fast Voxel Traversal Algorithm for Ray Tracing. In *Eurographics '87*, 3–10. Eurographics Association, 1987.
2. B. Cabral, N. Cam, and J. Foran. Accelerated volume rendering and tomographic reconstruction using texture mapping hardware. In *VVS '94: Proceedings of the 1994 symposium on Volume visualization*, 91–98, New York, NY, USA, 1994. ACM Press.
3. T. J. Cullip and U. Neumann. Accelerating Volume Reconstruction With 3D Texture Hardware. Technical report, University of North Carolina at Chapel Hill, 1994.
4. K. Engel, M. Kraus, and T. Ertl. High-quality pre-integrated volume rendering using hardware-accelerated pixel shading. In *Proceedings of the ACM SIGGRAPH/EUROGRAPHICS workshop on Graphics hardware*, 9–16. ACM New York, NY, USA, 2001.
5. M. Hadwiger, A. Kratz, C. Sigg, and K. Bühler. GPU-accelerated Deep Shadow Maps for Direct Volume Rendering. *Graphics Hardware*, 6:49–52, 2006.
6. M. Hadwiger, C. Sigg, H. Scharsach, K. Bühler, and M. Gross. Real-Time Ray-Casting and Advanced Shading of Discrete Isosurfaces. *Computer Graphics Forum*, 24(3):303–312, 2005.
7. E. Jurrus, M. Hardy, T. Tasdizen, P. Fletcher, P. Koshevoy, C. Chien, W. Denk, and R. Whitaker. Axon tracking in serial block-face scanning electron microscopy. *Medical Image Analysis*, 13(1):180–188, 2009.
8. J. Kniss, G. Kindlmann, and C. Hansen. Multidimensional Transfer Functions for Interactive Volume Rendering. *IEEE Transactions on Visualization and Computer Graphics*, 8(3):270–285, 2002.
9. J. Kniss, S. Premoze, M. Ikits, A. Lefohn, C. Hansen, and E. Praun. Gaussian transfer functions for multi-field volume visualization. In *Visualization 2003*, 497–504, Oct. 2003.
10. A. Knoll, Y. Hijazi, R. Westerteiger, M. Schott, C. Hansen, and H. Hagen. Volume ray casting with peak finding and differential sampling. *IEEE Transactions on Visualization and Computer Graphics (Proceedings of IEEE Visualization 2009)*, 15(6):1571–1578, Nov-Dec 2009.
11. J. Krüger and R. Westermann. Acceleration Techniques for GPU-based Volume Rendering. In *Proceedings IEEE Visualization 2003*, 2003.
12. C. Ledergerber, G. Guennebaud, M. Meyer, M. Bächer, and H. Pfister. Volume MLS Ray Casting. *IEEE Transactions on Visualization and Computer Graphics*, 14(6):1372–1379, 2008.
13. M. Levoy. Display of Surfaces from Volume Data. *IEEE Comput. Graph. Appl.*, 8(3):29–37, 1988.
14. W. E. Lorensen and H. E. Cline. Marching Cubes: A High Resolution 3D Surface Construction Algorithm. *Computer Graphics (Proceedings of ACM SIGGRAPH)*, 21(4):163–169, 1987.
15. E. Lum, B. Wilson, and K. Ma. High-quality lighting and efficient pre-integration for volume rendering. In *Proceedings Joint Eurographics-IEEE TVCG Symposium on Visualization 2004 (VisSym 04)*, 25–34. Citeseer, 2004.
16. G. Marmitt, H. Friedrich, A. Kleer, I. Wald, and P. Slusallek. Fast and Accurate Ray-Voxel Intersection Techniques for Iso-Surface Ray Tracing. In *Proceedings of Vision, Modeling, and Visualization (VMV)*, 429–435, 2004.
17. S. R. Marschner and R. J. Lobb. An evaluation of reconstruction filters for volume rendering. In *Proceedings of Visualization '94*, 100–107. IEEE Computer Society Press, 1994.
18. D. Mitchell and A. Netravali. Reconstruction filters in computer-graphics. *ACM Siggraph Computer Graphics*, 22(4):221–228, 1988.
19. T. M oller, R. Machiraju, K. Mueller, and R. Yagel. Evaluation and design of filters using a taylor series expansion. *IEEE Transactions on Visualization and Computer Graphics*, 3(2):184–199, 1997.
20. S. Parker, P. Shirley, Y. Livnat, C. Hansen, and P.-P. Sloan. Interactive Ray Tracing for Isosurface Rendering. In *IEEE Visualization '98*, 233–238, October 1998.

21. C. Rezk-Salama, M. Hadwiger, T. Ropinski, and P. Ljung. Advanced illumination techniques for gpu volume raycasting. In *ACM SIGGRAPH Courses Program*. ACM, 2009.
22. S. Roettger, S. Guthe, D. Weiskopf, T. Ertl, and W. Strasser. Smart hardware-accelerated volume rendering. In *VISSYM '03: Proceedings of the symposium on Data visualisation 2003*, 231–238, Aire-la-Ville, Switzerland, Switzerland, 2003. Eurographics Association.
23. S. Röttger, M. Kraus, and T. Ertl. Hardware-accelerated volume and isosurface rendering based on cell-projection. In *Proceedings of the conference on Visualization'00*, 109–116. IEEE Computer Society Press Los Alamitos, CA, USA, 2000.
24. J. Schreiner and C. Scheidegger. High-Quality Extraction of Isosurfaces from Regular and Irregular Grids. *IEEE Transactions on Visualization and Computer Graphics*, 12(5):1205–1212, 2006. Member-Claudio Silva.
25. C. Sigg and M. Hadwiger. Fast third-order texture filtering. *GPU Gems*, 2:313–329, 2005.
26. M. Sramek. Fast Surface Rendering from Raster Data by Voxel Traversal Using Chessboard Distance. *Proceedings of IEEE Visualization 1994*, 188–195, 1994.
27. J. Weickert. *Anisotropic diffusion in image processing*. ECMI Series, Teubner-Verlag, Stuttgart, Germany, 1998, 1998.

Part III
Volumes and Shapes

Vessel Visualization with Volume Rendering

Christoph Kubisch, Sylvia Glaßer, Mathias Neugebauer, and Bernhard Preim

Abstract Volume rendering allows the direct visualization of scanned volume data, and can reveal vessel abnormalities more faithfully. In this overview, we will present a pipeline model for direct volume rendering systems, which focus on vascular structures. We will cover the fields of data pre-processing, classification of the volume via transfer functions, and finally rendering the volume in 2D and 3D. For each stage in the pipeline, different techniques are discussed to support the diagnosis of vascular diseases. Next to various general methods we will present two case studies, in which the systems are optimized for two different medical issues. At the end, we discuss current trends in volume rendering and their implications for vessel visualization.

1 Introduction

For the diagnosis of vascular diseases, like stenosis, plaques, and other vessel wall abnormalities, contrast-enhanced image data, such as CT angiography or MR angiography data, are acquired. Direct volume rendering of these datasets is a viable option and complements slice-based viewing. The increased image intensity of the contrast-enhanced vessels is used to selectively emphasize them. Compared to threshold-based surface rendering or surface rendering of segmentation results, direct volume rendering (DVR) represents vascular structures, in particular small vessel abnormalities, more faithfully. This is achieved by avoiding the binary decision, which portions of the data belong to a surface. With carefully refined transfer functions (TF) different portions of the vessel wall, e.g. arease with plaque

C. Kubisch (✉) · S. Glaßer · M. Neugebauer · B. Preim
Institute of Simulation and Graphics, University of Magdeburg
e-mail: kubisch@isg.cs.uni-magdeburg.de; glasser@isg.cs.uni-magdeburg.de; neugebauer@isg.cs.uni-magdeburg.de; preim@isg.cs.uni-magdeburg.de

L. Linsen et al. (eds.), *Visualization in Medicine and Life Sciences II*, Mathematics and Visualization, DOI 10.1007/978-3-642-21608-4_7,

are displayed. A survey of surface-based vessel visualization techniques with applications in treatment planning and surgical training was presented in [1]. In contrast, this chapter is focussed on volume rendering and consequently diagnostic applications.

Data Quality. Ideally, the contrast agent is equally distributed in all vascular structures which are relevant for diagnosis. In clinical practice, the contrast agent spreads with a certain speed and thus cannot reach a complete vascular tree at a particular point in time. Moreover, the contrast agent diffuses in the surrounding and leads to (again irregularly) increased image intensity values. The non-uniform spatial distribution of the contrast agent is referred to as contrast agent inhomogeneity. A simple technique to compensate for this artefact is a so called background compensation [2], which estimates the different background values and adaptively corrects them.

Another general problem is the small scale of vascular structures. A large portion of the voxels belonging to a blood vessel are boundary voxels. These voxels comprise vessel portions and portions of other tissues resulting in a state that is generally referred to as the *partial volume effect*. This leads to an averaging of the image intensities. As a consequence, small side branches of vascular structures may be hidden and larger branches may appear smaller than they actually are.

Finally, the image intensity of contrast-enhanced vascular structures in CT data is in the same range as bony structures. In some regions of the body, these bony structures may be far away from the vascular structures and can be efficiently removed by applying clipping planes. However, in other regions, such as the skull, this is often not feasible due to a close neighborhood.

The remainder of this chapter is organized as follows. We describe a vessel visualization pipeline in Sec. 2. We go on and discuss the elements of that pipeline, namely data preprocessing (Sec. 3), classification (Sec. 4), and rendering (Sec. 5). Case studies, where the general principles are applied to specific diagnostic tasks, are presented in Sec. 6.

2 Vessel Visualization Pipeline

We define the visualization of vessels from volume data as a pipeline process. The common goal of this pipeline is to support the diagnosis of vascular diseases, such as stenoses (narrowings), aneurysms, and plaque formations. This goal is achieved by the use of isolated views and focus-and-context visualizations. The focus can be the vessels themselves, pathologic portions of a vessel, or their spatial relationship to other organs. The context is mostly provided by the surrounding tissue, which aids therapy planing by providing functional relationships. We will present the following stages of the pipeline:

1. *Data Preprocessing:* Additional anatomic structures next to the volume data are identified and delineated to aid the vessel identification. The volume may be filtered for noise removal.

2. *Classification with Transfer Functions:* The intensity values stored in the volume are mapped to optical properties, like color and opacity values.
3. *Rendering:* The volume data is rendered as 2D or 3D image and might be limited to single vessel paths.

Common visualization goals are the removal of obstructing tissue and highlighting vessel pathologies. When it comes to intervention planning, surrounding context structures gain more importance. The presented DVR pipeline focuses more on the vessels themselves.

3 Data Preprocessing

In a first stage, various preprocessing steps on the volume data can be carried out. Initially, filtering techniques may remove noise or compensate for contrast agent inhomogeneity [3]. Bone removal (Sect. 3.1), vessel segmentation (Sect. 3.2), and emphasis of elongated vessel-like structures (Sect. 3.3) are frequently performed to display the vessels. These methods can strongly improve the visual quality, since vessels are relative small structures inside the volume and thus can be easily obstructed.

3.1 Bone Removal

In most regions of the human body, (contrast-enhanced) vascular structures and skeletal structures cannot be discriminated reliably [4]. Bone removal is part of almost all modern radiology workstations. In most cases, it is sufficient if the user marks a connected skeletal component with one click of a pointing device to initiate a (more or less advanced) region growing method to roughly segment this structure (see Fig. 1).

3.2 Vessel Segmentation

Another common task is the segmentation of the vessel voxels. The segmentation of vascular structures allows to adjust a TF to the image intensities, in particular to the histogram of vessel voxels. This is essential since they occupy only a small fraction of the entire dataset ($\approx$2-5% in the coronary datasets used in Sect. 6.2) and thus are hardly recognizable in the global histogram. There exist various segmentation techniques. The conceptually most simple method is *region growing* [5]. From an initial set of seed points, more and more neighboring image voxels are included as long as they can be classified as vessels. Typically, this inclusion is bound to a global intensity threshold. The progressing boundary of the segmentation process can also

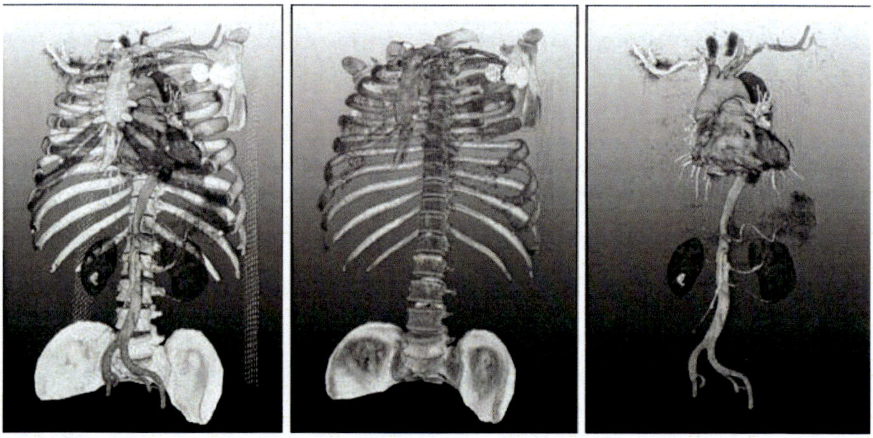

Fig. 1 DVR of vascular structures. Original scene (left), bones to be removed (middle) and resulting visualization restricted to the heart, the kidney and vascular structures (right). Compared to the initial scene, bone removal significantly improves the display of vascular structures. (Image courtesy of Johann Drexl, Florian Link and Horst Hahn, Fraunhofer MEVIS)

be interpreted as wave front that propagates through the segmented object. This interpretation forms the basis of the *level set* and *fast marching* methods [6]. Local image features, such as intensities, gradients, and textures, control this propagation adaptively. Therefore, the front is rapidly moved towards those regions, which are likely part of the vascular system, and is hindered from others.

After segmentation, the centerlines of vessels and the whole vessel tree can be extracted through *skeletonization*. This can be achieved by basic operations, which commonly use *thinning* to create a single voxel wide representation of the vessel tree. Thinning can be implemented by the morphological operator *erosion*, which has to consider the anisotropic character of medical image data (slice distances are usually larger than in-plane resolution). The centerline can be used for vessel specific rendering techniques (see Sect. 5.2) or for the analysis of vessel attributes, such as the diameter.

A variation of segmentation is *vessel tracking*, in which a path along the vessel is extracted from a given starting point. Various methods exist, some extract single paths, others can locate the entire vessel tree and correctly handle vessel branching. At the *Rotterdam Coronary Artery Algorithm Evaluation Framework* [7] many methods have been compared and analyzed with respect to accuracy, speed, and level of user interaction. For more details on vessel segmentation and skeletonization, we refer to Boskamp et al. [8].

3.2.1 Centerline Correction

An initial vessel path can be refined, so that it reflects the centerline more accurately. For each pathpoint M, a tangential plane T is generated, in which the point is moved

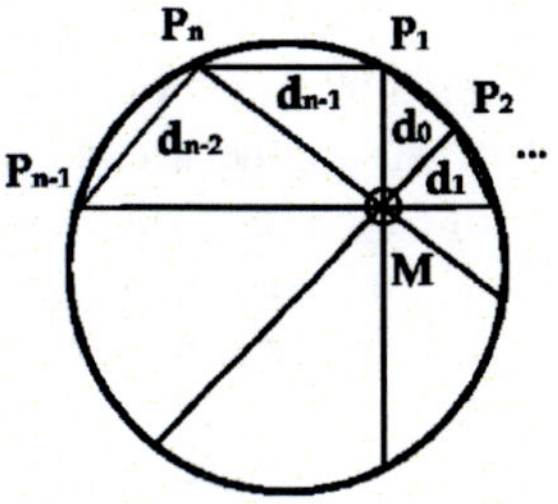

Fig. 2 The original center M is corrected by a weighted sum of the radial hitpoints P_i and the distances d_i between them

towards the center [9]. The center is approximated by casting radial rays within the plane, which generate a series of hitpoints P_i with the vessel wall (see Fig. 2). The hitpoint generation requires pre-defined thresholds for the vessel's inner and outer intensities. The distances d_i between adjacent hitpoints are used to approximate a new center point N:

$$N \approx \frac{\sum_{i=1}^{n} P_i \cdot (d_{i-1} + d_{(i)\text{mod}n})}{2 \cdot \sum_{i=0}^{n-1} d_i} \tag{1}$$

As vessels typically exhibit smooth curvature and no abrupt changes in flow, spline-interpolated pathpoints and tangents can be used. This is particularly useful when the segmentation yields a path with discrete voxel coordinates.

3.3 Filtering

Similar to segmentation, it is also possible to find voxels that are usually part of vessels with filtering, since they exhibit certain geometric features in their neighborhood. Whilst segmentation classifies voxels in a binary fashion, filtering assigns real probability values to each voxel, the *vesselness* factor. This factor is mostly related to finding elongated round structures. Filtering uses the local neighborhood of each voxel and treats all voxels in the same manner. Detection of the vesselness is challenging, because especially at branchings, the local neighborhood does not reveal elongated structures.

To compute the vesselness, Frangi et al. [10] used the Hessian matrix $\mathcal{H}$ to detect tubular structures. A common approach to analyze the local behavior of an image L is to consider its Taylor expansion in the neighborhood of a point x_o:

$$L(x_o + \delta x_o, s) \approx L(x_o, s) + \delta x_o^T \nabla_{o,s} + \delta x_o^T \mathcal{H}_{o,s} \delta x_o \tag{2}$$

This expansion approximates the structure of the image up to the second order. $\nabla_{o,s}$ and $\mathcal{H}_{o,s}$ are the gradient vector and the Hessian matrix of the image computed in x_o at scale s. Varying s will represent different vessel diameters, and the results of each filter run are combined using the maximum in the end. To calculate these differential operators of L, convolution with derivatives of Gaussians are used. Analysis of the second order derivatives information (Hessian) has an intuitive justification in the context of vessel detection. The second derivative of a Gaussian kernel at scale s generates a probe kernel that measures the contrast between the regions inside and outside the range $(-s, s)$ in the direction of the derivative.

At the core of the vessel filter lies an eigenvector decomposition of the Hessian matrix, which describes the local principal directions of the curvature. Given the 3D volume space, this yields three orthonormal vectors $(\lambda_1, \lambda_2, \lambda_3)$ due to the matrix's symmetry. The eigenvector with the smallest absolute eigenvalue corresponds to the direction that represents the least curvature, i.e. along the vessel direction. If the vectors are sorted by their magnitude $(|\lambda_1| < |\lambda_2| < |\lambda_3|)$, an ideal tubular structure in a 3D image would have the following attributes:

$$|\lambda_1| \approx 0 \tag{3}$$

$$|\lambda_1| \ll |\lambda_2| \tag{4}$$

$$|\lambda_2| \approx |\lambda_3| \tag{5}$$

These attributes only describe cylindrical structures, which means they are inappropriate at vessel branchings. The final vesselness is a product of different parameters that represent geometric details such as blob-like structures or plane- and line-like patterns. To refine the result, the user can control these parameters, which are based on the λ vectors, by custom weights. One of the parameters takes the intensity values into account to lower the impact of noisy background values. For further details on the entire filter construction, we refer to [10].

Because the vesselness is only defined for a single diameter, a multi-scale approach is employed. Since only tubular structures are determined by this filter, the detection of vessels in branching areas leaves room for improvement. The multi-scale approach is also limited in dealing with varying vessel diameters, due to its limitation to fixed diameters. These issues were dealt with by Joshi et al. [11], who used a mix of Hessian- and entropy-based filtering. Their entropy factor was generated from analyzing the polar profiles of each voxel. Figure 3 depicts such a profile in the 3D case. For each direction, the variance in intensities and average intensity are computed within the neighborhood. The likelihood of clusters within such profiles is used to calculate the vesselness.

Compared to the hessian-based technique, the profiles do not require multiple runs for different vessel diameters and especially the quality around branching points has been improved (see Fig. 4). However, the method's implementation in Matlab on a 3.2 GHz Pentium IV is very time-consuming, i.e. it resulted in a four hour computation, which is unfavorable in the clinical routine. With the use of per-voxel factors, either probability-based or binary from segmentation, the later rendering of volumetric vessel data is reduced or highlighted to the important vascular structures.

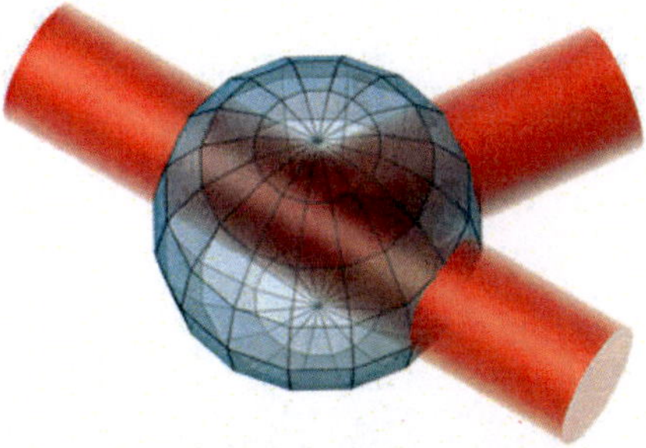
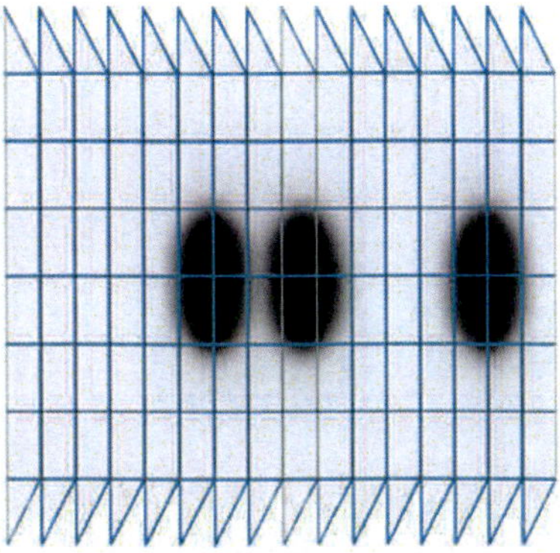

Fig. 3 A spherical profile (right) for branching vessels (left). The profile shows a phantom distribution of the intensities on the spherical map for a single radius. By using multiple radii the mean value intensities and variance of intensities along each solid angle can be derived. A cluster analysis on this data is used to contribute to the vesselness factor

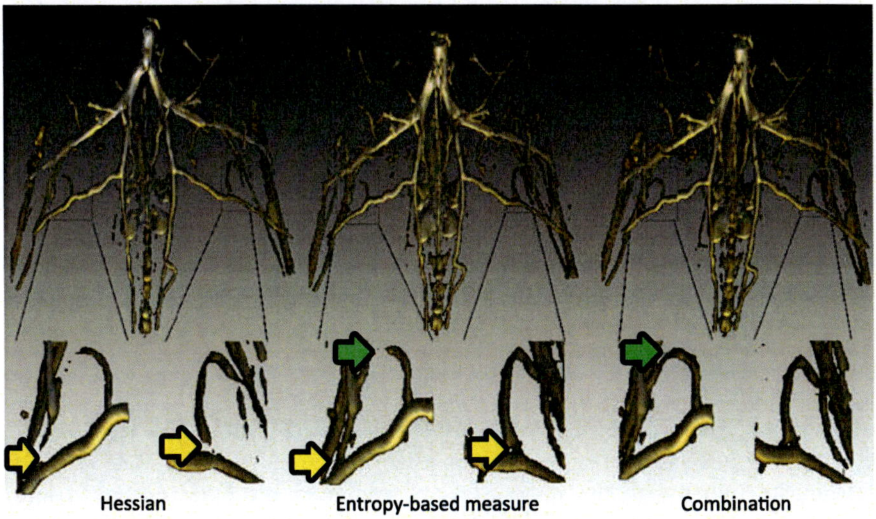

Fig. 4 Comparison of the Hessian-based (left) and entropy-based (middle), polar profile, vessel filters. The green arrows mark benefits of the earlier method, and the yellow arrows show improvements of the latter. The final (right) image shows a combination of both techniques. Image from [11]

4 Classification with Transfer Functions

Segmented data provides additional information, that can aid analysis of the volume data for classification. DVR requires a classification C of the measured intensity value i, which maps intensity values to colors and opacities:

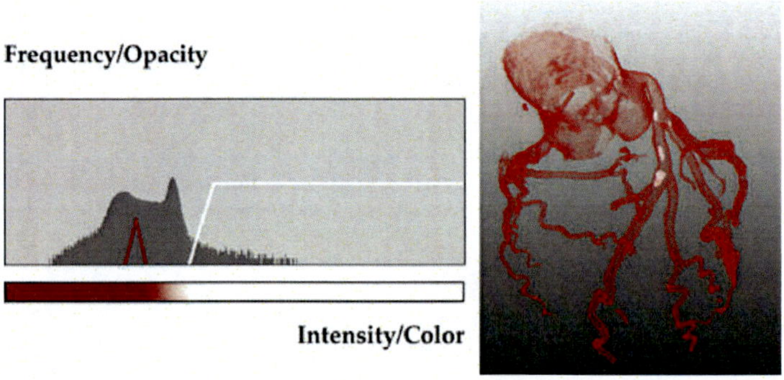

Fig. 5 TF specification with a piecewise linear function to emphasize two tissue types. The histogram is shown as context. The supporting points of the 1D transfer function can be arranged (left) to highlight only focus structures (right), in this case calcified plaques within the segmented coronary vessels

$$C(i) \rightarrow (r, g, b, \alpha) \tag{6}$$

This can be used to color-code anatomical structures differently and make them invisible or transparent. For faster processing, the classification is often precomputed and encoded into lookup-textures. It can be applied to each voxel before rendering (*pre-classifaction*), or during runtime sampling (*post-classification*). The latter yields much higher quality, because the interpolated intensity values between voxels are classified on their own.

A piecewise linear function can be defined through multiple supporting points that highlight certain intensity ranges and hide others (see Fig. 5). If the objects lie within certain intensity ranges, the histogram information aids the manual specification process.

The classification may not only depend on the intensity values. Kindlmann and Durkin [12] have shown how *multi-dimensional TFs (MDTF)* can be semi-automatically generated in order to emphasize transitions between materials. To discriminate values with the same intensity, additional attributes are taken into account. One typical attribute is the gradient magnitude: homogenous regions have a small gradient and larger gradients can be found at tissue transitions. With a higher-dimensional domain, the TF design process itself becomes harder, as arbitrary shapes and color gradients can now be used to classify the regions of interest in the multi-dimensional space. Several ways to deal with this complexity exist, either by automatizing the process, or by supporting the manual process through appropriate user interfaces [13].

In principle, MDTF allows to display vascular and skeletal structures simultaneously, as well as to discriminate them e.g. by using different colors, which was not possible with 1D TFs. Thus, expressive visualizations of complex spatial relations may be achieved, e.g. blood vessels close to the cranium [14] (see Fig. 6, where image intensity and gradient magnitude are employed).

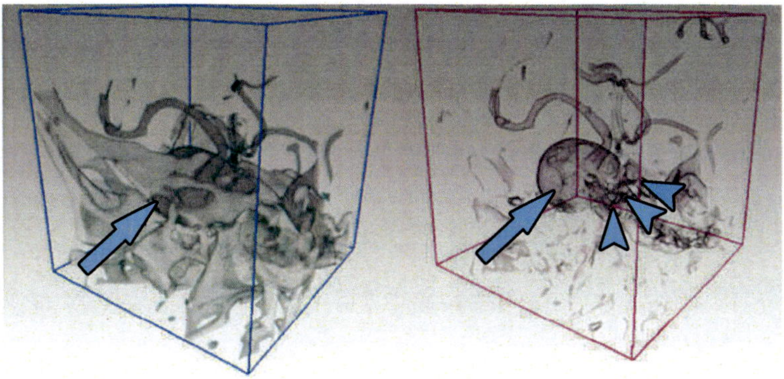

Fig. 6 Application of 2D TFs. Additionally to the intensity values in the 1D TF (left) the gradients are used for classification and therefore improve the image quality (right). The arrow shows the aneurysm, while the arrowheads point to the parent arteries, which were obstructed by skull tissue. Image from [14]

4.1 Histogram Analysis for Transfer Functions

A histogram analysis can automatize the TF specification for emphasizing dataset specific boundaries. Rezk-Salama et al. [15] used this approach for their *implicit segmentation*, in which anatomic structures appear as being (explicitly) segmented. A histogram-based method was also chosen by Vega Higuera et al. [16] to emphasize cerebral vasculature with specification of a 2D TF. The histogram is analyzed for gradient magnitudes and intensity values to only highlight blood vessel walls in datasets for diagnosing intracranial aneurysms (more detail in Sect. 6.1). This automatic method strongly speeds up the process, which otherwise requires a trained expert, due to the small size if the selected region in the 2D histogram. Based on a reference dataset and a reference classification, a non-rigid registration between examined and reference datasets' 2D histograms is performed. The reference TF is then smoothly deformed using bicubic B-Splines to the new dataset. The result of this method works very well and is presented in Fig. 6.

To detect peaks of smaller objects, *local histograms* can be determined in subregions, or along lines, also known as intensity profiles. Lundström et al. [17] partition the volume into regional neighborhoods for which they compute histograms. Because a clear visual segmentation is not always possible, animation techniques were introduced to represent the probability of the individual classification [18]. Instead of mixing the overlapping colors for each structure's intensity region, the pure colors are presented in an animated series proportional to their probability. Therefore, the information about the ambiguity is not lost and helps medical doctors becoming aware of uncertainties.

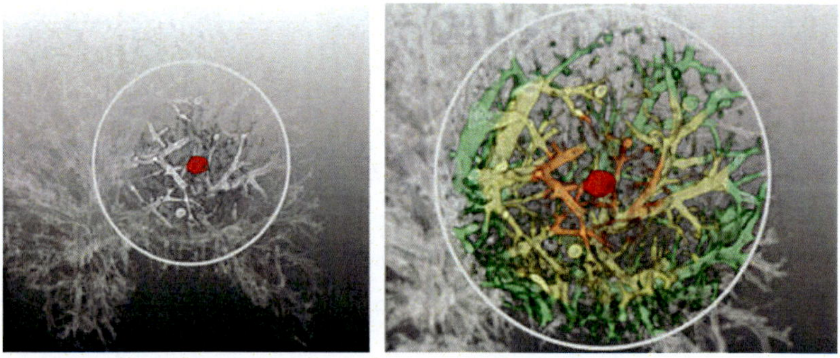

Fig. 7 Distance-based transfer function applied to abdominal CT data. In the left image the vessels within the security margin around the tumor (red) have higher opacity. In the right image different distance ranges are color-coded. The security margins around tumors are essential to assess operability. Image Courtesy of Andreas Tappenbeck, University of Magdeburg

4.2 Spatial Transfer Functions

The distance to focus structures can also be used as input to MDTFs [19] (see Fig. 7). Prior to the application, a distance field for the given key structure needs to be computed, which requires segmentation of that structure. The sign of the voxel in the distance field is used to discriminate between interior and exterior volume.

The local size of regions was used as secondary input to MDTFs by Correa et al. [20]. They did not use a segmentation, but based their relative size metric on a continuous scale-space analysis and a set of detection filters. Their algorithm works in three steps:

1. *Scale-space computation.* The user defines a set of interesting scales (similar to the vessel diameters in Hessian-based filtering in Sect. 3.3). Through forward Euler integration, this scale-field is generated based on diffusion using a custom conductivity function. The result of this process are various scale fields.
2. *Scale detection.* Within the scale fields a point set is created by running detection filters. The points are maxima in space and scale, referred to as Laplacians of Gaussians. In the authors' application, the point set defined spheres, which represent blob structures of various radii.
3. *Backprojection.* The relative size volume is created from the point set by scattered data interpolation. This was accomplished with Shepard's interpolation that uses a fourth-degree Wendland polynomial as common basis function.

In the end, the relative size is used for the classification. The authors have mapped size to color values and a combination of size and intensity to opacity and generated appealing illustrations of the vessels in the *visible human's* hand dataset.

5 Rendering

In the rendering stage the classification is finally applied. In medical workstations a combination of 3D and 2D views is used to provide the viewer with different spatial information about the vessels. We will mention some of the principle methods used in this stage. For the spatial context the 3D views, typically generated through ray-casting, are frequently used (Sect. 5.1), whilst reformation (Sect. 5.2) and unfolding techniques (Sect. 5.3) create 2D images for detailed vessel inspection. Kanitsar et al. [21] have performed extensive research on various 2D visualization techniques for diagnostic relevance.

5.1 3D View

The 3D view gives an overview of the basic branching pattern of the vascular tree, and – depending on the TF design – it gives clues to context structures surrounding the vessels. A wide-spread DVR option is the Maximum-Intensity Projection (MIP), a rendering mode, where the brightest voxel along each viewing ray is displayed independent from its position in 3D space. Thus, a single MIP image does not convey any depth cues. Therefore, it is essential to rotate such visualizations or to use predefined videos of such rotations in order to benefit from depth-cues resulting from the motion parallax. The image intensity of the selected voxels is usually linearly mapped to the brightness of a gray value. Given the intensities I along the viewing ray from s_{start} to s, the MIP is simply the maximum of all sampled intensities.

$$\mathsf{MIP}(s) = \max_{s_{\mathsf{start}}}^{s}(I(\tilde{s})) \tag{7}$$

For clinical users, the simplicity of MIP images is a great advantage: No parameter needs to be adjusted and the relation between image intensity in the final image and the underlying data is quite direct. The partial volume effect is often strongly disturbing, because it hides small vascular structures in front of larger vascular structures. As a remedy, the closest vessel projection (CVP) [22], also called Local MIP, has been introduced [23]. With this variant, instead of the global maximum along a viewing ray, the first local maximum above a certain threshold t is selected. CVP images thus provide correct depth information. However, the user has to specify t appropriately. As a rule of thumb, a certain percentage of the global maximum intensity of the whole dataset is usually appropriate and "suggested" as default value.

5.1.1 GPU-Ray-Casting

With the advancement of programmable graphics hardware *GPU-ray-casting* has become the state-of-the art technique also for the visualization of vascular structures.

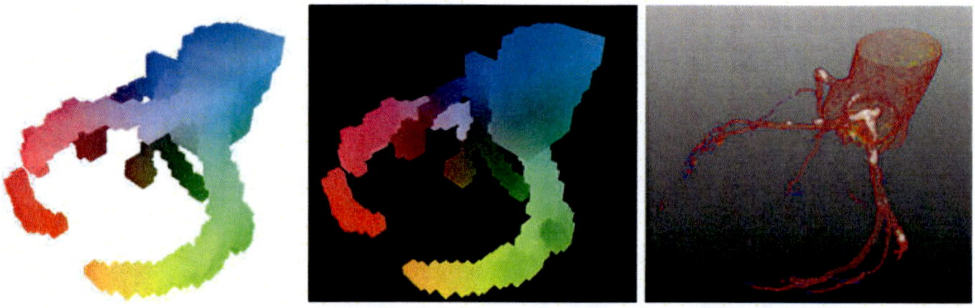

Fig. 8 GPU-ray-casting with position buffers applied to coronary vessels in CT data. An approximated mesh hull of the volume is rasterized into ray-start- (left) and ray-end-postion (middle) buffer. Both are used for the final volume rendering (right) through a viewport-spanning quad mesh

The approximated calculation of the *volume rendering integral* (I radiance, D end position, the radiance I_0 at the background position s_0, κ being the absorption and q being the emission of light):

$$I(D) = I_0 e^{-\int_{s_0}^{D} \kappa(t)\mathrm{d}t} + \int_{s_0}^{D} q(s) e^{-\int_{s}^{D} \kappa(t)\mathrm{d}t} \mathrm{d}s. \tag{8}$$

can be performed with custom loops in the pixel processing stage of the hardware, substantial flexibility and performance was gained. Scharsach et al. [24] provide a robust setup for perspective and orthogonal ray-casting. It uses acceleration strategies, such as empty space skipping and early ray termination. The key idea is to create two image space buffers for ray start and end positions (see Fig. 8 left and middle), which are used as input for the final pixel processing program. To accelerate empty space skipping, a hull mesh can be created at a coarser resolution than the actual volume. Its positions are rasterized into the ray position buffers. Special care must be taken to avoid artifacts resulting from clipping the hull mesh with the near plane. Empty space skipping is particularly useful, as the vascular structures typically cover only a small subset of the data.

When using *volume splatting*, such as presented by Vega Higuera et al. [25] for displaying neuro-vascular data, empty space skipping can be performed more accurately. As the ray position buffers only store entry and exit positions, the empty area inbetween cannot be encoded. Volume splatting, however, uses particles to approximate the filled parts of the volume. Compared to ray-casting, the splatting comes at the cost of rasterizing many small primitives on their own, creating many context switches for the hardware. Ray-casting does not suffer from these switches and its load can be spread over the GPU threads effectively,

The ray-casting approach also provides previous sample data in a more accessible fashion, which has lead to a technique called *opacity peeling* [26]. In opacity peeling, the ray sampling can skip regions along the view ray (assuming front-to-back traversal). These layers are created when the post-classified opacity O changes

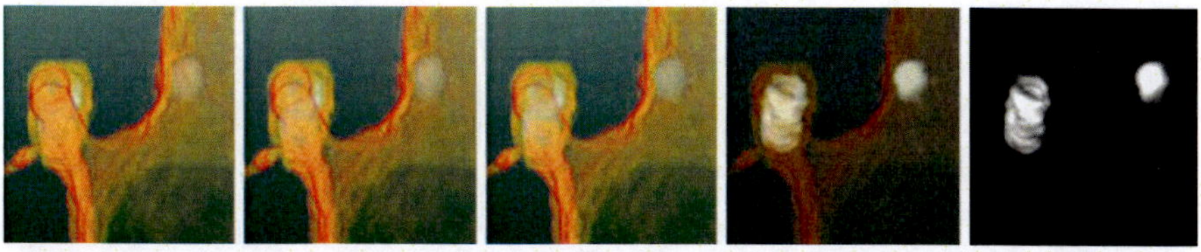

Fig. 9 Maximum intensity difference accumulation applied to coronary vessels in CT data. From left to right the MIDA control value γ is changed from -1 (left) to 0 (center) to 1 (right). This is equivalent to blending from DVR to MIDA to MIP. Compared to the original MIDA implementation, the opacity values for MIDA blending were different than for DVR to achieve a higher contrast

($c = O(s_{i+1}) - O(s_i)$) from one sampling point s_i to another s_{i+1}. Global or local thresholds can be used to define conditions that represent a layer toggle. For example, in a head CT data the skull can be peeled quickly: The ray will first start out in air, then hit skin and skull both giving $c \gg 0$, and finally reach the less dense inner tissue $c \ll 0$ again. Given this setup, it is easy to completely refine the visual results without segmentation, by specifying which layers shall be made invisible or rendered in a different style.

Bruckner and Gröller [27] introduced the *maximum intensity difference accumulation (MIDA)* to highlight possibly occluded structures. They altered the blending to favor classified samples that have a higher opacity than previous samples. Thus, similar to MIP, high opacity structures are not covered by the accumulation of many low opacity structures. Their γ control value allows a seamless transition from regular DVR to MIDA to MIP (see Fig. 9).

When 1D TFs are used, DVR greatly benefits from pre-integration tables. The quality of the approximation of the volume integral greatly depends on the step width that is used for ray traversal. This especially applies to small vessels when the classification of the vessel wall has other opacities than vessel interior and exterior. Increasing the amount of steps, and therefore lowering the distance between sample points, has a negative impact on the performance of ray-casting. Decreasing the steps, on the other hand, yields aliasing artifacts. However, the pre-integration tables store the integrals between two intensities using a finer sampling. Later at runtime, this pre-computed table T based on the classification C only needs to be sampled with front intensity s_f and back intensity s_b. The computation of T is as follows:

$$T(s_f, s_b) = \frac{1}{s_b - s_f}\left(\left(\int_0^{s_b} C(s)\mathrm{d}s\right) - \left(\int_0^{s_f} C(s)\mathrm{d}s\right)\right). \tag{9}$$

5.2 *Planar Reformation*

In addition to the 3D view, 2D views are crucial in vessel diagnosis. The profile of the vessel wall can give information about narrowings, which can hint at an

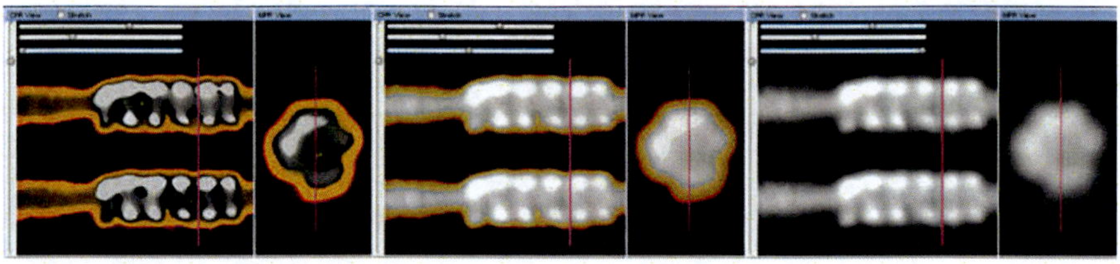

Fig. 10 Mulitplanar reformation of vessels. The vessel viewer shows the oppositing longitudinal sections on the left and a planar cross section on the right. Different rendering styles have been used to depict the stent placed inside the arteries. Far left shows isosurface style rendering, whilst far right a classic windowing transfer function. The middle section overlays colored TF with greyscale windowing TF

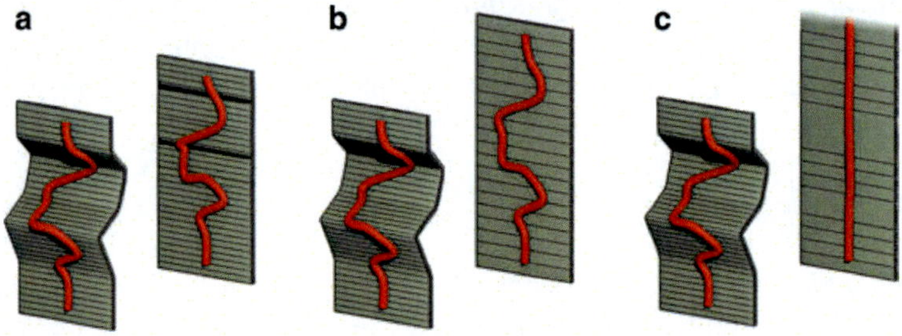

Fig. 11 Curved Planar Reformation. The unaltered projection perpendicular to the main direction of the vessel may cause overlaps (**a**). *Stretched* CPR (**b**) keeps true lengths between vessel path points. *Straight* CPR keeps true lengths and centers path points (**c**). Image from [30]

undersupply of tissues. Analyzing the vessel's diameter may also reveal signs of calcifications, also known as *plaques* inside the vessel, which also lower blood flow in this area. Wesarg et al. [28] determined vessel diameter changes during the segmentation and also used longitudinal cut views along the vessel's centerline. The cuts were defined by creating a virtual cylinder made of circles perpendicular to the vessel path. When all circles are stacked on top of each other, the vessel is effectively straightened and can be presented in 2D as longitudinal cut through the virtual stacked cylinder (see Fig. 10). To highlight the plaques, arrows are placed in the 3D and 2D views [28].

A similar strategy to depict the vessel along its path is called *curved planar reformation* (CPR). Next to the straightened reformation in 2D, a stretched version has been presented by Kanitsar et al. [29]. The path points are projected perpendicular to the vessel's main direction. When the projection is unaltered, the path can overlap itself and will not be isometric (Fig. 11 A). However, isometry is important for diagnosis, and therefore the following methods are able to provide it. *Stretched* CPR shifts the projected points along the longitudinal axis to account for original lengths between path points. *Straight* CPR eliminates cross-sectional movement, so that the diameter changes of the vessel are easier to read on the longitudinal axis.

Instead of generating a single CPR, the entire vessel tree can be spread into the image plane as well [30]. Iterative relaxation of the outgoing directions at joint points can assure that vessels do not overlap at branchings. Rotation around the projection plane is still possible, but will mean changes to the position of all subbranches from the root path.

Due to the focus on the vessels in the 2D views, the surrounding context is typically lost. Straka et al. [31] introduced the *vessel glyph* as a set of methods to combine advantages of both 3D and 2D views into a single visualization. They emphasize the vessels, e.g. by fading out or darkening adjacent structures. However, this only works well if the vessels do not exhibit high curvature. Otherwise, they would overlap themselves and create ambiguous images.

5.3 Vessel Unfolding

With the increasing resolution of MR and CT scanners, the interior of the vessel may be investigated. Due to the similarity of the geometry to much larger structures such as the colon, some techniques developed for the colon wall visualization can be adapted to the smaller vessels. One such technique is *virtual colon unfolding* [32]. It creates a 2D image of the entire colon wall. In a recent work, Ropinski et al. [33] used vessel unfolding and flattening to reveal measured data within the vessels. They have also aligned the unfolded structures to simplify comparison between different datasets.

Due to the aforementioned progress in scanning quality, it is expected that other visualization methods make this transition, e.g. virtual endoscopy turning into virtual angioscopy. Further reduction in noise and improved image resolution with more voxels shall also improve the output of histogram-based visualization methods (see Sect. 6.2, Fig. 17).

6 Case Studies

We have presented several basic and advanced DVR techniques to display vessel structures. In the following case studies, we will present two techniques that were customized to specific diagnostic tasks. The first is a simplified filtering technique for implicit segmentation of the main cerebral vessels, which provide context to simulation meshes. The second provides a refined local histogram analysis to create TFs, which highlight pathologies for diagnosis of the coronary artery disease (CAD).

6.1 Case Study 1: DVR of Cerebral Vasculature in MRA Data

Cerebral aneurysms develop from a congenital or acquired weakness of stabilizing parts of the cerebral arterial vessel wall. They bear a higher tendency of rupture, with often fatal consequences for the patient. To gain insight into the cause and evolution of cerebral aneurysms and to reduce the risk of surgical or endovascular treatment [34], a detailed characterization of morphology (size, shape), morphodynamic (pulsatile change of morphology), and hemodynamics (blood flow pattern) is important [35]. Neugebauer et al. [36] focus on the evaluation of the morphology with focus-and-context rendering. They propose a hybrid rendering: a polygonal representation of the mesh containing the flow data and a volume rendering of the contextual vasculature. A filtering is necessary to ensure that only relevant information are included into the volume rendering. In contrast to the complex filtering kernels mentioned in Sect. 3.3, they suggested a simpler series of image processing operations to detect vessels surrounding a cerebral aneurysm. As input, MR-Time of Flight (TOF) data is used. Due to a special setup, the TOF sequences yield a high signal from blood moving into the direction of the slice plane's normal (see Fig. 12c). By exploring ten datasets with varying intensity values and from different scanner devices, they have empirically created the following workflow (Fig. 13):

1. *Binary threshold.* Due to the high intensity values of vessels in MRA-TOF data, a simple threshold operation removes most of the disturbing data. The intensity histograms of the datasets are analyzed for common characteristics to determine a proper threshold. Because of individual varyings between datasets, the threshold is based on the equalized histogram, which compensates for image intensities and overall contrast between vessels and tissue. The mean value position p in the equalized image histogram is used to define the refined threshold as $T = p \cdot 0.5$, which is found to be close to $I_{max} \cdot 0.25$. After applying the threshold, the image contains all the arteries of interest, but also parts of high intensity brain tissue and skin, as well as some noise (see Fig. 13b).
2. *Connected component analysis.* Voxels belonging to arteries form large groups, whereas skin and tissue-related voxels split up into several small groups (see Fig. 13b). This observation motivated the use of a connected component analysis to distinguish between relevant (arteries) and irrelevant (skin, tissue) structures. In the following, the voxel-groups are referred to as components. The noise is removed by discarding all components smaller than 0.01% (empiric threshold) of the overall dataset volume. The components left are the main cerebral arteries and parts of the skin. Removal of the skin components is achieved by taking into account that the arteries are located along the center medial axis of the head (see Fig. 13c). The bounding boxes of all components are projected into the *transverse plane* (medical term). Since the arteries are close to the center and run mostly vertically, their bounding box centers c are clustered in this plane. In contrast, the skin components exhibit a lower clustering rate. The closest center to the heads vertical axis becomes the reference origin o. The distances $d_i = |c_i - o|$ of all bounding box centers are calculated relative to o and sorted

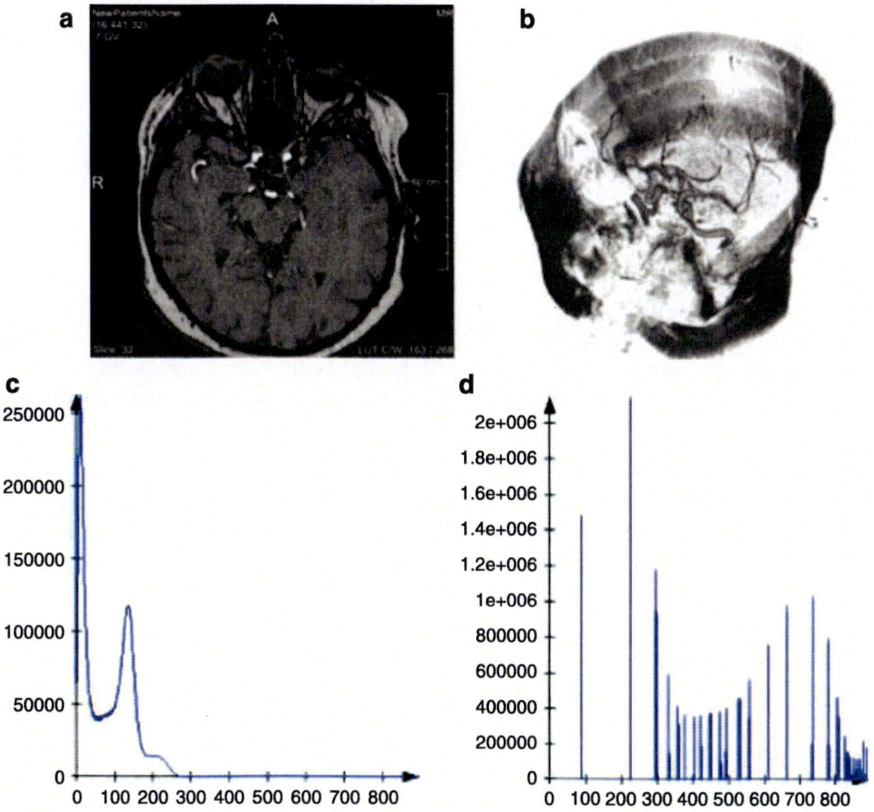

Fig. 12 Analysis of the dataset's intensity values: slice view (**a**), volume view (**b**), histogram (**c**), equalized histogram (**d**). The vessel and skin intensity values are within the upper half of the equalized histogram

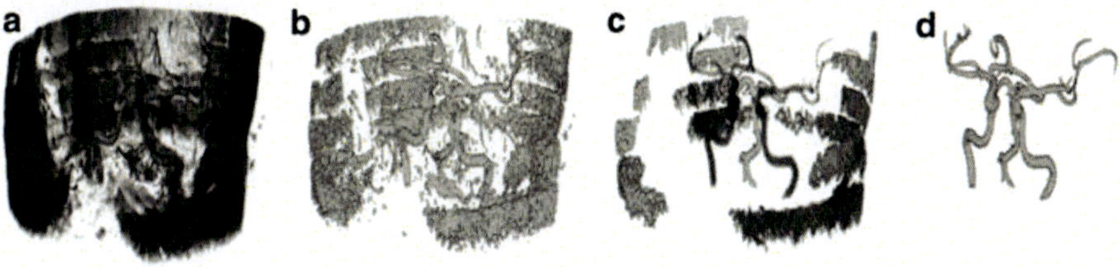

Fig. 13 The different stages of the filtering processes: original data (**a**), threshold applied (**b**), connected component analysis discarding small structures (**c**), removal of outer skin structures (**d**)

in ascending order. By sorting the distances, the aforementioned axial-center distance is included in all following calculations. Based on the ordered distance list, the mean m value of the distance intervals $D_i = d_{i+1} - d_i$ is calculated. This mean value is a simplified measure to identify a spatial clustering, the second of the aforementioned characteristics are utilized to distinguish between skin and arteries. All components that belong to an interval $D_i < m \cdot b$, are defined as components that represent arteries. Experiments on several datasets have shown that $b = 1.2$ is an adequate choice to describe the common cluster density of components belonging to cerebral arteries (see Fig. 13d). The parameter b describes the relative rate of distance-change and is therefore invariant with respect to resolution and size of the dataset.

3. *Proximity-based preservation of details.* Whereas the large arteries can be filtered in a robust manner by the connected component analysis, small vessels that potentially emanate from the aneurysm body, cannot be filtered explicitly. They have a small diameter (≈2–3 voxels), are often fragmented because of insufficient spatial resolution of the MR scan and exhibit low contrast differences with respect to the surrounding tissue due to the high relative impact of the partial volume effect. Thus, those vessels exhibit characteristics similar to noise, tissue and skin artifacts, and are consequently removed during the connected component analysis. Despite the vague representation, these small vessels are diagnostically important and should be included in the context visualization. An experienced radiologist is able to visually distinguish between noise and small vessels, if the view at the region containing the small vessels is not occluded. Hence, the region near the aneurysm is included in the context visualization. This region is chosen by proximity without any filtering. A smooth reduction of intensity with increasing distance to the aneurysm surface is applied to reduce the rate of occlusion. This is possible since the opacity of the rendered voxels will be linked to their intensity when the transfer function is applied within the final visualization.

   A weighted volume mask is created with an Euclidian distance transform (EDT) to show possible vessels near the voxelized aneurysm. It is crucial that the polygonal simulation mesh is registered properly with the volume data, so that its voxelisation matches the volume dataset. This should be taken into account when generating the original mesh. The distance field F is the result of the EDT and encodes scalar values between 0 and 1, where 0 describes the minimal distance to the aneurysm surface, and 1 is the maximum distance found at the border of the dataset. To create a smooth ramp mask covering the local surrounding of the aneurysm, first, F is inverted and then a rescaled distance field F' generated with:

$$F' = \max(0, (\mathrm{inv}(F)1 + (1/c)) \cdot c), \tag{10}$$

   whereas c is a constant. For all datasets, $c = 10$ led to a well-sized ramp mask for the local aneurysm surrounding.

4. *Final Mask Creation.* The result of the connected component analysis is a binary volume, on which a morphological dilation filter with a $3 \times 3 \times 3$ kernel is applied. Thus, it is ensured that the vessel surface is represented completely when masking the original data, as partial volume effect mainly affects the intensities of border voxels between vessel and surrounding tissue. This mask is combined with the EDT mask by applying the arithmetic image operation max. The result is a mask that will preserve the main arteries and low contrast information near

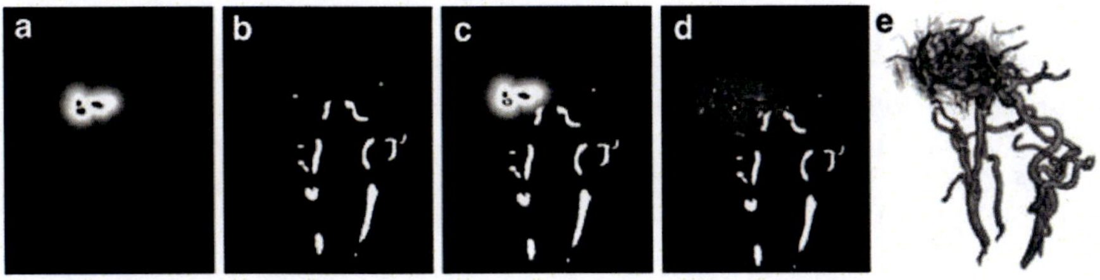

Fig. 14 To show small vessels surrounding the aneurysm, a weighted distance field (**a**) is generated from its voxelization. The mask is a combination of the main vessels (**b**) and the distance field using a maximum operator (**c**). Finally, it is weighted by volumes original intensity values (**d**) and used in the 3D view (**e**)

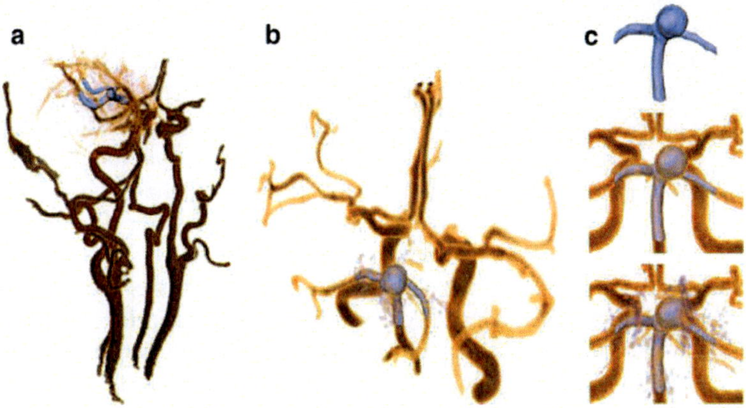

Fig. 15 The focus-and-context visualization shows both the polygonal simulation mesh and surrounding vessels via hybrid rendering. (**a**) and (**b**) show different 1D-transfer functions. The last image series illustrates the benefits of additional context information, and makes use of a grow distance field to reveal small vessels coming from the aneurysm

to the aneurysm surface, when multiplied with the original dataset (see Fig. 14d). In order to create this mask automatically, two defined constants are utilized: b and c. Both are independent variables and exhibit robust characteristics (for the tested datasets: stable results when altered +/–20%).

With the use of 1D TFs the final visualization is presented in Fig. 15. Hybrid rendering was used to integrate the polygonal simulation mesh into the final scene. To avoid artifacts around the mesh, its voxelized version was subtracted from the original volume data to avoid double representation of the same structure.

6.2 *Case Study 2: Emphasis of Plaques in Coronary Arteries*

In the second example, we present a local histogram-based approach to highlight atherosclerotic CAD. Atherosclerotic CAD is the result of accumulations in the coronary artery wall, so-called plaques. Plaque deposits are small structures with inhomogeneous densities inside the vessel wall and cannot be segmented directly in CT angiography (CTA) datasets. Therefore, a DVR visualization of the coronary artery tree should highlight the vessel wall and emphasize coronary artery plaques in CTA datasets by employing special TFs. Glasser et al. [37] developed a TF specification for DVR visualizations of the coronary arteries in CTA datasets. The TFs are automatically adapted to each dataset to account for varying CT values of the blood pool, due to the non-uniform spatial diffusion of the contrast agent. The presented workflow for TF specification is based on a coarse segmentation of the coronary artery tree including the segmentation's skeleton and contains the following steps:

1. *Approximation of the blood density values.* The segmented coronary artery tree primarily consists of voxels representing the contrast medium-enhanced blood and is analyzed for the approximation of the blood density distribution. The blood density is assumed to be a Gaussian distribution with parameters μ_{blood} and σ_{blood}. Because of other densities originating from surrounding tissue or interpolation issues, i.e. the partial volume effect, μ_{blood} and σ_{blood} could not be directly derived from the segmentation. Therefore, the approximation is carried out by means of an iterative reduction of a cost function. Costs are defined as differences between the approximated normal distribution and the distribution of all segmented voxels. The mean density of the blood (i.e. μ_{blood} and σ_{blood}) strongly differs for each dataset [37]. Therefore, a static threshold for a separation of hard plaques from blood was not applicable. Instead, a threshold t, similar to the threshold of the Agatston score [38], was introduced. To avoid overestimation, t is defined as

$$t = \mu_{blood} + 3\sigma_{blood}. \tag{11}$$

2. *Approximation of the vessel wall density values.* The approximation of the vessel wall density values is based on the evaluation of a coronary artery segment and its skeleton. The vessel wall of this segment is analyzed by means of intensity profile volumes (IPV), as described in [37]. The IPV calculation is carried out in four steps (see Fig. 16, left):

   a. Selection of a long, non-branching segment of the coronary artery tree. The selection is carried out by traversing the skeleton of the segmentation.
   b. The real vessel centerline is approximated by the segment skeleton.
   c. For each skeleton voxel, n rays perpendicular to the skeleton are casted.
   d. Along the rays, intensities are sampled and saved in a slice of the IPV.

   After the IPV extraction, the vessel wall densities can be approximated. For each skeleton voxel, the sampled intensities representing the vessel wall will appear as vertical structures in a slice of the IPV (see Fig. 16, right). The IPV is slicewise convoluted with a Gaussian and a Sobel filter by applying a 2D filter to each slice. The convolutions are followed by a search for vertical structures

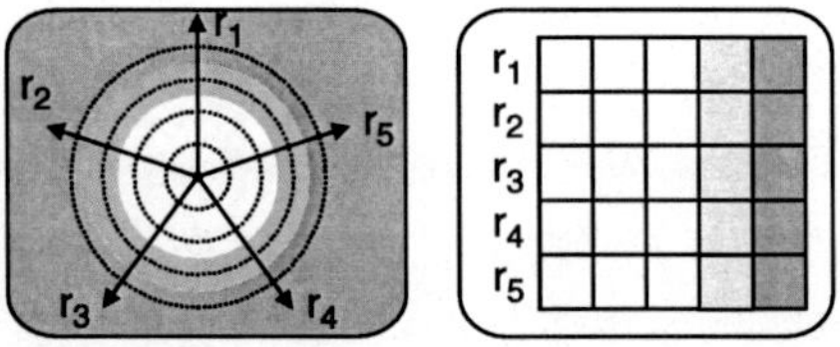

Fig. 16 Extraction of a slice of the intensity profile volume (IPV). For each voxel v of the skeleton of a vessel segment, n (e.g. 5) rays perpendicular to the centerline are casted (left). Along the rays, intensities are sampled and saved in a slice of the IPV (right). A cross-section of an ideal vessel (white) and the vessel wall (light gray) with surrounding tissue (dark gray) leads to vertical structures in the IPV slice for v. The IPV has as many slices as the number of centerline voxels

to obtain a sample set of vessel wall intensities. This sample comprises a set of intensities d_i, where a weight w_i is assigned to each d_i, see also [37]. Based on this sample, μ_{wall} and σ_{wall} are computed as weighted arithmetic mean and weighted arithmetic standard deviation, respectively:

$$\mu_{wall} = \frac{\sum w_i * d_i}{\sum w_i} \tag{12}$$

$$\sigma_{wall} = \sqrt{\sum w_i^2 * \sigma_i^2}, \tag{13}$$

where σ_i is the standard deviation for each density d_i.

3. *TF specification and DVR visualizations.* For the visualizations of the CTA datasets, TF_{2D} and TF_{3D} were determined for 2D and 3D visualizations. They solely depend on μ_{blood}, σ_{blood}, μ_{wall} and σ_{wall} and differ only in the interval size of the vessel wall densities. An interval of $\mu_{wall} \pm \sigma_{wall}$ is employed for the TF_{3D}, and an interval $\mu_{wall} \pm 2\sigma_{wall}$ for the TF_{2D}. This choice is motivated by a possible occlusion of inner structures within the vessel wall in 3D visualizations. Therefore, also smaller opacity values for the vessel wall are assigned for the TF_{3D} than for the TF_{2D}. In comparison, the vessel wall and hard plaques are mapped to higher opacity values, whereas absolute transparency is assigned to the surrounding tissue and the contrast medium-enhanced blood.

   A color scale from blue over red to green provides high contrasts for the visualization of different plaque deposits and thus different densities in the vessel wall. For hard plaques, the assigned colors range from beige to white, since these structures usually appear white or light gray in conventional CT displays. For a more intuitive view, the TF_{2D} is combined with the windowing TF. The user can choose interesting HU value intervals by manipulating the parameters of the windowing TF. The CPR and MPR view as well as the 3D spatial variation of the coronary artery tree are linked with each other. On the one hand, the user can traverse the CPR view with the corresponding MPR view being updated. On the other hand, it is possible to pick an interesting vessel part in the 3D view with the other views being updated accordingly.

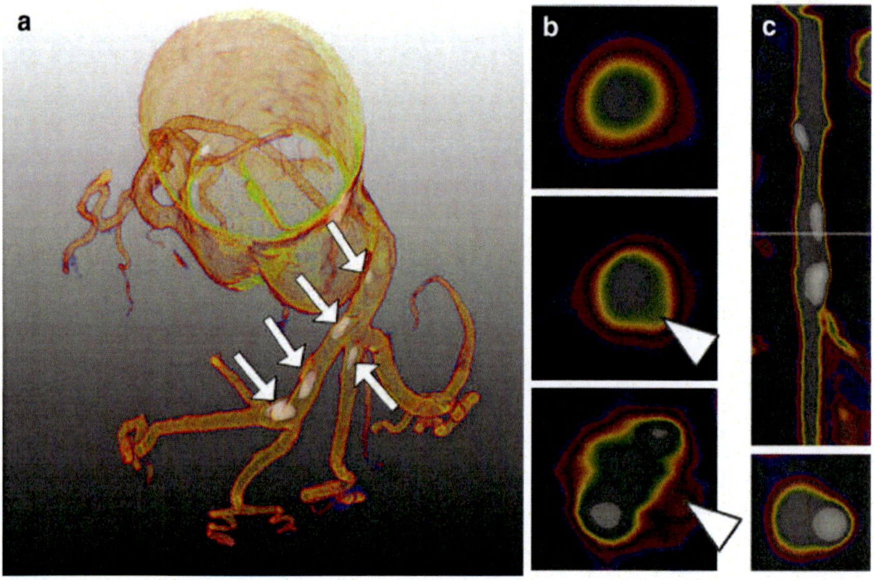

Fig. 17 Visualizations of coronary arteries with adapted TFs. (**a**) DVR visualization provides an overview about the spatial variation and of hard plaques (arrows). (**b**) MRP views provide cross-sectional views of the coronary artery walls. The type of plaque (arrowheads) may be inferred from the color coding: MPR view without plaques (top), greenish colors indicate fibrous plaques (middle) and pinkish colors indicate soft plaques, which are prone to rupture (bottom). (**c**) shows the combined CPR-MPR view. The transparent white bar indicates the current position in the CPR view (top) of the cross-sectional MPR view (bottom)

Discussions with two experienced radiologist indicate that the presented TFs highlight stenotic plaque deposits, and display smaller plaques without a significant stenosis. The 3D visualizations show the spatial variation of the coronary artery tree and indicate the patient's whole plaque burden by highlighting all hard plaques (see Fig. 17a). In the combined CPR-MPR view, the plaques are highlighted with different colors depending on the plaque type (see Fig. 17b). Comparisons between conventional views and the TF_{2D} and TF_{3D} color-coded visualizations indicate that the clinical expert can better detect hard plaques in the colored view. Especially smaller hard plaques, even with a size of one voxel, could be detected in the colored view. Furthermore, the color coding of the vessel wall with different colors for fibrous and soft plaque deposits indicates the plaque type, where an absolute identification is impossible due to overlapping density intervals of these types [39].

7 Concluding Remarks

The visualization of vascular structures is challenging due to their small size and complex topology. Various DVR techniques have been successfully employed to deal with these demands. The quality of vessel visualization benefits from preprocessing and filtering, which highlight elongated circular structures (vesselness). A histogram-based TF specification is particularly useful, when applied to local histograms, e.g. of segmented vessels. For a vessel disease diagnosis system (e.g. Vascu Vision by Boskamp et al. [40]) the combination of 2D and 3D views is essential, as only their combination can reveal the adequate detail (e.g. cross-sectional and longitudinal views to access the vessel's diameter and wall) and overview information (e.g. the patient individual spatial variation). In principle, automatically generated visualizations and an analysis of changes in the local vessel diameter might be used to emphasize potentially portions (see [28] for such attempts in cardiology). Compared to surface-based techniques, DVR may also display important changes of the vessel wall and not only its general shape. Preprocessing and detailed analysis of the data can help to create a customized workflow. To support diagnostic tasks, fast, reliable, and easy to setup techniques are favorable for clinical routine, but accuracy or visualization of uncertainty must not be compromised.

7.1 Future Work

The rising availability of memory and performance in consumer hardware, e.g. in many-core systems such as GPUs and future computing devices, will continue to aid volume analysis to enhance vessel exploration with DVR. However, not only analysis but also the rendering itself leaves room for improvements, especially when it comes to simulating advanced optical effects. Advanced illumination techniques, such as ambient occlusion (see Fig. 18 left), can improve the depth perception. Especially when many small vessels and many branchings exist, the structures may overlap each other extensively in a 2D image. Therefore, custom shading methods can give more clues about the spatial relationship among vessels and/or context structures. Other optical effects, like more sophisticated scattering and shadowing models, also raise the possibilities for visualizations (see Fig. 18 right, [41]). The major problem of these techniques remains complexity in time or resources, which limits them to high-end graphics cards or sacrificing interactivity. However, both technical and algorithmic advances continue to improve the situation in visual computing and future technology trends, such as cloud computing, will enhance the end-user experience.

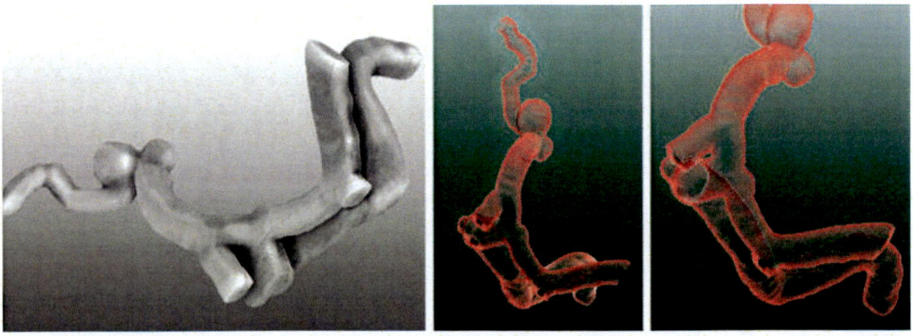

Fig. 18 Advanced illumination techniques. The left images shows parts of cerebral vessels. Through the use of self-shadowing, the geometric relationship between vessels is enhanced and fine-dotted lines show the occluded silhouettes. The flat planes at various branching points are the result of a limited region-of-interest. In the right advanced soft shadowing techniques were used on coronary vessels. (Right image taken from [41])

Acknowledgements The case studies were contributed to by O. Beuing, V. Diehl, R. Gasteiger, A. Mahnken, S. Oeltze and S. Wilhelmsen. We thank Dr. S. Achenbach (University of Erlangen-Nürnberg, Germany) for providing the coronary vessel CT image data and Fraunhofer MEVIS for providing advanced MeVisLab features. We thank P. Hastreiter, A. Joshi, S. Wesarg, A. Kanitsar and T. Ropinski for allowing us to use images from their publications and S. Quade and P. Greveson for their input.

References

1. Preim B, Oeltze S. Visualization in Medicine and Life Sciences. In: Linsen L, Hagen H, editors. 3D Visualization of Vasculature: An Overview. Springer Verlag; 2007. 19–39.
2. Selle D, Preim B, Schenk A, Heinz-Otto-Peitgen. Analysis of Vasculature for Liver Surgery Planning. IEEE Transactions on Medical Imaging. 2002 November;21(11):1344–1357.
3. Zerfowski D. Motion artifact compensation in CT. In: SPIE Medical Imaging 1998: Image Processing; 1998. p. 416–424.
4. Zheng L, Maksimov D, Stutzmann T. Bone removal in dual energy CT. In: CVII 2008: Computer Vision for Intravascular and Intracardiac Imaging; 2008. 120–127.
5. Blum H. Biological shape and visual science: Part I. J Theor Biology. 1973;38:205–283.
6. Sethian JA. Level Set Methods and Fast Marching Methods: Evolving Interfaces in Computational Geometry, Fluid Mechanics, Computer Vision, and Materials Science. 2nd ed. Cambridge University Press; 1999.
7. Schaap M, Metz CT, van Walsum T, van der Giessen AG, Weustink AC, Mollet NRA, et al. Standardized Evaluation Methodology and Reference Database for Evaluating Coronary Artery Centerline Extraction Algorithms. Medical Image Analysis. 2009;13/5:701–714.
8. Boskamp T, Hahn H, Hindennach M, Zidowitz S, Oeltze S, Preim B, et al. Geometrical and Structural Analysis of Vessel Systems in 3D Medical Image Datasets. In: Leondes CT, editor. Medical Imaging Systems: Technology und Applications. vol. V. World Scientific Press; 2005. 1–60.
9. Kanitsar A, Wegenkittl R, Felkel P, Fleischmann D, Sandner D, Gröler E. Computed Tomography Angiography:A Case Study of Peripheral Vessel Investigation. In: IEEE Visualization; 2001. 477–480.

10. Frangi AF, Frangi RF, Niessen WJ, Vincken KL, Viergever MA. Multiscale Vessel Enhancement Filtering. Springer-Verlag; 1998. 130–137.
11. Joshi A, Qian X, Dione D, Bulsara K, Breuer C, Sinusas A, et al. Effective visualization of complex vascular structures using a non-parametric vessel detection method. IEEE Transactions on Visualization and Computer Graphics. 2008;14(6):1603–1610.
12. Kindlmann G, Durkin JW. Semi-automatic generation of transfer functions for direct volume rendering. In: IEEE Symposium on Volume Visualization; 1998. 79–86.
13. Kniss J, Kindlmann G, Hansen C. Multidimensional Transfer Functions for Interactive Volume Rendering. IEEE Transactions on Visualization and Computer Graphics. 2002;8(3):270–285.
14. Vega Higuera F, Sauber N, Tomandl B, Nimsky C, Greiner G, Hastreiter P. Enhanced 3D-Visualization of Intracranial Aneurysms Involving the Skull Base. In: MICCAI (2); 2003. 256–263.
15. Rezk-Salama C, Hastreiter P, Scherer J, Greiner G. Automatic Adjustment of Transfer Functions for 3D Volume Visualization. In: VMV; 2000. 357–364.
16. Vega Higuera F, Sauber N, Tomandl B, Nimsky C, Greiner G, Hastreiter P. Automatic adjustment of bidimensional transfer functions for direct volume visualization of intracranial aneurysms. In: Proc. of Society of Photo-Optical Instrumentation Engineers (SPIE). vol. 5367; 2004. 275–284.
17. Lundström C, Ljung P, Ynnerman A. Local Histograms for Design of Transfer Functions in Direct Volume Rendering. IEEE Transactions on Visualization and Computer Graphics. 2006;12(6):1570–1579.
18. Lundström C, Ljung P, Persson A, Ynnerman A. Uncertainty Visualization in Medical Volume Rendering Using Probabilistic Animation. IEEE Transactions on Visualization and Computer Graphics. 2007;13(6):1648–1655.
19. Tappenbeck A, Preim B, Dicken V. Distance-Based Transfer Function Design: Specification Methods and Applications. In: Simulation und Visualisierung; 2006. 259–274.
20. Correa CD, Ma KL. Size-based Transfer Functions: A New Volume Exploration Technique. IEEE Transactions on Visualization and Computer Graphics. 2008;14(6):1380–1387.
21. Kanitsar A, Fleischmann D, Wegenkittl R, Gröller ME. Diagnostic Relevant Visualization of Vascular Structures. Favoritenstrasse 9-11/186, A-1040 Vienna, Austria; 2004. Human contact: technical-report@cg.tuwien.ac.at.
22. Zuiderveld KJ, Koning AHJ, Viergever MA. Techniques for speeding up high-quality perspective maximum intensity projection. Pattern Recogn Lett. 1994;15(5):507–517.
23. Sato Y NSTS Shiraga N, R K. LMIP: Local Maximum Intensity Projection: Comparison of Visualization Methods Using Abdominal CT Angiograpy. Journal of Computer Assisted Tomography. 1998;22(6):912–917.
24. Scharsach H, Hadwiger M, Neubauer A, Wolfsberger S, Bühler K. Perspective Isosurface and Direct Volume Rendering for Virtual Endoscopy Applications. In: EG/IEEE Eurovis; 2006. 315–322.
25. Vega Higuera F, Hastreiter P, Fahlbusch R, Greiner G. High Performance Volume Splatting for Visualization of Neurovascular Data. In: IEEE Visualization; 2005. 35–42.
26. Rezk-Salama C, Kolb A. Opacity Peeling for Direct Volume Rendering. Computer Graphics Forum (Proc Eurographics). 2006;25(3):597–606.
27. Bruckner S, Gröller ME. Instant Volume Visualization using Maximum Intensity Difference Accumulation. Comput Graph Forum. 2009;28(3):775–782.
28. Wesarg S, Khan MF, Firle E. Localizing Calcifications in Cardiac CT Data Sets Using a New Vessel Segmentation Approach. J Digital Imaging. 2006;19(3):249–257.
29. Kanitsar A, Fleischmann D, Wegenkittl R, Felkel P, Gröller ME. CPR: curved planar reformation. In: IEEE Visualization; 2002. 37–44.
30. Kanitsar A, Wegenkittl R, Fleischmann D, Groller ME. Advanced Curved Planar Reformation: Flattening of Vascular Structures. In: IEEE Visualization; 2003. 43–50.
31. Straka M, Cervenansky M, La Cruz A, Kochl A, Sramek M, Groller E, et al. The VesselGlyph: Focus & Context Visualization in CT-Angiography. In: IEEE Visualization; 2004. 385–392.

32. Bartrolí AV, Wegenkittl R, König A, Gröller E. Nonlinear virtual colon unfolding. In: IEEE Visualization; 2001. 411–420.
33. Ropinski T, Hermann S, Reich R, Schäfers M, Hinrichs KH. Multimodal Vessel Visualization of Mouse Aorta PET/CT Scans. IEEE Transactions on Visualization and Computer Graphics (TVCG) (Vis Conference Issue). 2009; 1515–1522.
34. Anxionnat R, Bracard S, Ducrocq X, Trousset Y, Launay L, Kerrien E, et al. Intracranial Aneurysms: Clinical Value of 3D Digital Subtraction Angiography in the Therapeutic Decision and Endovascular Treatment. Radiology. 2001; 799–808. Article dans revue scientifique avec comité de lecture.
35. Zhan C, Villa-Uriol MC, Craene MD, Pozo JM, Frangi AF. Morphodynamic Analysis of Cerebral Aneurysm Pulsation from Time-Resolved Rotational Angiography. MedImg. 2009 July;28(7):1105 – 1116.
36. Neugebauer M, Gasteiger R, Diehl V, Beuing O, Preim B. Automatic generation of context visualizations for cerebral aneurysms from MRA datasets. International Journal of Computer Assisted Radiology and Surgery (CARS). 2009 Juni;4 (Supplement 1):112–113.
37. Glaßer S, Oeltze S, Hennemuth A, Kubisch C, Mahnken A, Wilhelmsen S, et al. Automatic Transfer Function Specification for Visual Emphasis of Coronary Artery Plaque. Computer Graphics Forum. 2010;29(1):191–201.
38. Agatston AS, Janowitz WR, Hildner FJ, et al. Quantification of coronary artery calcium using ultrafast computed tomography. J Am Coll Cardiol. 1990;15(4):827–832.
39. Pohle K, Achenbach S, MacNeill B, et al. Characterization of non-calcified coronary atherosclerotic plaque by multi-detector row CT: Comparison to IVUS. Atherosclerosis. 2007;190:174–180.
40. Boskamp T, Rinck D, Link F, Kmmerlen B, Stamm G, Mildenberger P. New Vessel Analysis Tool for Morphometric Quantification and Visualization of Vessels in CT and MR Imaging Data Sets. Radiographics. 2004;24(1):287–297.
41. Ropinski T, Döring C, Rezk-Salama C. Advanced Volume Illumination with Unconstrained Light Source Positioning. IEEE Computer Graphics and Applications. 2010;Accepted for publication.

Efficient Selection of Representative Views and Navigation Paths for Volume Data Exploration

Eva Monclús, Pere-Pau Vázquez, and Isabel Navazo

Abstract The visualization of volumetric datasets, quite common in medical image processing, has started to receive attention from other communities such as scientific and engineering. The main reason is that it allows the scientists to gain important insights into the data. While the datasets are becoming larger and larger, the computational power does not always go hand to hand, because the requirements of using low-end PCs or mobile phones increase. As a consequence, the selection of an optimal viewpoint that improves user comprehension of the datasets is challenged with time consuming trial and error tasks. In order to facilitate the exploration process, informative viewpoints together with camera paths showing representative information on the model can be determined. In this paper we present a method for representative view selection and path construction, together with some accelerations that make this process extremely fast on a modern GPU.

1 Introduction

Optimal selection of viewpoints is an important task in volume rendering for improving the understanding of the inspected dataset. The amount of information revealed in an image is determined by the selected view and shading function. Although the visualization techniques are becoming faster and faster, the size of datasets also increases, and the time users are able to devote to a single analysis is limited. Automatic selection of informative views, as well as qualitative exploration paths are useful for reducing the costly manual trial-and-error good view selection. Moreover, informative views are also useful for book or articles

E. Monclús (✉) · P.-P. Vázquez · I. Navazo
Modeling, Visualization, and Graphics Interaction Group
Dep. LSI, Universitat Politècnica de Catalunya (UPC) / Spain
e-mail: emonclus@lsi.upc.edu; pere.pau@lsi.upc.edu; isabel@lsi.upc.edu

L. Linsen et al. (eds.), *Visualization in Medicine and Life Sciences II*,
Mathematics and Visualization, DOI 10.1007/978-3-642-21608-4_8,

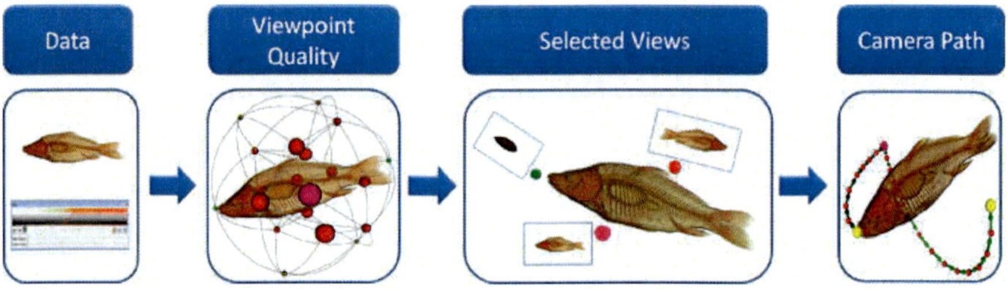

Fig. 1 Architecture of our application: Once the data is loaded and a transfer function is defined, first, an adaptive algorithm evaluates a set of views and determines the best view. With the analyzed images, we select a set of representative views that are used for a final exploration path definition

illustration, illustrating models libraries, fast model previsualization, and, in general, as a tool to optimize access to the interesting information.

In this paper we extend our previous approach [25] for representative views selection and exploration path construction. The input is a model classified through the definition of a transfer function (TF), and, optionally, the specification of a region of interest. The proposed algorithm consists of three different parts (see Fig. 1): (a) best view determination, (b) representative view selection, and (c) exploration path construction. Our method analyzes the views generated with a Direct Volume Rendering (DVR) algorithm and evaluates them based on the visible information. We use an entropy-based measure as view descriptor. Representative views are the ones that show perceivably different views of volume data. In order to determine them, we use a Kolmogorov complexity-based measure, the Normalized Compression Distance. Finally, these views are used for the generation of exploration paths around the models. The aspects of view and path selection we dealt in the previous paper are:

- The selection of a view descriptor for volume models and an adaptive algorithm for the fast selection of best views [25]. Best views can be chosen as a starting point for the exploration path for data inspection, thus reducing the human effort.
- A method based on Kolmogorov complexity that determines the set of visually different views of a volumetric model. These form the representative set of views of a model, usable to illustrate model libraries or to serve as key points for automatic exploration path construction [25].
- From the information gathered in the previous step, we build an exploration path around the model that reveals a high amount of information from it and passes through the visually different views of the model. This aids the user discovering the most important features of the model with a minimal effort.

Once the path is built, the users are given the possibility of modifying the resulting path, as well as changing the transfer function if required, in order to refine the inspection or to render the exploration path showing different information.

We improve the previous paper in various aspects:

- View descriptor evaluation: We compare our results with other viewpoint quality measures in literature. Concretely, we show that our view descriptor obtains similar best views for a set of models used by other authors. We also show the validity of the view descriptor for providing information about boundary regions [22], as views are also good if we use magnitude of gradient shading.
- Concerning the representative views selection, we provide insights on the features a compressor must fulfill in order to be used for similarity comparison and the most suitable image data format. Furthermore, we further accelerate the representative view selection by using a single color channel, as this allows to a higher reduction in file sizes, and show we maintain the quality of the results. The timings we obtain now are roughly half the ones we obtained with the previous approach.
- From the medical aspect, although it is not central to our approach, we assessed the validity of both the representative view selection and the video construction by asking two medical doctors. They believe the images generated can be useful for education materials creation. Moreover, one of the doctors also said the videos can help her for a fast diagnose of aneurysms in the aortic arch, and may serve in many cases for an initial evaluation done directly in her mobile phone.

The main advantage of our method is that it does not require any preprocess. Moreover, it gives the results in a time comparable to loading a model, and thus, the user can immediately begin the inspection starting from a good view of the model, or the resulting images or video can be quickly delivered to the medical doctor's mobile device for fast inspection.

Next section introduces previous work on view selection. Section 3 presents the visual quality evaluation measure we employ and our adaptive best view selection algorithm. In Sect. 4 we present the method for the selection of representative views based on Kolmogorov complexity. Section 5 gives details on how to build an exploration path around an object that reveals a high quantity of information from it. We discuss our results and summarize our work in Sect. 6.

2 Previous Work

Viewpoint selection has been an active research topic in many fields, such as object recognition, object modeling or cinematography, but has just recently received attention in computer graphics. Initial research focused on surface-based models, and later on, some techniques were developed for volume datasets.

In Computer Graphics, the problem of obtaining a representative view of an object is often addressed automatically, without prior knowledge on the geometry or orientation of the model. Facing the problem under these conditions is useful for many applications, such as the selection of thumbnails for objects databases, automatic camera positioning in CAD systems, automatic scene composition...

In surface-based approaches, the good viewpoints are selected taking into account the geometric information of the models. Usually, three parameters are taken into consideration: (a) the number of visible faces, (b) the area of the object or the visible faces' projections, and (c) the projected silhouette of the object. The analysis is focused under heuristic functions [16] or information theory-based approaches [24].

Sbert et al. [18] and Polonsky et al. [17] have analyzed some of the most relevant view descriptors. The main conclusion is that none of them can be coined as *universal*. On the contrary, one can always find some object whose best view is *missed* by the developed metrics. One of the relevant ones is *viewpoint entropy*, due to Vázquez et al. [24]. It is based on Shannon's entropy. The main idea is to assume the relative area of a projected polygon is a probability, and apply Shannon's formula to obtain the view that provides a higher amount of information according to this entropy.

Previous approaches for best view selection of triangular models are not amenable to direct volume rendering because the rendering result is not a set of polygons, and because the usual rendering techniques produce images where, for each pixel, more than a single (iso)value contributes to its final color. Thus, some methods have been developed in order to analyze volumetric information, such as the ones by Bordoloi and Shen [3], Ji and Shen [8], and Takahashi et al. [21]. These measures developed by these authors are referred to in literature as: voxel entropy ([3]), opacity entropy ([8]), and surface area entropy ([21]), respectively. They mainly focus on global features of interest. Tao et al. [22] present an algorithm that uses the bilateral filter for measuring global details, and combine its results with a view descriptor tailored to measure local details. Wang et al. use a two viewpoint entropy-derived measures for evaluating the quality of a viewpoint for volume rendering [27]: the total entropy and the important entropy, that rely on the visibility and importance of the voxels as captured from a certain viewpoint.

These methods do not measure quality directly on color images generated the rendered views, and thus, in some cases different structures might be treated separately although they could produce a uniform colored region on screen. In contrast to these, our approach works on the generated image, with the objective of measuring only the information that will be effectively seen by the user.

Mühler et al. [14] focus on intervention planning. They preprocess a set of viewpoints placed at 4096 positions on a bounding sphere. At each point, a set of parameter maps is computed that indicate the influence of the current quality parameter settings on the viewpoint. From this information, by using weighted parameter maps that are application dependent, the authors get the information needed to generate cutaway views of the objects of interest. Viola et al. [26] have also developed a preprocess method which takes into account, not only a viewpoint quality metric, but also information on focus of attention. Our objective is somewhat complementary to those, as we focus on the fast selection of representative views and exploration paths of models, without the need of segmenting the dataset and minimizing the manual interaction, which is possible thanks to the viewpoint quality measure we selected. However, we may also analyze selected regions of the model.

3 Efficient Best View Selection

In contrast to other approaches, we evaluate the viewpoint quality by analyzing the information that finally arrives at the user given a certain transfer function.

A lot of research has been carried out in the analysis of the information contents in an image. For instance, some techniques have been developed for the measurement of information [20], or automatic selection of lighting configuration either from an inverse engineering fashion [15], or with direct image measurement processes [6, 19, 23]. Those approaches take into account the final image, and measure the information using perceptual or information-theoretic measures. For the view quality evaluation of volume models, we use the approach by [23], that analyzes the rendered images with a multiresolution entropy measure, due to its simplicity and its robustness to noise. We will show that this metric, when applied to volumetric data renditions, leads to good views and it is robust to changes in the resolution of the rendered view.

3.1 *Multi Scale Entropy*

Multi Scale Entropy is a metric that measures the information of an image as the sum of the information at different resolution levels. The core idea behind that is that most information content metrics rely on histogram analysis which, unfortunately, are insensitive to the correlation of pixels. Multi scale entropy was introduced by Starck et al. in order to overcome these problems [20]. The idea is to use the wavelet transform to generate a set of resolutions of the image and measure the entropy of the wavelet coefficients in all the resolutions. Information is measured using Shannon entropy:

$$H(X) = -\sum_{i=1}^{N} p_i \log p_i, \tag{1}$$

where $X = \{X_1, X_2, \cdots, X_N\}$ is an image containing integer values and N is the number of different values of a pixel and p_i are the values obtained from the histogram of X, that is, the probabilities of each histogram entry. The logarithms are taken in base 2 and $0 \log 0 = 0$ for continuity; $-\log p_i$ represents the *information* associated with the result X_i. The unit of information is called a *bit*. Observe that this entropy will be 0 if all the pixels have the same value and maximum when all pixels have a different value. The entropy gives the average *information* or the *uncertainty* of a random variable

By measuring the information of a wavelet transform W of an image (measured using the wavelet coefficients $w_{l,k}$ for a fixed set of levels), the metric is less sensitive to noise (present in astronomical images, objective of the initial measure) and captures the correlation between pixels. If the number of levels is high enough, the remaining information can be considered background. We adapted this measure for the use with RGB images, and use it to determine the amount of information in a rendered view. Thus, the amount of information contained in an image generated through DVR can be measured by analyzing the Shannon entropy of wavelet coefficients of each color channel at each level:

$$H_W(X) = -\sum_{l=1}^{L}\sum_{k=0}^{N_l} h_{RGB}(w_{l,k}), \tag{2}$$

where $h_{RGB}(w_{l,k})$ is $-p(w_{(l,k)})\log p(w_{(l,k)})$, with $p(w_{(l,k)})$ being the relative number of coefficients of the channel with value k in level l. Hence, h_{RGB} means that the entropy is measured separately for each RGB channel. We have applied this method and found that it is suitable for volume rendering analysis because it reasonably analyzes the structures given a TF, as shown next. In our initial implementation we use the Haar wavelet transform, over RGB images encoded in 8 bits per color channel. As shown in [23], four levels of wavelet decomposition is usually enough.

The described measure gives good results, as shown in Fig. 2 for the hip and the feet models. Here, the quality has been measured for a dense set of viewpoints around a model, and we can observe how it computes higher values (warmer and larger spheres, being the pink sphere the best viewpoint) where more details are provided from the object.

3.2 *Comparison with Other Methods in Literature*

We also compare the best and worst views as generated by our measure with some typical models that have been analyzed by other authors when dealing with best view descriptors. The results are shown in Fig. 3. Note how the best views we select are roughly the same than the ones selected by other authors. For instance, the tooth model is also analyzed in [3, 8, 22] and they obtain results for best and worst view very similar to ours ([22] only shows the best view for the shape view descriptor for this model) . The same happens with the daisy pollen grain, where the best view selected by our technique is similar to the one by [22], as well as the hydrogen molecule (again with the view shape descriptor).

Our viewpoint quality measure does not require parameter tweaking, such as the method Ji and Shen [8]. We see this as an advantage. Moreover, we only take into account the visible information, that depends on the transfer function. Therefore, our viewpoint quality measure can also be used with images rendering the magnitude gradient. As shown in Fig. 4, we still obtain views that show information on the boundary regions of the object.

Unfortunately, for the purpose of fast model previsualization, a brute-force approach is not practical, because the analysis of a sufficient set of views takes roughly half a minute. Next, we propose some acceleration techniques in order to reduce the cost of the computation while maintaining the quality of the results.

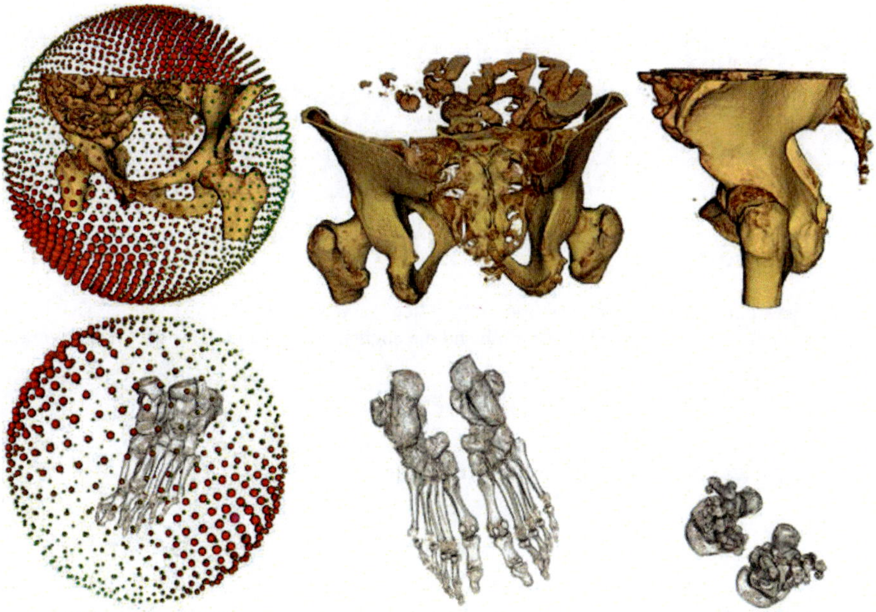

(a) Viewpoint Quality (b) Best View (c) Worst View

Fig. 2 View quality for a dense set of views around the hip and the feet models. (**a**) Viewpoint quality is encoded in both the color and the node size (the higher the value the warmer the color and larger the size). The best view is painted pink. (**b**) and (**c**) show the best and worst views

(a) Tooth (b) Daisy pollen grain (c) Hydrogen

Fig. 3 Top row shows the best views as selected by our algorithm for some typical models also analyzed by other researchers ([3, 8, 22]). Our views almost coincide exactly with the ones selected by other methods. Bottom row shows the worst views

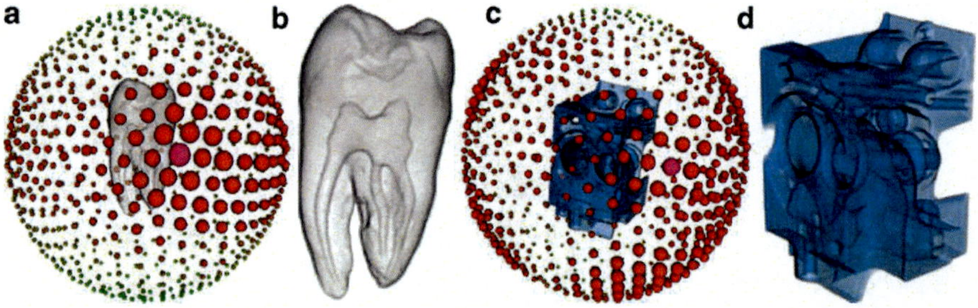

Fig. 4 Viewpoint selection for magnitude of gradient rendering. Images (**a**) and (**c**) show the viewpoint quality for the models of the tooth and the engine, and (**b**) and (**d**) show the best views for these models

3.3 Interactive Good Viewpoint Selection

As we said, analyzing a high number of views using a brute-force algorithm may be time consuming even for a modern machine. As the loading times of the models take several seconds (see Table 1) on a fast machine (a QuadCore PC with 8GB of memory with a 280 GTX card), it seems acceptable to have a good viewpoint selection algorithm that takes at most the loading time. Note (cf. Table 1) that all timings depend on the dataset dimension, number of analyzed views, and the transfer function applied (which determines the number of processed voxels), that is why a smaller model may have a worse timing than a larger one.

In order to achieve better efficiency, we proposed two improvements that greatly improve good view selection cost while preserving the quality of the results:

Viewport size reduction: View quality evaluation depends on the viewport size, reducing it will lead to improve time. Nonetheless, we have to ensure that smaller images still maintain enough information not to skew the measured quality.

Adaptive Search: In order to achieve interactive rates, we also perform the best view search in an adaptive fashion, starting from a coarse sampling of the bounding sphere and adaptively subdividing it according to an estimator of the entropy.

The adaptive approach is similar to the method by Gumhold [6]. Initially, the algorithm builds a spherical triangle mesh with the vertices placed at the vertices of an icosahedron. For each vertex i, the quality measure H_i and the total maximum H_{max} is evaluated and stored. We treat the quality function as Lipchitz-continuous in a neighborhood of the analyzed points and seek for the Lipschitz constant L of the mesh. This value gives an estimation of the maximum quality variation around a viewpoint.

Our algorithm consists of two steps: First, for each triangle of the current mesh, the maximum reachable entropy point at each edge ($He_1 \cdots He_3$) is estimated using L (see Fig. 5a). Then, we estimate the maximum value inside the triangle He_{max} by

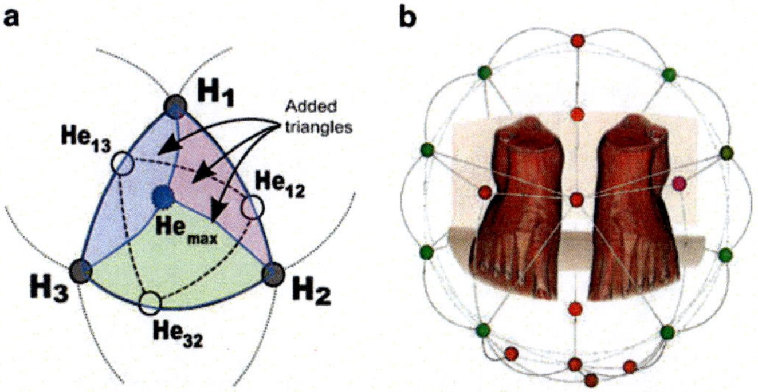

Fig. 5 Adaptive subdivision algorithm. Next subdivision point is selected by estimating the maximum reachable entropy inside the triangle. Right image shows the mesh subdivision after two steps of our algorithm for the feet model

Table 1 Comparison of the computation times for different models compared to the loading time. Fourth column shows the best brute algorithm using a set of 162 positions. The fifth and sixth columns shows the time needed to obtain the best view with our adaptive method with 256^2 and 128^2 resolution

Model	Slices	load time (s)	Best view 512^2 / 256^2 / 128^2	Adaptive 256^2 #views / time	Adaptive 128^2 #views / time
Hydrogen	128 × 128 × 128	0.0623	16.13 / 3.787 / 1.24	14 / 0.10	15 / 0.125
Daisy	192 × 180 × 168	0.148	15.91 / 3.75 / 1.41	13 / 0.30	17 / 0.155
Tooth	256 × 256 × 161	1.35	16.91 / 4.89 / 2.4	31 / 0.94	32 / 0.46
Engine	256 × 256 × 256	0.5	19.53 /5.49 / 2.40	34/ 1.226	30 / 0.42
Carp	256 × 256 × 512	6.66	19.17 / 6.36 / 2.33	22 / 0.79	32 / 0.43
Feet	512 × 512 × 282	4.1	31.69 / 11.73 / 4.3	39 / 2.76	34 / 0.84
Head	512 × 512 × 486	6.9	39.47 / 16.799 / 6.19	30 / 3.24	33 / 1.29

interpolating the computed values (He_1 to He_3). If $He_{max} + K_{safe}$ is higher than H_{max}, the actual value is measured by rendering a new viewpoint from this position and adding the subdivided triangle (see Fig. 5*a*). We add a safety constant K_{safe} in order to be conservative and avoid missing any maximum. We found experimentally that this constant is only necessary when the range of entropy values of the initial subdivision is not very large, which makes the Lipschitz constant small (below 1). Therefore, we add a $K_{safe} = 0.02 * H_{max}$ when $L < 1$. Each time a new viewpoint is added, the Lipschitz constant is recalculated. Our algorithm stops when none of the estimated entropy values is higher than H_{max}, or when those views are too close (i. e. 5 degrees) to existing analyzed positions. Figure 5*b* shows an example of the subdivision produced for the feet model.

This method obtains similar maximum values than the ones obtained with the brute-force method, but at a fraction of the time (see Table 1). Although we obtain good results when comparing with dense sets of views (we tested up to 2562 for some models), usually a lower number of regularly placed viewpoints is usually enough, being 162 a good compromise, and therefore this is the number of views we use in the timings shown in Table 1 for a fair comparison with the brute force approach. The time needed to compute the best view using our adaptive method is depicted in Table 1 for different models. Note how, in all cases, we may compute the best view in, roughly, $1/4$th of the time needed to load a model, and with low resolution viewports (i.e. 128^2) even faster.

4 Representative Views Selection

When providing images for models databases a single view may be insufficient. This is especially true for complex models, because important details of the object may still be missing or because it does not provide enough information from the 3D relationship of the structures. Selecting a set of views of high entropy may be not adequate because many objects present some level of symmetry that may lead to show similar views if quality is the only measure we consider. Moreover, our brain is very efficient in inferring symmetries, and therefore, it is more valuable to get a set of views that show information not captured in the best view. This approach is intended to *complete* the information provided by the initial view, as this could provide a better understanding of the relationships between the different structures of the model.

Previous approaches require the knowledge of structures present in the dataset, which are obtained via a segmentation process. However, we would like to avoid this preprocess. In order to do that, we propose an approach that consists in directly comparing the images obtained with the execution of the algorithm explained in Sect. 3.3. The view set that adequately represents a model is determined with a greedy scheme:

1. Select the best view B_0 with the adaptive algorithm.
2. Measure the distances from B_0 to the remaining views.
3. Next representer view B_1 is the one at the highest distance from B_0.

The algorithm cost is linear with the number of views, being the comparing process the most costly one. In order to determine the distance between two views ($d(X, Y)$), we use the Normalized Compression Distance as explained next.

Once we have the two initial representative views, if we want to gather the missed information by these two, we can proceed the same way: We compute the distances from the remaining views against B_1 and choose as new view X the one that maximizes the geometric average of the distances to B_0 and B_1. This process can be repeated several times, but three or four are usually enough for most models. In order to find the optimal set of M views, a global maximization could be applied, but would have a quadratic cost in the number of views and we obtain similar results with our approach.

At this point, the remaining work is to select a robust view similarity measure. Bordoloi and Shen [3] use the Jensen-Shannon divergence over the visible voxels, and do not take advantage of potential symmetry of the dataset, which is desirable. Nonetheless, we cannot compare views by using simple image metrics such as Mean

Square Error, as those metrics are sensitive to transformations of the image such as rotation or translation, which may become a problem when comparing two views automatically generated for model inspection. In order to solve these problems, we propose a conceptually simple solution with foundations in algorithmic complexity, concretely, we evaluate view likelihood with the *Normalized Compression Distance*.

4.1 Normalized Compression Distance

Normalized Compression Distance is a universal metric of distance between sequences. It has its roots in Kolmogorov complexity (also known as algorithmic complexity). We briefly detail here some concepts of algorithmic complexity. The interested reader can refer to Li and Vitányi's book [11] for a deeper and more theoretical introduction.

The **Kolmogorov complexity** ($K(x)$) of a string x is the length of the shortest binary program to compute x on a universal computer (such as a universal Turing Machine). Thus, $K(x)$ denotes the number of bits of information from which x can be computationally retrieved. As a consequence, strings presenting recurring patterns have low complexity, while random strings have a complexity that almost equals their own length. Hence, $K(x)$ is the lower-bound of what a real-world compressor can possibly achieve. The **conditional Kolmogorov complexity** $K(x|y)$ of x relative to y is the length of a shortest program to compute x if y is provided as an auxiliary input. Both Kolmogorov complexity and conditional Kolmogorov complexity are machine independent up to an additive constant.

Bennet et al. [2] define the **information distance** between two, not necessarily equal length binary strings as the length of the shortest program that can transform either string into the other one, both ways. The information distance is a metric. Li et al. [10] present a normalized version of information distance dubbed *similarity metric*, defined as:

$$d(x, y) = \frac{max\{K(y|x), K(x|y)\}}{max\{K(y), K(y)\}} \tag{3}$$

The authors also prove that this metric is universal (two files of whatever type similar with respect to a certain metric are also similar with respect to the similarity metric). Being Kolmogorov complexity not computable, it may be approximated with the use of a real-world compressor, leading to the **Normalized Compression Distance (NCD)**:

$$NCD(x, y) = \frac{C(xy) - min\{C(x), C(y)\}}{max\{C(x), C(y)\}} \tag{4}$$

where function $C(F)$ is the size of the compression of a certain file F, and xy is the concatenation of files x and y. Although the similarity metric has values in [0..1], NCD values are usually in the range of [0..1.1], due to compressor imperfections. NCD has been used for applications such as language classification and handwriting recognition [5]. Cilibrasi and Vitányi also analyze the conditions that compressors must fulfill in order to be used for computing the Normalized Compression Distance:

- *Idempotency*: For a repetition of a string, the compressor should be able to detect the repetitions and thus compress the file to a similar size than the original string compression $C(x, x) = C(x)$, and $C(\lambda) = 0$ where λ is the empty string.
- *Monotonicity*: The concatenation of two strings should yield to a less compressible file than taking a single string alone, up to a certain precision: $C(x, y) \geq C(x)$.
- *Symmetry*: $C(x, y) = C(y, x)$. Compression should be symmetric, that is changing the order of the concatenated strings should no affect the length of the compression.
- *Distributivity*: $C(x, y) + C(z) \leq C(x, z) + C(y, z)$. Real-world compressors seem to satisfy this property.

The data compressors with these properties are dubbed *normal compressors*. Most real-world compressors do fulfill those properties, at least to a point where they are usable for NCD computation. Some of the candidates are: stream-based (*zlib*), block-based (*bzip*), and statistical (*PPMZ*) compressors. Note that we are not searching for the best compression, as reducing the file size does not imply a reduction in NCD computation ([5]). Even with tested compressors, their behavior may change with the size of the file. As a consequence, while stream-based compressors (such as the based on Lempel-Ziv) will probably improve their behavior with respect to symmetry, because they adapt to the regularities of the string throughout the compression process, large files may affect the efficiency in NCD computation for block-based compressors. As studied by Cebrián et al. [4], in the case of *bzip2*, the *best* option works properly for files up to 900KB before being compressed. Larger sizes make the comparison processes less effective. Other predictive compressors, such as the ones belonging to the PPM family (i. e. PPMZ) are not precisely symmetric, for long strings, they are close to symmetry. If symmetry is not fulfilled to a great extent, the compressor should not be used at all.

Although bzip2 seems to be the obvious choice, the fact that we are dealing with images easily suggests other compression algorithms especially tailored for image compression. Thus, we carried out some experiments using JPEG. A simple study let us see that one of the properties that was not fulfilled by JPEG compression was monotonicity, as in some cases $C(x, y)$ was smaller than $C(x)$. As a consequence, there were some negative values for the NCD. The degree up to which monotonicity was not fulfilled was noticeable. Therefore, we sticked with *bzip2*, as the original authors of the NCD paper argue it works properly.

4.2 Applying Normalized Compression Distance to Images

To the author's knowledge, the first attempt to use Normalized Compression Distance with images is due to [9]. The authors use NCD for gray-scale image classification. In their case, a PPM-based compressor is used for object recognition. They show that the measure performs better than histogram-based approaches. A further work by Li and Zu ([12]) improves the optimization task by using a Lempel-Ziv encoding of the data and using either the dictionary or the compressed patterns for measuring the image similarity.

Bardera et al. [1] use Normalized Compression Distance for image registration. In order to do this, they select a *window* of pixels in one image and another one in the other reference image. Then, pixels are interleaved forming a new image where the red channel holds the pixels of reference image 1 and green channel the ones of reference image 2. These images are then compressed using JPEG 2000 and the compressed size is used as $C(x, y)$ in (4). Thus, the *concatenated file* is not exactly the concatenation of both images. They also present a second approach where the gray-scale values are treated as elements of a string, and bzip2 is used to compress the resulting string. Again, the values of both images are interleaved. In our case, we deal with RGB color values and the images are not interleaved, but concatenated. Note that, as commented by Macedonas et al.[13], images contain 2D information that is not trivially changed to a 1D string. They use a new similarity measure based on the dictionary generated by the compressor, and they test image comparison both taking images row by row and column by column. In their case, the row-by-row approach yields better. Our approach also takes images row-by-row.

A second question to take into account is image representation. There are many image formats that we might use: PPM, JPEG, PNG, BMP, TIFF, and so on. As we are going to use a normal compressor for view similarity and treat the images as strings, we require the image format to contain the minimum amount of information that encodes an image. This is fulfilled by PPM either in its ascii or its binary form to a great extent, as only comments (that can be removed) and the size of the images is added apart from the colors. Other formats such as the compressed ones might encode some sort of dictionary that will vary from image to image. Therefore, the simple concatenation of both files does not necessarily generate a string with redundant information that the compressor may deal with. This is exactly what happens with JPEG. If used as image format, the NCD does not work properly. Currently, no study has been done on what is the best image format and compressor. Nonetheless, our empirical results show that PPM files with bzip2 fulfill our needs. We will see later that the color images can even be replaced for gray-scale values in order to reduce bzip2 compression time, with the same results.

4.3 NCD-Based Representative View Selection

As studied by Cebrián et al. [4], the use of compressor for comparison purposes requires some issues to be taken into account, for instance, the size of the compressed files limits the efficiency of the comparison processes. In the case of *bzip2*, the *best* option works properly for files up to 900KB before being compressed. Larger sizes make the comparison processes less effective. Based on this, our initial

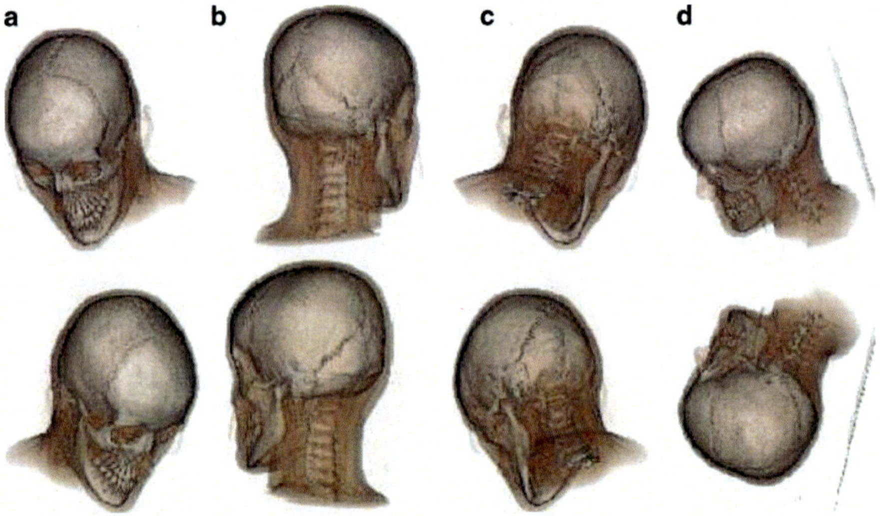

Fig. 6 Top row shows four random views of the head model, and bottom line shows view whose distance is minimum to the one on the top. Note how our similarity measure is robust with respect to rotation and symmetry. Images are measured as is, that is, without any rotation to align them

approach was to use *bzip2* compressor on PPM images. If we compare images of size 256×256 or 128×128, the size of the concatenated file never exceeds 900KB.

In consequence, to determine the distance between two images ($d(X,Y)$), we concatenate them, and then we compress the original and the concatenated files. Then, the distance is measured using (4). In Fig. 6 we show, from a set of views around the head model, the four pairs whose distance is the smallest. Note how symmetric views are correctly ranked as being very similar. An example of selected representative views is shown in Fig. 7, where the three selected views for the engine and hip models are shown. Observe that these views show substantially visually different information.

Representative views computation is quite fast (see Table 2), concatenation and compression are the most costly parts. However, in most cases we still require a total time (adaptive best view selection + representative view selection) smaller than the loading time. Moreover, we may perform the whole process in roughly one second for most of the models tested if we restrict ourselves to offscreen-viewports of 128^2. Because we measure *NCD* over rendered views, our representative view selection method yields views that *look different*. For most models, this will be a key issue, as selecting views according to the visible voxels does not guarantee that the final rendered information will enhance our knowledge of the model (it may be symmetric).

For a further acceleration on representative view selection, there is a simplification we can do. The most important cost is incurred in the compression of the images. We already reduced their size by reducing the viewport, while maintaining the quality of results. What we will do now is to reduce the amount of information

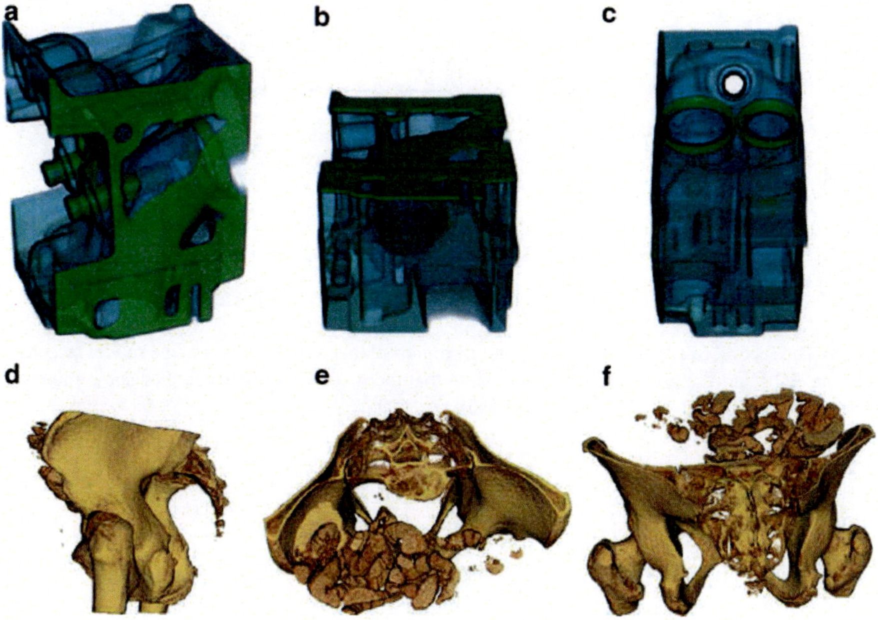

Fig. 7 The 3 representative views of the engine and the hip model. Note that they properly represent different appearances of the models

Table 2 Comparison of the representative views selection for different models compared to the loading time. All times are in seconds. Fourth and fifth column shows the time required for the representative view selection using RGB and gray-scale images, respectively

Model	Slices	load time	Represent. views RGB 256^2 / 128^2	Represent. views 1 channel 256^2 / 128^2
Hydrogen	128 × 128 × 128	0.062	0.11 / 0.12	0.06 / 0.06
Daisy	192 × 180 × 168	0.148	0.60 / 0.17	0.23 / 0.07
Tooth	256 × 256 × 161	1.35	1.00 / 0.25	0.56 / 0.14
Engine	256 × 256 × 256	0.5	2.16 / 0.37	0.61 / 0.14
Carp	256 × 256 × 512	6.66	1.14 / 0.32	0.14 / 0.14
Feet	512 × 512 × 282	4.1	2.05 / 0.44	0.69 / 0.15
Head	512 × 512 × 486	6.9	1.64 / 0.49	0.51 / 0.14

per pixel. Instead of storing a RGB value, for the representative view selection, we will work on gray-scale images. This will slightly change the results, but in all the cases we tested, the results were perfectly acceptable. We can see a comparison for the hip and the engine models. Note that the second image is now a view placed at the opposite side with respect to the second representer (cf. Figs. 7 and 8), but still shows substantially different information than the first view. Moreover, this reduces the computation time to the half, as shown in Table 2, so it might be appropriate if speed is an issue.

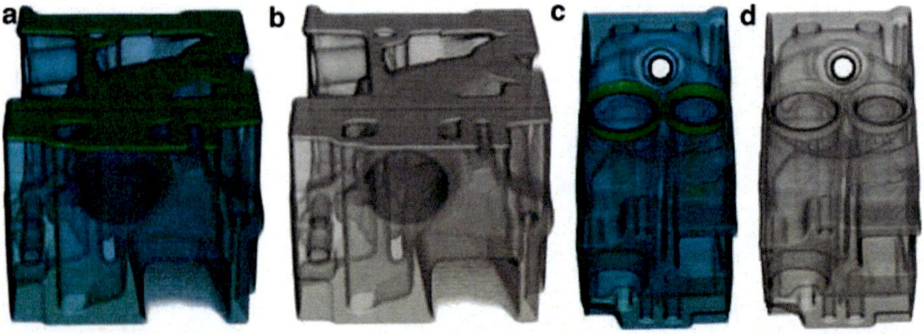

Fig. 8 The second (**a** and **b**) and third (**c** and **d**) representative views of the engine model selected with the RGB and the greyscale strategies. Note that there are not differences. For some models, some views might change a bit, but they are still acceptable

5 Exploration Path Construction

For complex objects, the construction of an inspection walkthrough may be very useful, as a set of views may be insufficient for providing a global vision of the model. Therefore, we propose an algorithm for generating exploration paths for model inspection. This path consists on a set of positions located on a bounding sphere of the object or region of interest. We use as key points the important viewpoints selected in the previous section and ensure visiting all of them. The viewpoint with the highest entropy is visited in second place. This allows the camera to visit its surroundings that, intuitively, should be more informative than the rest.

The path construction algorithm determines the minimal length path that passes through all the representative views and ensures we are providing a high amount of information. As we have only three to four candidates, an exhaustive search of the shortest path that ensures we pass through the best view in second position is computed instantaneously. We call this the *simple exploration* algorithm.

In order to maximize the information gathered through the exploration, we introduce an *improvement*: the camera may deviate from the short path up to a certain distance at each step. For each step, we analyze the entropy of three candidates, which are placed toward the destination point and separated by ± 4 to 6 degrees. The one with the highest entropy is chosen. The allowed total deviation is limited and reduced as long as we get closer to the next key point. This ensures we are visiting all keypoints. Furthermore, the speed of the camera is reduced when visiting points nearby the best one (B_0) because, intuitively, they will show a higher amount of valuable information. In Fig. 9 we can see the results with the *simple path* and the *improved path* for the exploration of the carp and the tooth models. The improved approach gathers a higher amount of details than the simple method. The *simple path* calculation time is negligible, but the *improved path* requires the evaluation of entropy at each step, and therefore, the time will depend on the number of steps we want to produce.

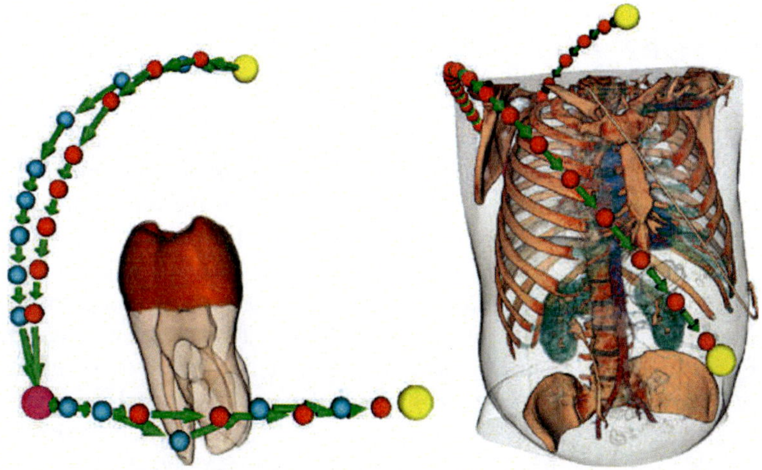

Fig. 9 Exploration paths around a tooth and the torax model

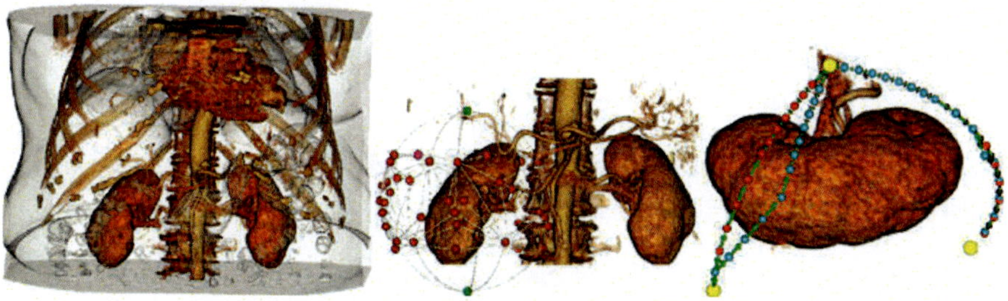

Fig. 10 Analysis of a region of interest around the kidney

An example of an exploration path built around a selected part of an model is shown in Fig. 10. In this case, the model is a torax of a human. Our region selection tool allows us to determine a spherical region around a kidney, and the analysis is performed only using this region.

6 Conclusions

Throughout the paper we have analyzed and commented upon the efficiency and quality of the results obtained applying our techniques either for the good view selection and the exploration path construction. Its fast response, ease of use, and lack of parameter definition, facilitates its incorporation in daily work by i.e radiologists at almost null cost. Its simplicity makes it easy to incorporate in a scientific volume visualization package but may also be applied to other rendering metaphors. We may also apply our technique to a region of interest defined by the user, who also defines the proper TF, and then, our algorithm computes an exploration path around the region of interest. The result is shown in Fig. 10. Once the exploration path is obtained, we can modify the TF in order to emphasize different anatomic structures and help in understanding their spatial relationship.

In this paper we have extended our previous results on viewpoint selection for volume models by providing further acceleration to the representative view selection algorithm. We have also extended the discussion on which image format to use and the properties a compressor should fulfill in order to be suitable for direct image comparison. Our method is useful, not only for volume model inspection, but also easy to incorporate to medical visualization tools, document illustration support, and automatic labeling of model libraries. For the concrete case of medical use, we asked two medical doctors: a maxillofacial surgeon and a cardiologist. Both of them agreed that the resulting views were useful for the preparation of medical education materials. Moreover, the cardiologist suggested to have the prepared videos delivered onto her iPhone in order to perform a fast diagnosis. Concretely, she believes that aneurisms at the aortha arch can be easily detected, and, in some cases, other aneurisms may be initially diagnosed. This conforms to previous research on video creation for diagnosis [7], although our video generation algorithm is more general, in the sense that can be applied to any models, and cardiologic diagnostic may be an useful result, but was not our goal. Our method is adequate for fast previsualization of models, and can be used to select initial inspection views for limited capacity devices such as iPhones or low-end PCs.

Acknowledgements Supported by TIN2007-67982-C02-01 of the Spanish Government. Thanks to "The Volume Library" at www9.informatik.uni-erlangen.de/External/vollib/ for the datasets.

References

1. Bardera, A., Feixas, M., Boada, I., Sbert, M.: Compression-based image registration. In: Proc. of IEEE International Conference on Information Theory. IEEE (2006)
2. Bennett, C., Gacs, P., Li, M., Vitanyi, P., Zurek, W.: Information distance. IEEETIT: IEEE Transactions on Information Theory **44** (1998)
3. Bordoloi, U., Shen, H.W.: View selection for volume rendering. In: IEEE Visualization, 487–494 (2005)
4. Cebrián, M., Alfonseca, M., Ortega, A.: The normalized compression distance is resistant to noise. IEEE Transactions on Information Theory **53**(5), 1895–1900 (2007)
5. Cilibrasi, R., Vitanyi, P.: Clustering by compression. IEEE Trans. Information Theory **51**(4), 1523–1545 (2005)
6. Gumhold, S.: Maximum entropy light source placement. In: Proc. of the Visualization 2002 Conference, 275–282. IEEE Computer Society Press (2002)
7. Iserhardt-Bauer, S., Hastreiter, P., Tom, B., Kötner, N., Schempershofe, M., Nissen, U., Ertl, T.: Standardized analysis of intracranial aneurysms using digital video sequences. In: In Proceedings Medical Image Computing and Computer Assisted Intervention, 411–418. MICCAI, Springer (2002)

8. Ji, G., Shen, H.W.: Dynamic view selection for time-varying volumes. IEEE Transactions on Visualization and Computer Graphics **12**(5), 1109–1116 (2006)
9. Lan, Y., Harvey, R.: Image classification using compression distance. In: Proceedings of the 2nd International Conference on Vision, Video and Graphics, 173–180 (2005)
10. Li, M., Chen, X., Li, X., Ma, B., Vitanyi, P.: The similarity metric. IEEE Transactions Informmation Theory **50**(12), 3250–3264 (2004)
11. Li, M., Vitanyi, P.M.: An Introduction to Kolmogorov Complexity and Its Applications. Springer-Verlag, Berlin (1993)
12. Li, M., Zhu, Y.: Image classification via lz78 based string kernel: A comparative study. In: PAKDD, 704–712 (2006)
13. Macedonas, A., Besiris, D., Economou, G., Fotopoulos, S.: Dictionary based color image retrieval. J. Vis. Comun. Image Represent. **19**(7), 464–470 (2008)
14. Mühler, K., Neugebauer, M., Tietjen, C., Preim, B.: Viewpoint selection for intervention planning. In: EG/ IEEE-VGTC Symposium on Visualization, 267–274 (2007)
15. Patow, G., Pueyo, X.: A survey on inverse rendering problems. Computer Graphics Forum **22**(4), 663–687 (2003)
16. Plemenos, D., Benayada, M.: Intelligent display in scene modeling. new techniques to automatically compute good views. In: Proc. International Conference GRAPHICON'96
17. Polonsky, O., Patanè, G., Biasotti, S., Gotsman, C., Spagnuolo, M.: What's in an image? The Visual Computer **21**(8-10), 840–847 (2005)
18. Sbert, M., Plemenos, D., Feixas, M., Gonzalez, F.: Viewpoint quality: Measures and applications. In: L. Neumann, M. Sbert, B. Gooch, W. Purgathofer (eds.) Computational Aesthetics in Graphics, Visualization and Imaging, 185–192. EuroGraphics Digital Library (2005)
19. Shacked, R., Lischinski, D.: Automatic lighting design using a perceptual quality metric. Computer Graphics Forum (Proceedings of Eurographics 2001) **20**(3), C–215–226
20. Starck, J., Murtagh, F., Pirenne, B., Albrecht, M.: Astronomical image compression based on noise suppression. Publications of the Astronomical Society of the Pacific **108**, 446–455 (1998)
21. Takahashi, S., Fujishiro, I., Takeshima, Y., Nishita, T.: A feature-driven approach to locating optimal viewpoints for volume visualization. In: IEEE Visualization, 495–502 (2005)
22. Tao, Y., Lin, H., Bao, H., Dong, F., Clapworthy, G.: Structure-aware viewpoint selection for volume visualization. Visualization Symposium, IEEE Pacific **0**, 193–200 (2009)
23. Vázquez, P.: Automatic light source placement for maximum illumination information recovery. Computer Graphics Forum **26**(2), 143–156 (2007)
24. Vázquez, P.P., Feixas, M., Sbert, M., Heidrich, W.: Viewpoint selection using viewpoint entropy. In: Proceedings of the Vision Modeling and Visualization Conference (VMV-01), 273–280. Stuttgart (2001)
25. Vázquez, P.P., Monclús, E., Navazo, I.: Representative views and paths for volume models. In: SG '08: Proceedings of the 9th international symposium on Smart Graphics, 106–117. Springer-Verlag, Berlin, Heidelberg (2008)
26. Viola, I., Feixas, M., Sbert, M., Gröller, M.E.: Importance-driven focus of attention. IEEE Transactions on Visualization and Computer Graphics **12**(5), 933–940 (2006)
27. Wang, Y., Zhou, D., Zheng, Y., Wang, K., Yang, T.: Viewpoint selection using PSO algorithms for volume rendering. In: IMSCCS '07: Proceedings of the Second International Multi-Symposiums on Computer and Computational Sciences, 286–291. IEEE Computer Society, Washington, DC, USA (2007)

Feature Preserving Smoothing of Shapes Using Saliency Skeletons

Alexandru Telea

Abstract We present a novel method that uses shape skeletons, and associated quantities, for feature-preserving smoothing of digital (black-and-white) binary shapes. We preserve, or smooth out, features based on a saliency measure that relates feature size to local object size, both computed using the shape's skeleton. Low-saliency convex features (cusps) are smoothed out, and low-saliency concave features (dents) are filled in, respectively, by inflating simplified versions of the shape's foreground and background skeletons. The method is simple to implement, works in real time, and robustly removes large-scale contour and binary speckle noise while preserving salient features. We demonstrate the method with several examples on datasets from the shape analysis application domain.

1 Introduction

Noisy shapes occur in a wide variety of applications and domains, such as image acquisition and processing in medical imaging and computer vision, object retrieval and matching from shape databases, image compression, and contour simplification in computer graphics. Considerable work has been invested in developing processing methods that are able to remove certain characteristics of a given shape, regarded to be noise from the perspective of the application at hand, and keep (or even enhance) other characteristics, known as *features*. In the following, we shall refer to such methods as feature-preserving smoothing methods.

Numerous feature-preserving smoothing techniques have been developed. They differ in several respects, e.g. the definition of what are features (to preserve) and noise (to be removed), shape representation (implicit or explicit), shape

A. Telea (✉)
Department of Computer Science, University of Groningen, the Netherlands
e-mail: a.c.telea@rug.nl

L. Linsen et al. (eds.), *Visualization in Medicine and Life Sciences II*,
Mathematics and Visualization, DOI 10.1007/978-3-642-21608-4_9,

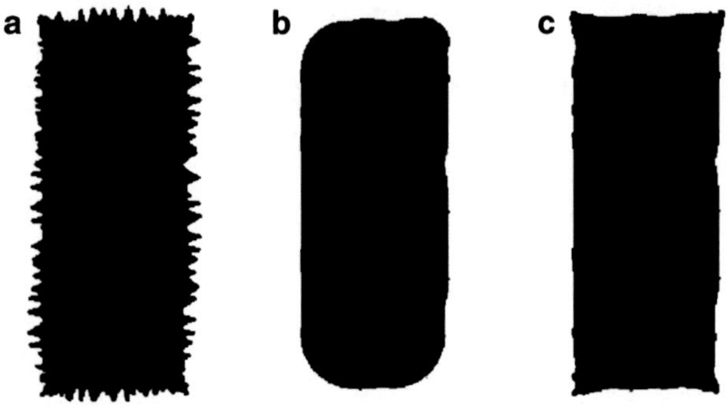

Fig. 1 (**a**) Noisy shape. (**b**) Local smoothing affects the corners. (**c**) Feature-preserving smoothing removes noise but keeps corners sharp

dimensionality (typically 2D or 3D), and space discretization (based on a fixed, uniform, pixel or voxel spatial grid or an explicit unstructured contour representation). If we consider shapes $\Omega \subset \mathbb{R}^n, n \in \{2, 3\}$ as having (closed) orientable and locally differentiable boundaries $\partial\Omega$, then a common definition of features and noise is based on the analysis of local perturbations of the surface $\partial\Omega$, typically in normal direction.

Traditionally, such perturbations are measured using various instruments based on higher-order surface derivatives, such as gradients, principal components or moments. Although a wealth of such techniques exists, curvature estimation on noisy shapes, discretized or not, is still a delicate process. This type of smoothing typically implies some form of signal filtering. By e.g. inversely correlating the filter width with the strength of the feature signal, smoothing can achieve a certain degree of feature preservation. Filtering can be applied on several scales, thereby smoothing the shape at several levels of detail. However, filtering of discrete signals usually cannot avoid some finite amount of undesired smoothing, *e.g.* in regions where one wants to preserve features. Figure 1 illustrates the idea: Typical curvature-based smoothing will produce the smooth shape (b) from the noisy shape (a), thereby removing cusps and dents but also smoothing out the perceptually important rectangle corners. The method proposed here produces image (c), which smooths out the noise but keeps the corners sharp at pixel level.

In this paper, we approach the goal of smoothing shapes in a feature-preserving manner from a different angle. Our scope is digital, or binary, shapes, e.g. black-and-white images obtained from grayscale segmentation. Such shapes admit a clear, unambiguous, definition of foreground, background, and boundary. Given such shapes, we proceed as follows. First, we characterize both features and noise on a shape's boundary $\partial\Omega$ by a new saliency metric computed on the shape's skeleton or medial axis $S(\Omega)$. The saliency metric relates the size of a boundary perturbation, encoded by the skeleton's so-called importance metric, to the local object size, encoded by the shape's distance transform. Secondly, we prune the saliency-attributed skeleton by simple upper thresholding. Due to the properties of our saliency metric on the skeleton $S(\Omega)$, this effectively removes all features below a given saliency value but fully preserves features above that value, a property which is relatively difficult to achieve when using purely local techniques. Thirdly, we reconstruct the smoothed shape by inflating the pruned skeleton. By considering both the foreground and background skeletons of a digital image, we can smooth out cusps (protrusions) and fill dents (concavities) respectively.

The paper is structured as follows. Section 2 overviews related work in the area of feature-preserving shape smoothing and related medial axis methods. Section 3 introduces our new saliency metric. Section 4 describes the feature-preserving smoothing algorithm. Section 5 presents implementation details. Section 6 presents several results obtained on different datasets related to medical and life sciences, and discusses our method. Section 7 discusses various aspects of the presented method. Finally, Sect. 8 concludes the paper.

2 Related Work

Shape smoothing methods can be classified along the way in which features and noise are defined, measured, and represented, as follows.

2.1 Local Methods

A first class of methods models features and noise locally, using the surface curvature tensor. In this way, corners of 2D shapes and corners and edges of 3D shapes can be detected.Many methods exist for curvature evaluation on discrete surfaces, e.g. [5, 7, 15, 18]. Besides curvature, surface classifiers can be based on related integral quantities, such as moments [6]. Gaussian scale-space representations of progressively smoothed shapes can be obtained by successive convolutions [14].

Distinction between features and noise are typically taken by analyzing the magnitude of the detector signal. Smoothing attempts then to preserve the former, and remove the latter, by e.g. convolving the classifier signal with filters correlated with the signal's strength [5,31,32]. When normals are present in the input data, they can be used to efficiently perform feature-preserving filtering [10, 17]. However, if normals lack, their computation from position data involves a finite amount of smoothing, similar to curvature estimation. All in all, since differential classifiers decide whether a certain signal variation at point $x \in \partial\Omega$ is a feature or noise based on the analysis of a small neighborhood $N(x) \subset \partial\Omega$ of size, or diameter, δ, around x, they will typically smooth both features and noise details below scale δ. Also, discrete curvature computations need to be regularized, typically by local integration over a neighborhood of finite size smaller or equal to the noise size δ, which introduces some artificial smoothing. An early example of the challenges involved in curvature-based salient feature detection in a multiresolution setting is offered by [9]. A good overview of the above-mentioned challenges involved in discrete curve curvature computations is given in [12].

2.2 Global Methods

A second class of smoothing methods models features and noise in a more global manner, using the so-called skeleton, or medial axis, of a shape, defined as follows. For a shape boundary $\partial\Omega$, we first define the distance transform $D : \Omega \rightarrow \mathbb{R}_+$ and the feature transform $F : \Omega \rightarrow \mathscr{P}(\partial\Omega)$ as

$$D(x \in \Omega) = \min_{y \in \partial\Omega} \|x - y\| \tag{1}$$

$$F(x \in \Omega) = \{y \in \partial\Omega | \, \|x - y\| = D(x)\} \tag{2}$$

The skeleton of $\partial\Omega$ is defined as the locus of centers of maximally inscribed balls. Since these balls touch $\partial\Omega$ in at least two feature points [13], we have

$$S(\Omega) = \{x \in \Omega | \quad |F(x)| \geq 2\} \tag{3}$$

It is well known that the terminations of the skeleton branches (which are curves in 2D and manifolds in 3D) map, via the feature transform, to curvature maxima on $\partial\Omega$ [13]. Several methods exploit this property to smooth a shape by pruning, or regularizing, its skeleton and then inflating it back. Pruning a so-called skeleton terminal branch $b \subset S$, i.e. one branch which captures a boundary curvature maximum, effectively corresponds to replacing the points $F(x \in b) \subset \partial\Omega$ by a circle arc (in 2D) or spherical sector (in 3D). Skeleton-based shape simplification is intimately related to differential shape processing, e.g. curvature flow: The so-called skeleton scale-space obtained by computing skeletons of increasingly smoothed versions of a given shape Ω, corresponds to increasingly pruning $S(\Omega)$ from its endpoints to its center [4, 16, 19].

Hisada et al. detect salient features (edges) of polygonal surfaces by extracting the 3D skeleton, detecting the terminations (edges) of the separate 3D skeletal sheets, and mapping these back to the shape [11]. Since both the shape and skeleton are represented as a non-uniformly sampled point set, problems arise with the density and continuity of the detected salient features. The skeleton computation, based on Voronoi techniques and related to the power crust [2], is extremely noise sensitive. Robustness is achieved by Laplacian smoothing of both the surface and it skeleton, but this actually removes salient details one wants to find.

Computing robust, exact, and connected skeletons is, however, perfectly doable for noisy 2D and 3D shapes. Among others, Telea et al. achieved this by defining the importance $\rho(x)$ of each $x \in S$ as the longest shortest-path length between any two feature points of x, i.e. $\max_{a,b \in F(x)}(\arg\min_{P \subset \partial\Omega} \|P(a,b)\|)$ where $P(a,b)$ is a path, or boundary segment, on $\partial\Omega$ between the feature points a, b of x.

Intuitively, ρ is the boundary length subtended by a skeleton point's features. Efficient implementations are provided for 2D images [30] and 3D voxel volumes [22]. A key observation is that ρ is minimal at skeleton endpoints and increases monotonically towards the skeleton's center or root [16, 22]. Hence, a regularized robust skeleton S_ρ can be obtained by upper-thresholding ρ with a desired value ρ_{min}. Using this property, Reniers et al. proposed a skeleton-based classifier for 3D shapes, which detects salient features such as valleys and ridges simply by computing the image of $F(x \in S, \rho(x) > \rho_{min})$ for a given saliency value ρ_{min}. This classifier showed to be more noise-resistant than curvature-based classifiers e.g. [27]. However, the classifier is not further used to smooth the shape.

Tek and Kimia presented probably the earliest result in boundary smoothing using skeletons [28]. They iteratively smooth a 2D shape by applying a so-called splice transform, which removes terminal branches from both the inner and outer skeletons. Removals are ordered by saliency measure, which is defined as the area difference between the smoothed and original shapes divided by skeleton branch length. The splice transform, however, needs to manipulate a relatively complex graph representation of the skeleton, and maintain explicit connections between explicit (non pixel-aligned) curve representations of boundary segments and fixed pixel-grid-aligned representations of the distance transform.

Our proposal is related to [28] as follows. We exploit the same principle of smoothing a shape by pruning its internal and external skeletons. However, our saliency metric is different (see Sect. 3), and so is the skeleton pruning order, algorithm, and results (Sects. 5 and 7). We work fully in a pixel-based setting, without the need of maintaining a skeleton graph representation or to explicitly manipulate boundary curve segments. In particular, we obtain the reconstructed (smoothed) shape by inflating the regularized skeleton using its distance transform, rather than explicitly editing the boundary to replace fragments by circular arcs. This yields a much simpler overall implementation with arguably lower complexity, which enables us to smooth considerably more complex shapes (Sect. 6).

3 Skeleton-Based Saliency Metric

Our general aim is similar to that of Tek and Kimia [28]: We want to build a multiscale shape representation so that perceptually salient features, such as sharp corners, are retained on coarse scales. Consider Fig. 1: On a coarse scale, we arguably see a rectangle with sharp corners but no noise on the edges (Fig. 1c), and not a rectangle with rounded corners (Fig. 1b). Hence, the saliency of a feature relates not just to its size, but to whether that feature is relevant for the local interpretation of the object (see also [8]). In the following, we consider first convex salient features. Concave features are treated similarly (see Sect. 4.4).

The first step in our saliency metric design is to measure the size of a feature. For this, we use its boundary length, which is exactly the importance metric ρ in Sect. 2 [30]. Consider the case of the noisy rectangle in Fig. 1. Figure 2 shows ρ using a blue-to-red colormap. Terminal skeleton branch pixels have low ρ values (blue), whereas pixels closer to the skeleton center have high values (red).

Let us compare the skeleton branches corresponding to the upper-right rectangle corner, respectively the neighboring small-scale noise in this figure. Along the

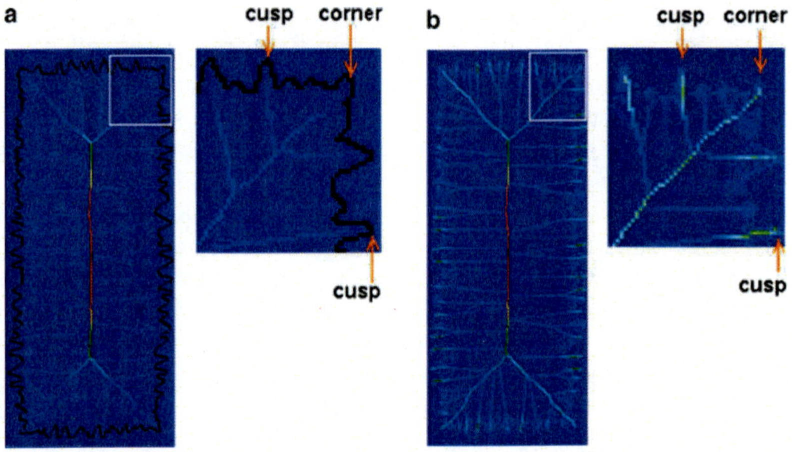

Fig. 2 Comparison of skeleton importance ρ (**a**) with saliency metric σ (**b**), both shown with a blue-to-red colormap. While ρ stays constant along ligature branches, σ decreases markedly: the signal ρ is low (dark blue) along the branch going the rectangle corner, while σ is high over the *entire* branch going to the corner but low along ligature portions of branches going to cusps

corner branch, ρ increases steadily. Along the two cusp branches, ρ increases until the branches leave the cusp and enter the rump of the shape. After that, ρ stays constant on that branch, indicating the presence of so-called ligature points [3], i.e. skeleton points that connect a branch fragment, corresponding to a small protrusion, with the skeleton's main part.

By upper thresholding ρ with some desired value ρ_{min}, and inflating back the regularized skeleton S_ρ, we achieve smoothing which essentially replaces all boundary features shorter than ρ with circle arcs (Fig. 1). However, what we want is to keep the rectangle's corners sharp and smooth out the cusp.

We first make two observations concerning the perceptual saliency of a shape detail:

- Saliency is proportional with *size*, which can be measured by boundary length. Longer features are more salient than shorter ones [16].
- Saliency is inversely proportional with local object thickness. A feature located on a thick object is less salient than the same feature located on a thin object [28].

Hence, we can define a saliency metric σ on the skeleton as

$$\sigma(x \in S(\Omega)) = \frac{\rho(x)}{D(x)} \tag{4}$$

where $\rho(x)$ is the skeleton importance defined as in [30] and $D(x)$ is the distance transform (1).

Figure 2b shows the saliency computed for the noisy rectangle shape. We see that σ stays constant along the branches corresponding to the rectangle corners, but decreases rapidly along ligature portions of the small-scale cusp branches, as indicated from the light-blue to dark-blue color transitions. Following (4), σ is 0 for all non-skeletal points, has a constant value of $2/\tan(\alpha)$ for points along the skeleton branch of an angular cusp of aperture angle α, and has a theoretical maximal value of $|\partial\Omega|/\Phi$, where $|\partial\Omega|$ is the shape's perimeter and $F = \min_{x \in S(\Omega)} D(x)$ is the shape's minimal local thickness.

4 Saliency-Preserving Smoothing

We can now use σ to selectively smooth out low saliency features. In the following, we consider low-saliency convex features, or cusps (see e.g. Fig. 1). Concave features are treated analogously (see Sect. 4.4). The entire smoothing pipeline, illustrated in Fig. 3, is described below.

4.1 Saliency-Based Pruning

The first smoothing step is to prune the skeleton S to S_σ retaining all points where σ exceeds a minimal saliency σ_{min}

$$S_\sigma = \{x \in S | \sigma(x) \geq \sigma_{min}\} \tag{5}$$

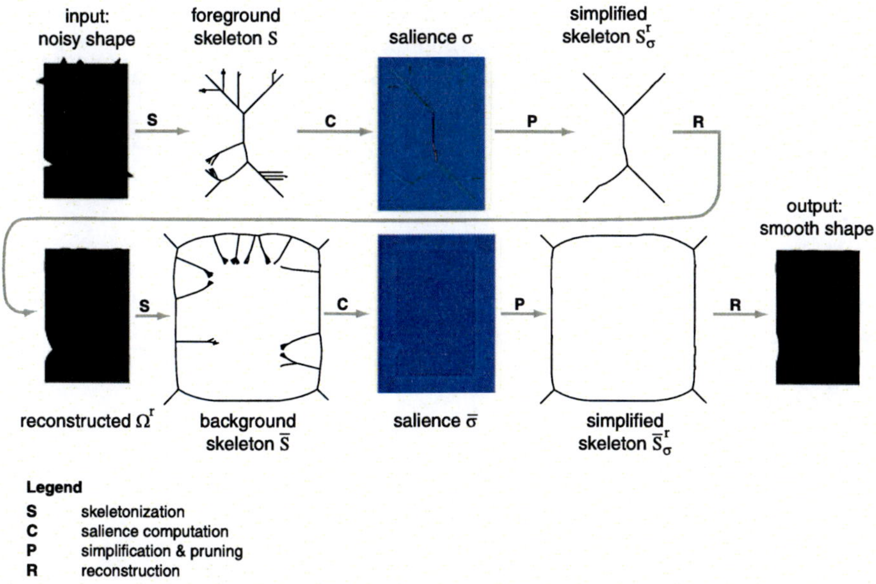

Fig. 3 Feature-preserving smoothing pipeline. See Sect. 4 for details

In (5), S denotes the *full* skeleton of a given shape Ω. In practical terms of the implementation used (see Sect. 5), this means all image points where $\rho > 1$. Using the full skeleton, thus having a method able to compute it with pixel precision, is essential, as we later inflate this skeleton and wish to exactly reconstruct salient features (Sect. 4.3).

4.2 Connected Component Detection

Since σ is not monotonic over Ω, S_σ will be a collection of disconnected, compact, skeletal components. For 2D shapes, disconnection immediately follows the thresholding in (5), since the skeleton S computed by the AFMM is always one pixel thick [21]. Note, however, that this is not always valid for 3D shapes (Sect. 7). These skeletal components correspond to multiple small fragments located inside low-saliency features, and one large *rump* fragment. Note that this holds also for higher-genus shapes, i.e. shapes with holes.

The second step isolates the skeleton rump component, $S_\sigma^r \subset S_\sigma$, defined as the connected component of S_σ which passes through $\max_{x \in \Omega} \rho(x)$. This is easily done e.g. by performing a 8-connected flood-fill on S_σ from the maximum of ρ. S_σ^r is a different regularization of S than S_ρ: While both eliminate ligature branches corresponding to small-scale noise, S_ρ also shortens branches corresponding to salient features by the same factor ρ. In contrast, S_σ^r does not touch salient feature branches.

The fact that the above process eliminates *entire* ligature branches, but does not touch salient branches, is thus essential to our method. If this were not the case, reconstruction of the shape from the simplified skeleton would not just remove small-scale noise, but also smooth out salient details, as shown in the example in Fig. 1.

4.3 Reconstruction

The third and last step inflates S_σ^r back to reconstruct a smoothed shape Ω^r. For this, we use the well-known Fast Marching Method [24]: We evolve each point $x \in S_\sigma^r$ until it reaches its distance transform value $D(x)$. To simplify the stopping criterion, we actually solve

$$|\nabla T| = 1 \tag{6}$$

$$T|_\Omega = -D \tag{7}$$

on Ω, and obtain the reconstructed shape as the level set $T(0)$. This effectively removes all shape details which correspond to the pruned skeleton branches.

4.4 Concave Features

The result of the three-step procedure outlined above effectively eliminates small scale *convex* features (cusps), while it preserves the coarse-scale convex features intact. However, we would like to treat *concave* features (dents) similarly. For this, we cannot use the shape's skeleton directly. While convexities map to terminal branches, which we can easily prune as described in Sect. 4.1, concavities map to so-called inner skeleton branches [3], which cannot be edited with the same ease, as they also describe other non-noise shape points.

An elegant solution is to use the background skeleton $\bar{S}$, i.e. the skeleton of all points $\bar{\Omega}$ located outside Ω. While S describes (and allows the selective smoothing of) the shape's convexities, $\bar{S}$ describes, and allows the selective smoothing of, the shape's concavities. Figure 3 (bottom row) illustrates this by showing the entire process described in Sects. 4.1–4.3 for the background of the rectangular shape in the image. The background skeleton $\bar{S}$, saliency $\bar{\sigma}$ and simplified skeleton $\bar{S}^r$ are computed using the same settings as their foreground counterparts. Note that the background salience image in Fig. 3 is almost everywhere dark blue. This is correct, since $\bar{\sigma}$ is very low everywhere for this shape. Indeed, the shape has no salient *dents* which are to be preserved, like it would be the case, for example, with the four concave corners of a cross shape.

4.5 Convex and Concave Smoothing Combination

If we simply apply the pipeline described above separately to the foreground, respectively background, of a given image, we obtain two smoothed shapes Ω^r and $\bar{\Omega}^r$ from which the cusps, respectively dents of the original shape Ω have been removed. However, how can we combine these two shapes in one final result? Using e.g. a distance-based interpolation between Ω^r and $\bar{\Omega}^r$ would be doable, but it would have the undesired effect of reducing the elimination of the cusps and dents to roughly half.

A better solution is to apply the concave smoothing pass (Sect. 4.4) on the background of the output of the convex smoothing pass, i.e. on $\bar{\Omega}^r$ (Sect. 4.3) rather than on the background of the original image Ω. This effectively combines the dent and cusp removal in one single result. Intuitively, the foreground skeleton simplification pulls cusps inwards, whereas the background skeleton simplification pulls dents outwards. Figure 3 bottom-right shows the final result. Note that the foreground skeleton simplification can also affect branches which describe the convex borders of a dent. For example, the smoothed Ω^r in the same figure shows a rounded dent on the left side. However, this is not a problem: This dent will be further removed by the background skeleton simplification pass.

Conversely, it is possible to apply the background skeleton simplification on the original Ω and then the foreground simplification on the result thereof, i.e. Ω^r.

We have tried this method on the images shown in this paper and obtained almost pixel-identical results with the previous solution. Again, this is due to the fact that foreground simplification does not make dents deeper (since S branches only model cusps) and a background simplification does not make cusps shallower (since $\bar{S}$ branches only model dents). The small, maximally two-pixel, differences between the results of different convex-concave pass orders are due to our current skeletonization code which models boundaries slightly asymetrically, i.e. Ω boundaries as foreground pixels having background neighbors and $\bar{\Omega}$ boundaries as background pixels having foreground neighbors.

Performance-wise, the pass ordering is also not relevant. Removing a cusp decreases the computational effort for $\bar{S}$ by two branches corresponding to the two convexities surrounding the cusp in $\bar{\Omega}$, and removing a dent decreases the computational effort for S by two branches corresponding to the two convexities surrounding the dent in Ω.

5 Implementation

Computing σ for any pixel or voxel-based shape is immediate. Here, we use the AFMMStar implementation [21,30] which delivers us the skeleton S, importance ρ, and distance and feature transforms D and F. The AFMMStar implementation is available at [29].

To prevent division by zero in (4), we only compute σ, in a pixel representation of Ω, only for the pixels strictly inside Ω and not on $\partial\Omega$ where D is by definition zero. For the inside pixels, D is by definition at least one.

The entire smoothing method is implemented in under 300 lines of C++. As expected, the method is very efficient: Images of roughly 1000^2 pixels can be processed in under one second on a 2.5 GHz PC computer. For an image of N pixels, the complexity of the algorithm is $N \log N$, determined by the underlying Fast Marching Method [24].

6 Results

Figure 4 shows feature-preserving smoothing applied to several datasets, details as follows.

The first two rows (a-h) show pairs of noisy shapes and their respective smoothings. The shapes are used as test cases for several papers in the image processing community (e.g. [12, 16]). Small-scale noise is eliminated, and sharp coarse-scale corners are kept unsmoothed, e.g. the tails of the leafs. The method is able to handle relatively large-scale noise, as illustrated by the star figure.

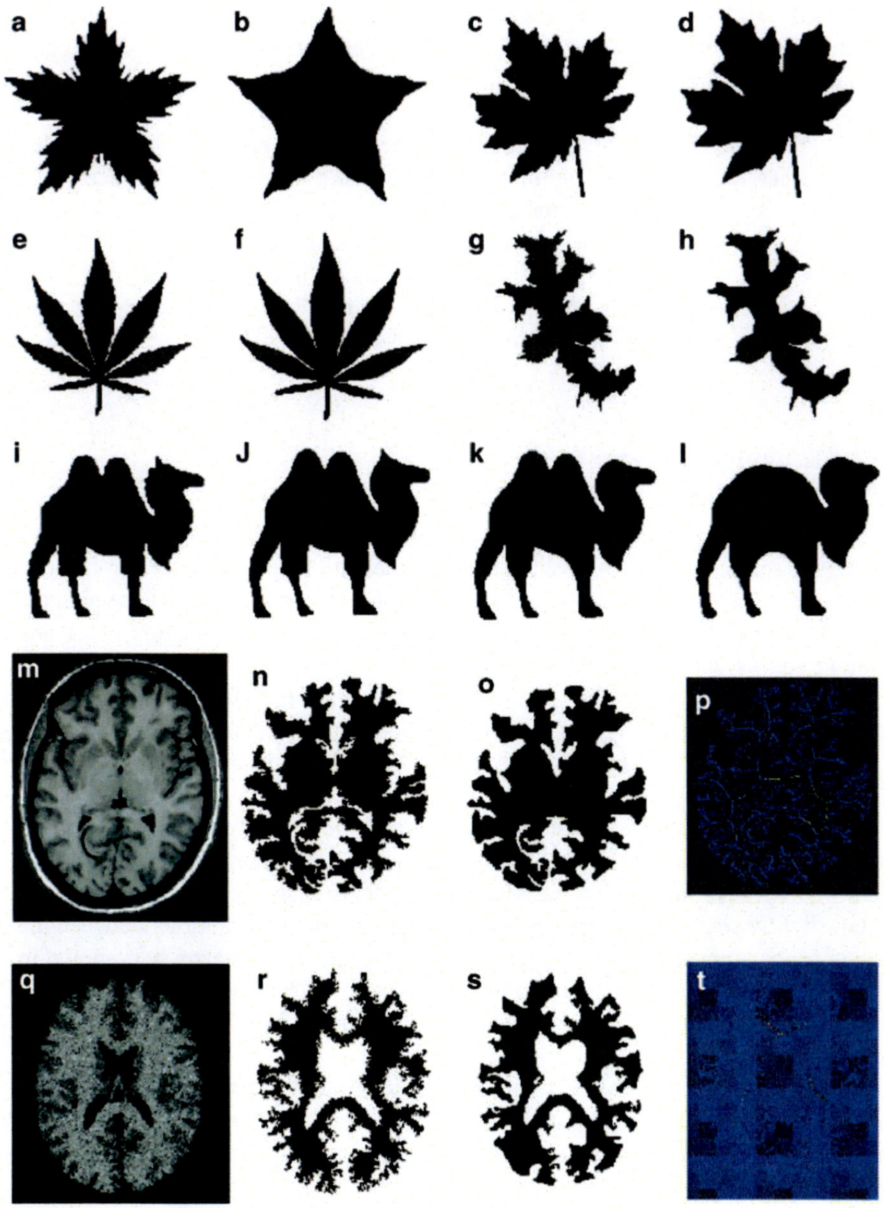

Fig. 4 Feature-preserving shape smoothing examples. See Sect. 6 for explanations

6.1 Multiscale Smoothing

By successively increasing the salience threshold σ_{min} , increasingly more sharp features are removed, as illustrated by Fig. 4i–l, where σ_{min} was successively increased by 10%. The notion of scale corresponds to the definition of σ (4), i.e. feature size as a fraction of local object thickness. Fine scales capture small wiggles at thick object parts, coarse scales capture large boundary perturbations at thinner object parts.

The simplification of the foreground and background skeletons by increasing σ_{min} successively removes terminal branches, and hence only replaces (bot does not add) sharp shape features by circle arcs with radii higher than $\min_{p \in \partial\Omega, q \in S_\sigma} \|p - q\|$. As such, the so-called "scale-space causality" is respected, so the space of progressively simplified shapes under σ has the necessary properties of a scale space. A similar observation was made by Tek and Kimia for their skeleton-based simplification method.

6.2 *Complex Shapes*

We have tested our method also on more complex shapes. Figure 4 m–p and 4 q–t show the smoothing of two value-based segmentations of white matter in MRI grayscale images, taken from [20, 25]. Since no preprocessing of the segmented data was done, apart from a simple morphological pass to fill in tiny holes and remove specks, the shapes are quite noisy and complex (Fig. 4 n,r). Note that this preprocessing step is necessary, as our method cannot remove small-scale *round* specks and/or holes, since these elements are not cusps and dents along a boundary. However, this is not a problem, since well-known morphological operators based on area can deal with specks and holes.

The remaining shapes after preprocessing exhibit both segmentation noise, visible as fine-level wiggles and filament-like structures along the boundary, but also actual shape information, visible as ondulations of the boundary at a level of detail slightly coarser than the noise. In certain areas, drawing a border between the actual shape boundary and noise is quite challenging. The saliency σ is shown in Fig. 4p,t using a blue (low) to red (high) colormap. The final results show that small-scale noise has been eliminated, but the boundary has not been excessively rounded and smoothed (Fig. 4o,s).

As a final example of the ability of the presented method to handle highly noisy shapes, we show a smoothing of a value-based segmentation of the MRI image from Fig. 4q *prior* to performing morphological hole filling and small-scale island removal. The input image (Fig. 5a), which is actually a set of many high-genus shapes with complex boundaries, is smoothed to yield Fig. 5b. The simplification of the foreground and background skeletons achieves similar effects to the morphological operators used to create Fig. 4r, at no added cost. The explanation resides in the definition of ρ: small-scale islands in the image have a perimeter, thus can be removed by low ρ thresholds. Note that this is the original skeleton simplification, and not the salience-based threshold.

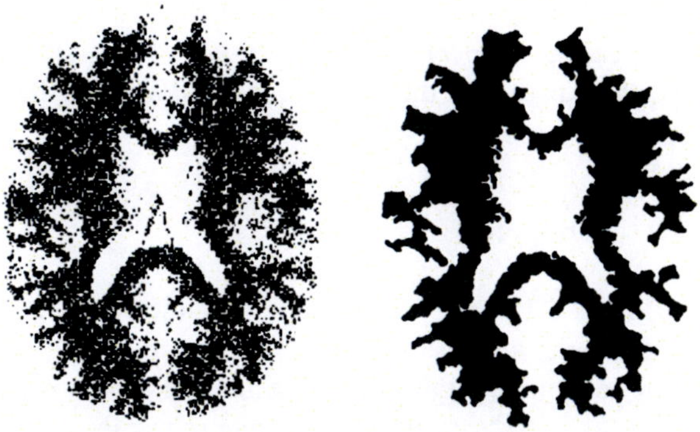

Fig. 5 Smoothing of highly noisy images achieves the effect of morphological erosion and dilation operators

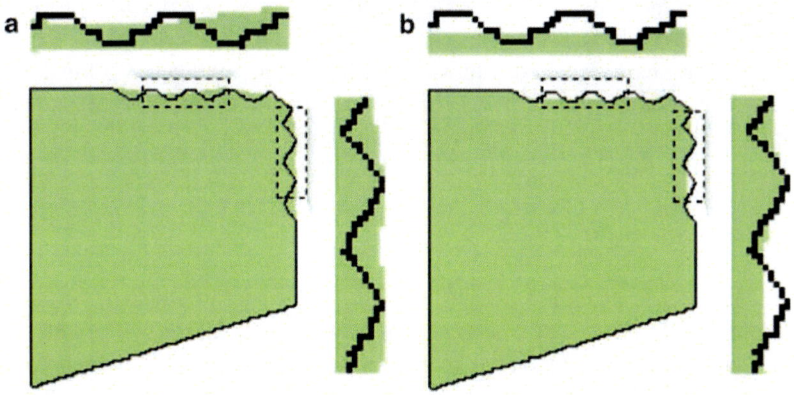

Fig. 6 Feature-preserving shape smoothing. Tek and Kimia (**a**) and current method (**b**). Original shape in black outline. Smoothed shape is shown filled. Insets show selected zoom-in details

7 Discussion

Below we discuss a number of aspects of our proposed method.

7.1 Comparison with Tek and Kimia

Figure 6 compares our method with the one proposed in [28]. The image is taken from the respective publication. The initial noisy image is shown in black outline; the smoothed result is filled green. Two insets show zoomed-in details. Looking at these, we see that our method constructs the smoothed edge roughly *between* the cusp and dent extrema, thereby behaving like an approximator of these points. In contrast, the method of Tek and Kimia tends to pull the smoothed contour outwards in most places, thereby preferring to fill the dents rather than remove the cusps. This is also visible in other images from the respective paper. The explanation lies in the different type of saliency metric used and performed optimization: While we consider feature size related to object thickness, Tek and Kimia remove branches in increasing order of affecting the overall shape's area. Also, we apply the dent smoothing pass after the cusp filling pass, whereas Tek and Kimia interlace the two passes.

7.1.1 Parameter Setting

The method has one dimensional parameter: σ_{min}, with a geometric meaning. For example, setting σ_{min} to 0.1 means removing all wiggles smaller than 10% of the local object thickness. The images in this paper have been produced with $\sigma_{min} \in [0.2, 0.3]$. Setting σ_{min} is quite robust to small variations: In contrast to e.g. curvature-based methods, skeleton-based removal of noise is a discrete process, as noise details go out in a successive sequence as skeleton branches get pruned out. For all shapes we experimented with, we noticed there are only a few critical values of σ_{min} where notable removal events happen. Similar observation were made for a different type of shape descriptor, shock graphs, by [23, 26].

7.1.2 Saliency Measure

The saliency metric proposed here essentially favors sharp corners which are far away from locally thick shape parts. As such, smoothed shapes will exhibit long straight edges, without wiggles, but keep convex or concave corners which connect coarse-scale edge-like structures. This is in line with the perceptual saliency model proposed by [1]. If desired, other saliency measures can be easily incorporated. For example, given a skeleton point $x \in S(\Omega)$, the area A of the feature corresponding to x can be computed by cutting away the feature from the shape's rump using the line segment delimited by the furthest two feature points of x (Fig. 7)

$$\{p_1, p_2\} = \underset{p_1, p_2 \in F(x)}{\arg\max} \|p_1 - p_2\| \tag{8}$$

Since $x \in S$, it is the center of a maximal ball C touching $\partial\Omega$ at p_1 and p_2, so $p_1 p_2 \subset \Omega$. Defining a saliency based on the so-called feature aperture angle $\alpha = \widehat{p_1 x p_2}$ can be done analogously. The segment length $||p_1 p_2||$ represents the thickness of the feature at its base, i.e. where it joins the main shape rump. If we want to smooth features based on their thickness, one metric to use is $\epsilon = \alpha ||p_1 p_2||$. Figure 8a–d illustrates this possibility. Here, the noisy brain segmentation contour (Fig. 4n) was smoothed progressively for increasing values of ϵ. Comparing with Fig. 4s, we see now that features which are thin at the base and have a relatively small length, are removed. Thereby, relatively narrow and short dents and cusps disappear more rapidly, but without the effect of excessive overall rounding.

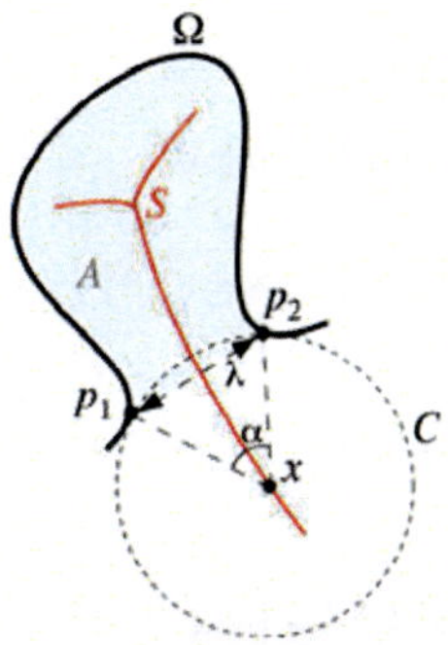

Fig. 7 Definition of additional skeleton-based saliency measures (see Sect. 7)

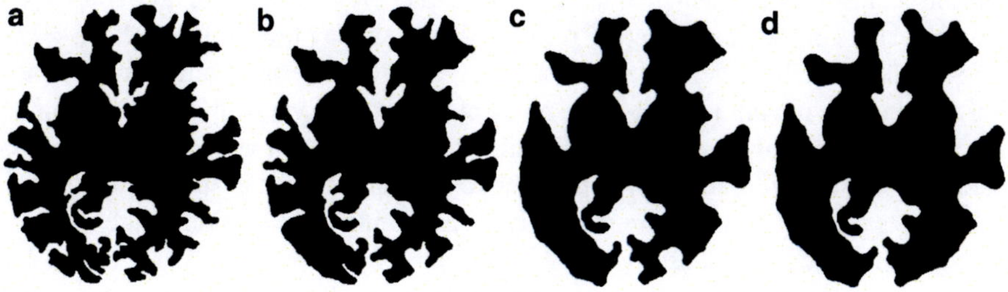

Fig. 8 Progressive smoothing based on feature saliency and thickness (Sect. 7)

7.1.3 Combined Smoothing

In this paper, we have used the full, unsimplified, skeleton as a starting point for the saliency computation. This guarantees that sharp corners are precisely kept. However, if desired, one can combine our saliency-based smoothing, based on the σ metric, with classical skeleton pruning based on the importance ρ, e.g. by first performing the ρ-based pruning, followed by our σ-based smoothing. This allows combining smoothing of all corners with removal of noise details. Studying the scale-space properties of the two-dimensional space created by ρ and σ is a potentially interesting avenue.

7.1.4 Topological Stability

The salience-based smoothing is guaranteed to keep the genus of the processed shape. Indeed, the convex and concave smoothing passes only remove terminal skeleton branches corresponding to shape cusps and dents, thus do not change the genus of the skeleton itself. This is a desirable property since topology change-inducing processing, e.g. hole or speck removal, can thus be done independently, by separate methods like morphological operators (see Sect. 6).

7.1.5 Selective Processing

Since cusps and dents are detected separately, they can be processed accordingly with no additional effort. For example, one can imagine applications where cusps are to be smoothed more aggressively than dents, or conversely.

7.1.6 3D Extension

Computing the skeletons, distance and feature transforms, and ρ and σ metrics is immediate for 3D shapes, using e.g. the approach in [22]. The only step of our smoothing method which does not immediately generalize to 3D shapes is the connected component detection following the saliency thresholding (Sect. 4.2). Thresholding σ will, in general, not disconnect the 3D skeleton. Think for example of a cubic shape having a small-scale noisy cusp that extends from the middle of a face to one of the cube's edges. In 2D, this problem does not appear: Removing a pixel from a skeleton branch corresponding to an arbitrary size cusp or dent always disconnects the skeleton. We are currently investigating an extension of the current analysis to handle 3D shapes.

8 Conclusions

We have presented a method for feature-preserving smoothing of digital shapes using shape skeletons. A saliency metric that relates the importance of a shape feature to its size and the local object thickness enables us to smooth relatively large-scale boundary noise and fully preserve sharp convex and concave corners. Another application is creating compact shapes from images corrupted by high amounts of binary speckle noise. The method uses skeletons at various places: to compute feature sizes, local object thickness, and to perform the actual feature smoothing. A simple and efficient implementation is presented here that can robustly handle complex shapes.

In the future, we aim to extend this method to efficiently and effectively handle 3D shapes, as outlined in Sect. 7. Another extension is the ability to handle full grayscale images, e.g. by representing these as a stack of skeletons for different isophotes.

References

1. T. Alter and R. Basri. Extracting salient curves from images: An analysis of the saliency network. *Intl. J. of Computer Vision*, 27(1):51–69, 1995.
2. N. Amenta, S. Choi, and K. Kolluri. The power crust. In *Proc. Solid Modeling*, 249–260. IEEE, 2001.

3. J. August, K. Siddiqi, and S. W. Zucker. Ligature instabilities in the perceptual organization of shape. *CVIU*, 76(3):231–249, 1999.
4. X. Bai, J. Latecki, and W. Y. Liu. Skeleton pruning by contour partitioning with discrete curve evolution. *IEEE TPAMI*, 29(3):449–462, 2007.
5. U. Clarenz, U. Diewald, and M. Rumpf. Processing textured surfaces via anisotropic geometric diffusion. *IEEE Trans. Image Proc.*, 13(2):248261, 2004.
6. U. Clarenz, M. Rumpf, and A. Telea. Robust feature detection and local classification of surfaces based on moment analysis. *IEEE TVCG*, 10(5):516–524, 2004.
7. Mathieu Desbrun, Mark Meyer, Peter Schröder, and Alan H. Barr. Implicit fairing of irregular meshes using diffusion and curvature flow. In *Proc. ACM SIGGRAPH*, 317–324, 1999.
8. G. Dudek and J. K. Tsotsos. Shape representation and recognition from multiscale curvatures. *CVIU*, 68(4):170189, 1997.
9. C. Fernmuller and W. Kropatsch. Multiresolution shape description by corners. In *Proc. CVPR*, page 271276. IEEE, 1992.
10. S. Fleishman, I. Drori, and D. Cohen-Or. Bilateral mesh denoising. *ACM TOG*, 22(3):950953, 2003.
11. M. Hisada, A. Belyaev, and L. Kunii. A skeleton-based approach for detection of perceptually salient features on polygonal surfaces. *Computer Graphics Forum*, 21(4):689700, 2002.
12. A. Jalba, M. Wilkinson, and J. Roerdink. Shape representation and recognition through morphological scale spaces. *IEEE Trans. Image Proc.*, 15(2):331341, 2006.
13. R. Kimmel, D. Shaked, and N. Kiryati. Skeletonization via distance maps and level sets. *CVIU*, 62(3):382391, 1995.
14. J. J. Koenderink. The structure of images. *Biological Cybernetics*, 50:363370, 1984.
15. H. Moreton and C. Séquin. Functional optimization for fair surface design. In *Proc. ACM SIGGRAPH*, 167–176, 1992.
16. R. L. Ogniewicz and O. Kübler. Hierarchic Voronoi skeletons. *Pattern Recognition*, 28(3):343–359, 1995.
17. S. Osher and J. Sethian. Fronts propagating with curvature-dependent speed. *J. of Computational Physics*, 79:1249, 1988.
18. J. Peng, V. Strela, and D. Zorin. A simple algorithm for surface denoising. In *Proc. IEEE Visualization*, page 107112, 2001.
19. S. M. Pizer, W. R. Oliver, and S. H. Bloomberg. Hierarchical shape description via the multiresolution symmetric axis transform. *IEEE TPAMI*, 9(4):505511, 1987.
20. K. Rehm, K. Schaper, J. Anderson, R. Woods, S. Stoltzner, and D. Rottenberg. Putting our heads together: a consensus approach to brain/non-brain segmentation in T1-weighted MR volumes. *Neuroimage*, 22(3):12621270, 2004.
21. D. Reniers and A. Telea. Tolerance-based distance transforms. In *Advances in Computer Graphics and Computer Vision (eds. J. Braz, A. Ranchordas, H. Araujo, J. Jorge)*, page 187200. Springer LNCS, 2007.
22. Dennie Reniers, Jarke Van Wijk, and Alexandru Telea. Computing multiscale curve and surface skeletons of genus 0 shapes using a global importance measure. *IEEE TVCG*, 14(2):355–368, 2008.
23. T. B. Sebastian, P. N. Klein, and B. Kimia. Recognition of shapes by editing their shock graphs. *IEEE TPAMI*, 26(5):550571, 2004.
24. J. A. Sethian. *Level set methods and fast marching methods*. Cambridge University Press, 2nd edition, 1999.
25. M. Shah. MRI brain image segmentation, 2009. www.81bones.net/mri/mri_segmentation.pdf.
26. K. Siddiqi, A. Shokoufandeh, S. Dickinson, and S. W. Zucker. Shock graphs and shape matching. *Intl. J. of Computer Vision*, 35(1):1332, 2004.
27. G. Taubin. Estimating the tensor of curvature of a surface from a polyhedral approximation. In *Proc. IEEE ICCV*, 992–997, 1995.
28. H. Tek and B. Kimia. Boundary smoothing via symmetry transforms. *J. of Math. Imaging and Vision*, 14:211223, 2001.

29. A. Telea. AFMMStar software, 2009. www.cs.rug.nl/~alext/SOFTWARE/AFMM/afmmstarvdt.zip.
30. A. Telea and J. J. Van Wijk. An augmented fast marching method for computing skeletons and centerlines. In *Proc. Symp. on Data Visualisation (VisSym)*, 251–259, 2002.
31. B. ter Haar Romeny. *Geometric-Driven Diffusion in Computer Vision*. Kluwer, 1994.
32. J. Weickert. A review of nonlinear diffusion filtering. In *Proc. Intl. Conf. on Scale Space*, page 328. Utrecht, the Netherlands, 1997.

Part IV
Tensor Visualization

Enhanced DTI Tracking with Adaptive Tensor Interpolation

Alessandro Crippa, Andrei C. Jalba, and Jos B. T. M. Roerdink

Abstract A novel tensor interpolation method is introduced that allows Diffusion Tensor Imaging (DTI) streamlining to overcome low-anisotropy regions and permits branching of trajectories using information gathered from the neighbourhood of low-anisotropy voxels met during the tracking. The interpolation method is performed in *Log-Euclidean space* and collects directional information in a spherical neighbourhood of the voxel in order to reconstruct a tensor with a higher linear diffusion coefficient than the original. The weight of the contribution of a certain neighbouring voxel is proportional to its linear diffusion coefficient and inversely proportional to a power of the spatial Euclidean distance between the two voxels. This inverse power law provides our method with robustness against noise. In order to resolve multiple fiber orientations, we divide the neighbourhood of a low-anisotropy voxel in sectors, and compute an interpolated tensor in each sector. The tracking then continues along the main eigenvector of the reconstructed tensors.

We test our method on artificial, phantom and brain data, and compare it with (a) standard streamline tracking, (b) the Tensorlines method, (c) streamline tracking after an interpolation method based on bilateral filtering, and (d) streamline tracking using moving least square regularisation. It is shown that the new method compares favourably with these methods in artificial datasets. The proposed approach gives the possibility to explore a DTI dataset to locate singularities as well as to enhance deterministic tractography techniques. In this way it allows to immediately obtain results more similar to those provided by more powerful but computationally much

A. Crippa · J.B.T.M. Roerdink (✉)
Johann Bernoulli Institute for Mathematics and Computer Science, University of Groningen, P.O. Box 407, 9700 AK Groningen, The Netherlands
e-mail: a.crippa@rug.nl; j.b.t.m.roerdink@rug.nl

A. C. Jalba
Institute for Mathematics and Computing Science, Eindhoven University of Technology, P.O. Box 513, 5600 MB, The Netherlands
e-mail: a.c.jalba@tue.nl

L. Linsen et al. (eds.), *Visualization in Medicine and Life Sciences II*, Mathematics and Visualization, DOI 10.1007/978-3-642-21608-4_10,

more demanding methods that are intrinsically able to solve crossing fibers, such as probabilistic tracking or high angular resolution diffusion imaging.

1 Introduction

Diffusion Tensor Imaging (DTI) is a Magnetic Resonance (MR) technique that allows quantitative measurement of three-dimensional diffusion of water molecules in biological tissues [5, 35]. DTI has found many applications in medicine and neuroscience [25, 45] and nowadays is the only method enabling *in vivo* and non-invasive exploration of architectonic organisation of human brain (among other organs), particularly of the cerebral white matter. Self-diffusion of water molecules in white matter is more probable to occur along neural fibers than perpendicular to them [8], leading to anisotropic diffusion of water which reflects the fibrous structure and the directional arrangements of bundles of axons [7]. Although axon myelination seems not to be essential for diffusion anisotropy [52], it is widely assumed to be the major barrier to water diffusion in brain white matter.

The DTI method approximates the diffusion process in each voxel of a DTI dataset by modelling the probability density function of water displacement via a second order diffusion tensor that represents the covariance matrix of a Gaussian diffusion process. This tensor is expressed as a three-by-three symmetric non-negative definite matrix. For each voxel, the tensor's eigensystem is used to describe the local diffusion process; in particular, the principal eigenvector (corresponding to the largest eigenvalue) shows the direction of main diffusivity.

DTI provides an estimation of anatomical connectivity patterns and reconstructs fiber pathways within the human brain [35]. Several techniques have been proposed in the literature: deterministic tractography (or fiber tracking) [4, 14, 24, 30] is one of the most commonly used techniques and relies on the assumption that the principal eigenvector of a tensor accurately describes the orientation of the underlying fiber bundles. This information is used to perform streamlining (or other line propagation techniques) in the vector field induced by the principal eigenvectors of all voxels in the dataset. Fiber tracking has been shown to be effective in many brain regions.

A major limitation of DTI is the fact that local diffusion information is not always sufficient to determine the underlying fiber direction. In the case of low anisotropy, i.e., when the two biggest eigenvalues (or even all three) have comparable magnitude, the principal eigenvector of the tensor does not necessarily correspond to the main diffusion direction. In this case a tensor representation can sometimes provide an approximated average of the multiple compartments present within a voxel [51] and thus the streamlining may suffer from cumulative tracking errors that could lead to erroneous results [28,41].

Fractional anisotropy (FA) is a measure of the anisotropy of a tensor. Voxels with a small FA mainly occur because of partial volume effects at locations where fiber crossing, fiber branching, or fiber kissing occur when two fibers meet and depart from each other within a voxel [1, 22] (cf. Fig. 1). The presence of low-anisotropy voxels is also related to the resolution of DTI data (being roughly $2 \times 2 \times 2$ mm^3, while axons have a diameter in the order of μm) and to the high susceptibility of DTI to noise during data acquisition [35, 37]. The inability to deal with areas of low anisotropy and to distinguish among singularities (i.e., crossing, branching, or kissing fibers) is considered to be the biggest problem of DTI [6, 30, 46]. Thus a tensor representation may not always be adequate to describe the underlying white matter structure. Possible solutions are to perform MR acquisition with a large number of gradient directions, as in High Angular Resolution Diffusion Imaging (HARDI) [44] or Q-ball imaging [43], or process the data using probabilistic tractography with multiple fiber orientations [9]. However, these methods require longer scanning times and longer preprocessing steps (computing the directional probability distributions needed for probabilistic tractography is very time consuming) and are not always available in current clinical environments.

The method proposed in this paper aims to improve deterministic DTI tracking by adding the ability to solve singularities without introducing a different diffusion model. We present a tensor interpolation method which can achieve noise reduction and resolve singularities to enhance subsequent streamline tractography in areas of low anisotropy. The method improves deterministic fiber tracking by interpolating voxels with low anisotropy reached during the tracking process. In such voxels directional information is gathered from the neighbourhood and the track is split to follow both crossing or kissing fibers. Our approach provides a good alternative when more involved approaches are not available or when the emphasis is more on speed than quality: while probabilistic tractography requires hours of preprocessing, our method detects and resolves fiber crossings in a few seconds.

2 Related Work

The issue of characterising fiber orientation in voxels with a population of more than one fiber has been addressed in several ways in the literature [23]. Model-based approaches include for instance multiple tensor fitting [11,40,44], probabilistic techniques [9,10,27,29,31], DTI tracking based on front propagation algorithms [32,33, 38], and higher order tensor models [20]. Model free approaches include Diffusion Spectrum Imaging (DSI) [47], HARDI [44] and Q-ball Imaging [43]. For a review of tracking methods we refer to [2].

The importance of considering singularities in fiber tracking algorithms has been shown by Behrens et al. [9]. Nevertheless, depending on the quality of the data and the complexity of the tissue, model parsimony measures are required to determine when more complex models are justified [23]. For instance, probabilistic and front propagation algorithms can intrinsically resolve singularities but have the major drawback not to output a connection path but values describing the likelihood to find a connection between two brain regions. Hence the user has to heuristically set a threshold on these values and decide how reliable the result is.

The method proposed in this paper has been inspired by three different papers: Hamarneh & Hradsky's bilateral filtering of tensors [19], the surface reconstruction method via Coulomb potentials [21], and the Tensorlines propagation method by Weinstein et al. [48]. Bilateral filtering of DTI data is a technique that tries to reduce noise by identifying and delineating areas with similar diffusion properties. Several papers on interpolation of tensor data or of expressions derived from tensors (such as eigenvalues, eigenvectors, or FA) have been published [16, 18, 36, 42]. For instance, Westin and Knutson [50] have shown how normalised convolution can be used as regularisation of tensor fields. Welk et al. [49] proposed median filtering, and Castano-Moraga et al. [12] proposed anisotropic interpolation of DT-MRI data. Hamarneh's method is the application of bilateral filtering to DTI images and seems to perform well in detecting edges in tensor fields. It can also handle, as a special case, tensor interpolation; see the Appendix for more information on this method.

The surface reconstruction method proposed in [21] is not related to DTI; it is an approach that tries to gather the necessary information from the whole dataset, weighting the contribution of each sample by its distance from the area to be interpolated. Using the whole dataset instead of only a certain neighbourhood of the surface gives this method a better resistance to noise. We incorporate an adapted version of this distance weighting in our method.

The Tensorlines method [48] does not address the issue of estimating a smooth tensor but it is a technique that adaptively interacts with streamline propagation using the whole local tensor information to deflect the incoming tracking direction. In this method the tracking direction in a voxel is defined by the collinearity between the local main eigenvector and the main eigenvector of the previously visited voxel. This method does not solve singularities but can perform tracking in low-anisotropy regions and can produce longer tracks than standard streamlining. A drawback of this method is that the integration step for tracking the fibers is equal to the voxel size, while standard streamlining usually uses smaller integration steps. Furthermore, this tracking technique seems to be very sensitive both to noise and to the parameters set by the user, who has to choose the relative weight of the incoming direction for the local tracking.

Another interpolation method that adaptively interpolates tensors along streamlines is the technique proposed by Zhukov *et al.* [53]. This method is based on a moving least square regularisation of the tensors along the streamline. After reconstructing a continuous tensor field in the volume through trilinear interpolation, the method finds a polynomial that fits the data, in a least square sense, in a region around the tensor to be interpolated. The fitting depends on the location, the orientation of the streamline at the point, and the history of motion along the streamline. This method was successfully applied to brain as well as heart diffusion data [53, 54].

3 Methods

Our method is meant to enhance deterministic tracking techniques by interpolating diffusion information from neighbourhoods of voxels with low anisotropy. Inspired by [48] and [53], we do not apply interpolation to every voxel of the dataset but only to those voxels reached during tracking. When a voxel with low anisotropy is reached, regional information is gathered from the surrounding voxels and this is used to find the direction, or directions, in which to continue the tracking. This prevents, in contrast to global interpolation, that the interpolation affects high anisotropy as well as low-anisotropy voxels.

The basic steps of our algorithm are the following:

1. Choose a starting voxel in the dataset and start tracking a fiber.
2. Continue the tracking until a voxel with low FA is reached (FA< 0.3).
3. If such a voxel is encountered, interpolate the neighbouring voxels to find an interpolated tensor with higher FA than the original. If this is possible, continue the tracking along the main eigenvector of this tensor.
4. Visualise the resulting fiber.

3.1 *Tensor Interpolation*

Diffusion tensors do not form a vector space and special attention must be paid when performing calculations on them [34]. The value of the determinant of a tensor is in fact a measure of the dispersion of the diffusion process [3] and Euclidean averaging of tensors has been shown to lead tensor swelling effects [13,17,42]. This could lead to a decrease of FA and to a possible stopping of the tracking algorithm. Therefore tensor averaging is performed in *Log-Euclidean space* as proposed by Arsigny [3]. Performing computations in this space prevents the increase of the determinant of the averaged tensor [15].

We recall here that the logarithm of a tensor T is defined as

$$\log_m(T) = R^T \log(D)\, R,$$

where D is the diagonal matrix of the eigenvalues of T and R is the matrix of its eigenvectors. The formula for tensor exponentiation $\exp_m$ is analogous.

Given a (low-anisotropy) voxel at a certain position x, the corresponding tensor $T(x)$ is interpolated by gathering information from a neighbourhood of x containing N voxels. The weight of the contribution of a certain neighbouring voxel at position ξ_i is set proportional to its linear diffusion coefficient $C_L(\xi_i)$ and, like in [21], inversely proportional to a power n of the spatial Euclidean distance between the two voxels x and ξ_i:

$$w(x, \xi_i) = C_L(\xi_i) d(x, \xi_i)^{-n} \tag{1}$$

The linear diffusion coefficient C_L is a rotational invariant of a diffusion tensor [26] that measures the amount of linear diffusion in comparison to the overall diffusion. It is defined as

$$C_L = \frac{\lambda_1 - \lambda_2}{\lambda_1 + \lambda_2 + \lambda_3} \tag{2}$$

where $\lambda_1, \lambda_2, \lambda_3$ are the three eigenvalues of the tensor, ordered from the largest to the smallest.

For a low-anisotropy voxel x an interpolated tensor $\tilde{T}(x)$ is thus computed according to the following formula:

$$\tilde{T}(x) = \exp_m \left(k \cdot \log_m(T(x)) + (1-k) \cdot \sum_{i=1}^{N} \frac{w(x, \xi_i)}{W(x)} \log_m(T(\xi_i)) \right) \tag{3}$$

where

$$W(x) = \sum_{i=1}^{N} w(x, \xi_i) \tag{4}$$

is the total weight at voxel x. The parameter k weights the influence of the tensor $T(x)$ and the influence of the neighbouring tensors $T(\xi_i)$. It allows us to divide the local and regional information into two complementary components. As explained above, the tensor logarithm and tensor exponentiation used in this equation are obtained by computing the eigenvalues and eigenvectors of the tensor field.

The weighting provided by C_L ensures that voxels with predominantly linear diffusion will provide more information than voxels with planar or isotropic diffusion. We prefer to use the linear coefficient instead of the more commonly used FA because we eventually aim to eigensolve the interpolated tensor and find proper directions to follow during tracking. For this reason we want to give the same weight (given equal spatial distance from the voxel to be interpolated) to voxels with planar diffusion and to voxels with isotropic diffusion; using FA would have given more weight to voxels with planar diffusion.

The inverse power law dependence of the weight on the distance extends the area of influence so that even voxels further away will contribute to the interpolation. This approach allows us, by taking the exponent $n > 1$, to consider a wider area of influence than using a weighting that is linearly proportional to the inverse of the distance. As shown in [21], this provides our method with more robustness against noise. In contrast to the approach of [21], we do not extend the radius of influence to the whole dataset, which would not be physically plausible for DTI brain data, but restrict it to a spherical neighbourhood of the voxel under consideration. In the artificial dataset of Fig. 3, where the low FA area is very large, a neighbourhood of 10 voxels was used for tracking; a neighbourhood of only 5 voxels was used for the tracking in the synthetic volume and in the brain volume.

3.2 Extended DTI Model

Because low-anisotropy voxels are mainly due to partial volume effects, the tensor model cannot distinguish among cases in which crossing, kissing or branching fibers occur within a single voxel (cf. Fig. 1). Thus, interpolation techniques may fail or lead to erroneous results and it is not possible to tell if the tracking is following the correct direction once it encounters a low FA voxel.

The second ingredient of our approach is meant to tackle this problem as follows. Whenever the tracking enters a low anisotropy voxel x, its 3D neighbourhood is divided in 26 sectors S_i, one per direct neighbour (on the voxel grid, using 26-connectivity). An interpolated tensor $\tilde{T}_i(x)$ is computed for each sector S_i, $i = 1, \ldots, 26$ according to (3), by interpolating the tensors of sector S_i (cf. Fig. 2). For each sector S_i, centered at a voxel x, a likelihood value $l_i(x)$ is computed by

$$l_i(x) = (\tilde{C}_L^i(x)\mathbf{v}_i(x)) \cdot \mathbf{u}_i \quad (5)$$

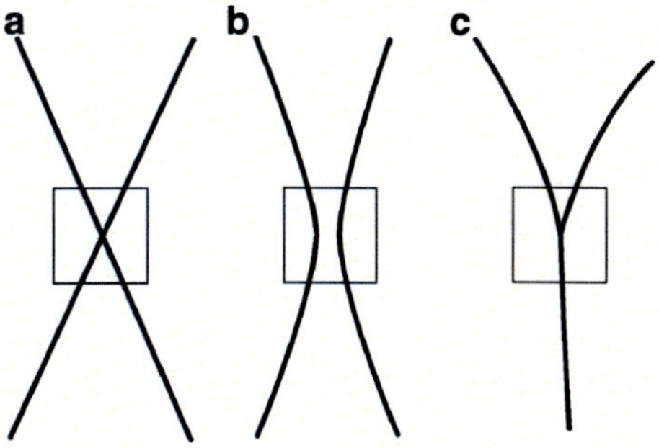

Fig. 1 Schematics of possible scenarios in a local fiber configuration. Within a voxel the fibers can (**a**) cross each other, (**b**) kiss each other or (**c**) a fiber can split into two branches

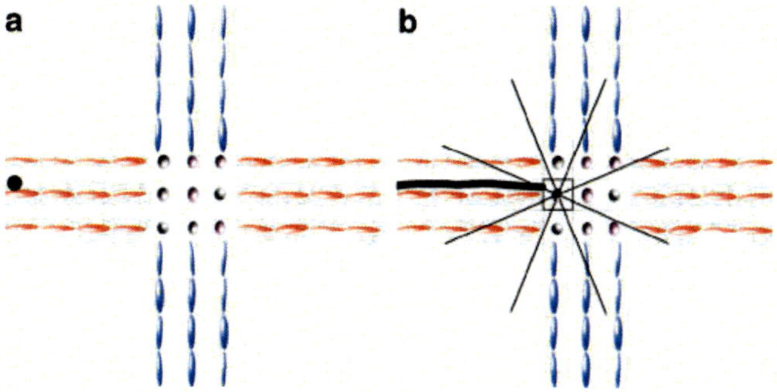

Fig. 2 Dataset subdivision when a low FA voxel is reached by the tracking (2D case; the 3D case is a straightforward extension). (**a**) The figure shows two fibers meeting in the center of an example dataset. Tensors are displayed as ellipsoids whose major axis corresponds to the main eigenvector and whose eccentricity depicts the amount of anisotropy; the colour indicates the direction of main diffusivity (red: horizontal, blue: vertical). A seed is positioned in the black spot. (**b**) When the tracking (black line) reaches a low FA voxel (square) the dataset is uniformly subdivided into eight sectors; each sector is centered at a direct neighbour of the low-FA voxel. Tensors are computed in each slice according to (3)

where $\mathbf{v}_i(x)$ is the principal eigenvector of the interpolated tensor $\tilde{T}_i(x)$ and $\mathbf{u}_i$ is the unit vector bisecting sector S_i (here $\mathbf{u} \cdot \mathbf{v}$ denotes the dot product of two vectors $\mathbf{u}$ and $\mathbf{v}$). $\tilde{C}_L^i(x)$ is the average distance-weighted linear diffusion coefficient in sector S_i,

$$\tilde{C}_L^i(x) = \sum_{j \in S_i} C_L(\xi_j) d(x, \xi_j)^{-n}.$$

The values $l_i(x)$ are used as the likelihood for each sector S_i that the tracking should continue along the direction $\mathbf{u}_i(x)$ (cf. Fig. 2). Because of the impossibility of differentiating between crossing, kissing or branching fibers, we continue tracking along the bisections of the two sectors with the highest likelihood value. Sectors whose unit vector forms an angle bigger than 80 degrees with the incoming tracking direction are discarded (according to the literature, a fiber bundle should not bend more than 80 degrees within a single voxel). Doing so we both spot the singularity and continue tracking along the correct direction.

Although it would in principle be possible to follow more directions, we chose to continue the tracking only along two directions, resolving in this way the crossing of not more than two fibers. It is not possible to determine the number of fibers that may produce the partial volume effect in a voxel, and two is obviously the lower bound.

4 Results

In this section we present results of our method for artificial data, phantom data and DTI brain scans; the method is also compared with existing techniques.

4.1 *Artificial Data*

4.1.1 High-Anisotropy Area with a Low-Anisotropy Core

The first artificial dataset we used consists of a 20×20×20 three-dimensional tensor field containing high-anisotropy voxels (FA = 0.85) but with a cubic (7×7×7 voxels) low-anisotropy core (FA = 0.1), cf. Fig. 3, left upper picture. Tensors in the high-anisotropy area are aligned to the vertical direction, while tensors in the low-anisotropy area are aligned to the horizontal direction. This dataset is used to compare our method (streamline tracking combined with the new interpolation method) with (a) standard streamline tracking, (b) the Tensorlines method, (c) streamline tracking after an interpolation based on bilateral filtering of tensors (see the Appendix) and (d) streamline tracking using moving least square regularisation. The tests conducted on this dataset explore the ability of the different approaches to reconstruct fibers passing through the low-anisotropy area.

White Gaussian noise was added to each tensor component. Figure 3 shows the central slice of the dataset and a comparison among standard streamline tracking,

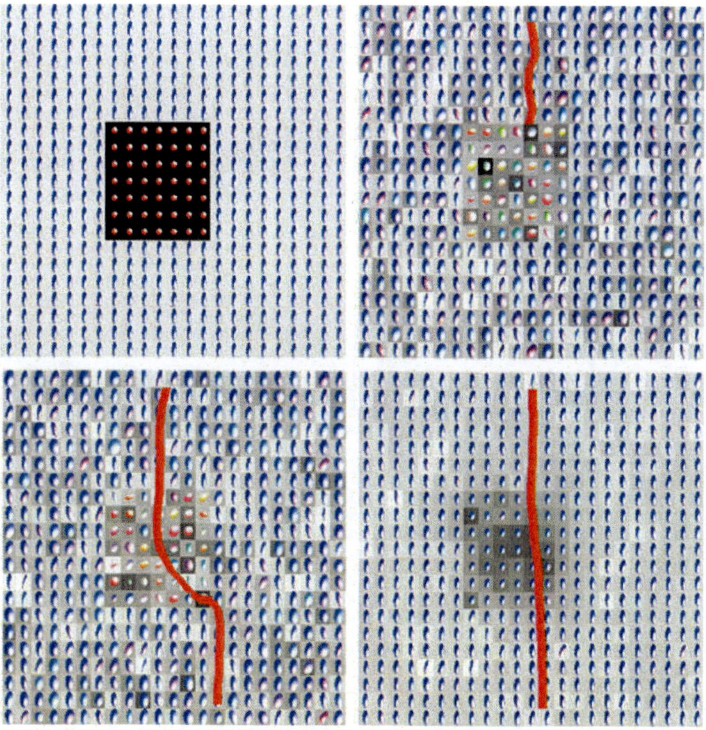

Fig. 3 Top left: the artificial dataset is a 20×20×20 three-dimensional tensor field consisting of highly anisotropic tensors (FA = 0.85) with a 7×7×7 low-anisotropy area (FA = 0.1) in the middle. The picture shows the central slice of the dataset. Background colour represents FA (black: 0, white: 1); the colour of the ellipsoids represents direction (blue: vertical, red: horizontal). Top right: white Gaussian noise ($\sigma = 0.02$) was added to each tensor component. Streamline tracking (red line) was seeded in a single voxel in the top of the picture. Bottom left: the Tensorlines technique. Bottom right: streamline tracking after our new tensor interpolation

Tensorlines and our new method. Deterministic tractography was achieved using Euler integration. The top right figure shows that standard tracking stops as soon as it reaches the low-FA area. Tensorlines tracking (bottom left) gets distorted already with a low noise level ($\sigma = 0.02$), while our method (bottom right) is able to reconstruct the connection between the upper part and the lower part of the dataset. For illustration purposes, the interpolation was applied to the whole dataset (and not only to the voxels reached by the tracking) to show its effects on directionality and on FA values.

Figure 4 shows a comparison among our method, interpolation based on bilateral filtering, and Zhukhov's moving least square regularisation. Since these methods were all able to track through the low-anisotropy region of Fig. 3, we measured the effectiveness of the three interpolation techniques on three different voxels in the dataset: the voxel in the middle of the low-anisotropy area; a voxel with low FA, directly on the border of the inner low-FA core, and a high-FA voxel on the border

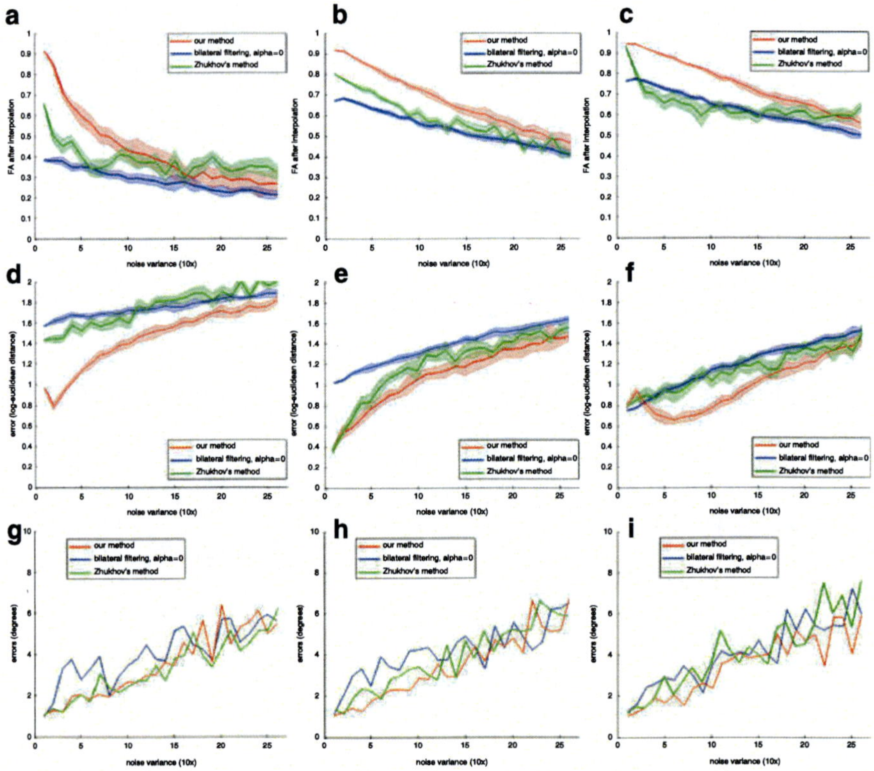

Fig. 4 Comparison between results achieved by our method (red lines), by the interpolation based on bilateral filtering (blue lines), and by the moving least square regularisation method of Zhukov *et al.* (green lines). Error measures were computed for three different voxels in the dataset shown in Fig. 3: a voxel in the middle of the low-anisotropy area (column 1), a voxel with low FA, directly on the border of the inner low-FA core (column 2), and a high-FA voxel on the border of the inner core (column 3). Row 1: FA value of the interpolated voxel. Row 2: Log-Euclidean distance between the interpolated voxel and a voxel in the high-FA area. Row 3: angular error of the interpolated voxel. All error measures are shown as a function of noise level σ. Each line represents the average of 25 runs and the shaded area indicates the corresponding standard deviation

of the inner core. Three different measures were used to compare the techniques. First, the FA value of the interpolated voxel; this shows the ability of each technique to gather anisotropy from the neighbourhood. Second, the Log-Euclidean distance between the interpolated voxel and a voxel in the high-FA area (before the noise addition); this indicates the ability to reconstruct the tensors. Third, the angular difference, in degrees, between the main eigenvector of the reconstructed tensor and the main eigenvector of a voxel in the high-FA area (before the noise addition); this indicates the ability to perform correct tracking after interpolation.

The experiments were conducted at several noise levels. Gaussian white noise was added individually to each tensor component with standard deviation σ varying between 0.001 and 0.025. For each noise level we repeated the experiment 25 times.

When adding Gaussian noise, we maintained the non-negative definiteness of the diffusion tensor. This prevents to generate tensors with non positive eigenvalues which are difficult to interpret physically (since they describe non positive diffusion factors) [3].

As seen in Fig. 4, the interpolated tensors present an FA always bigger than 0.3, even at relatively high noise levels (a threshold of FA = 0.2 is usually used as a stopping criterion for streamline tractography). Our method is more effective in restoring high-FA values than the interpolation based on bilateral filtering, and it is also more effective than moving least square regularisation, except for high noise levels. The Log-Euclidean distance between the interpolated tensors and the "expected" tensor is quite small in all three cases. Again, our results show lower Log-Euclidean distances than those achieved by the interpolation based on bilateral filtering, except for a small region below $\sigma = 0.02$ in Fig. 4f. Compared with moving least square regularisation, our method produces lower Log-Euclidean distances especially for the voxel the center of the low FA region (Fig. 4d). Regarding the angular error of the three interpolation techniques we see no significant difference, see Fig. 4(g, h, i). The average angular error grows roughly linearly with the amount of noise, and it does not exceed 6 degrees (at $\sigma = 0.025$).

4.1.2 Crossing Fibers

The second artificial dataset consisted of three fibers that meet each other in the center of a 20×20×20 dataset. Each fiber is represented by a strip of high-FA voxels (FA = 0.85). The dataset was used to evaluate the ability of our method to recognise directional information in different sectors of the dataset. Figure 5(left) shows a slice of the dataset where the three fibers cross. Tracking was seeded at the top of the figure and the track did split when it reached the low-FA area. White Gaussian noise was added to each tensor of the dataset and the tracking was repeated 25 times per noise level; the standard deviation of the noise varied from 0.001 to 0.25. Figure 5(right) shows statistics on the ability to find the correct fibers as a function of noise level. The blue line represents the probability to find all five segments that meet in the low-FA area (the incoming direction was not considered as a possible solution). The blue shaded area represents its standard deviation. The red line indicates the probability to find the two fibers that the algorithm should detect, i.e., the only two fibers forming an angle smaller than 80 degrees with the incoming direction. The red shaded area is its standard deviation (Fig. 6).

4.2 *Phantom Data*

The next dataset considered was a physical phantom DTI dataset representing two 90 deg crossing fibers. The dataset was made of Dyneema® fibers. The fibres were grouped in parallel bundles of 780 filaments which were crossed, surrounded by a shrinking tube, and immersed in water (Courtesy of E. Fieremans, NYU Medical

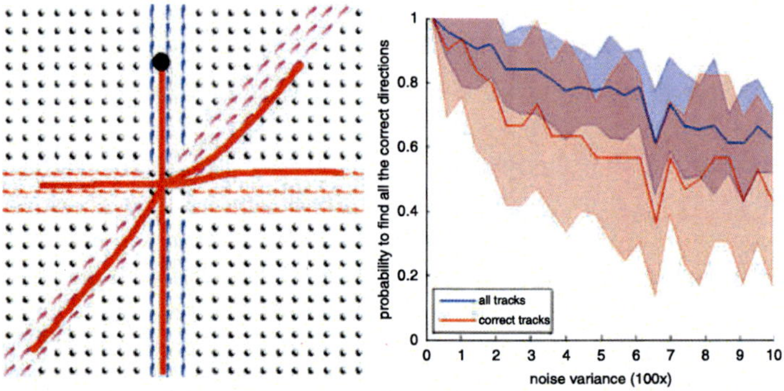

Fig. 5 Left: artificial dataset with three fibers crossing in the middle. The figure shows the central slice of the dataset. Tracking was seeded in the black spot. The tracking follows the fiber until the low-anisotropy area in the middle and then detects all other fibers. Right: The blue line shows the probability, as a function of noise level, to detect all five fibers meeting in the center (the incoming direction was not considered as a solution). The red line indicates the probability to find the two fibers that the algorithm should detect (the only two fibers forming an angle smaller than 80 degrees with the incoming direction)

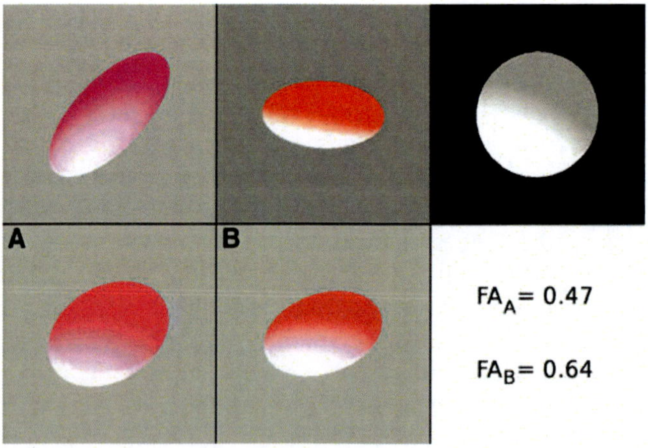

Fig. 6 Comparison between averaging of tensors in Euclidean versus Log-Euclidean space. First row: three input tensors. Second row: tensor A is obtained by Euclidean averaging, tensor B by log-Euclidean averaging. The tensor in A shows a lower FA compared to the tensor in B

Center, and J. Sijbers, Univ. of Antwerp). Figure 7 shows a comparison among standard deterministic tractography, our method and the results of probabilistic tracking performed with FSL [39]. The white disc indicates the position where the tracking was seeded. No constraints were set on the angles between the incoming and the outgoing directions. Standard deterministic tractography was not able to

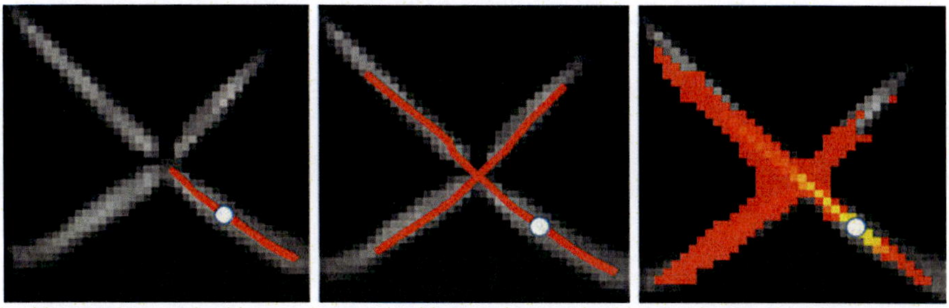

Fig. 7 Tracking performed in a physical phantom DTI dataset depicting two crossing fibers. The pictures show a slice of the dataset. The white disc indicates the position where the tracking was seeded. Left: tracking results of standard deterministic tractography; the tracking stopped in the crossing area. Middle: results of our algorithm; crossing fibers were detected and all branches were found. Right: tracking by probabilistic tractography with FSL

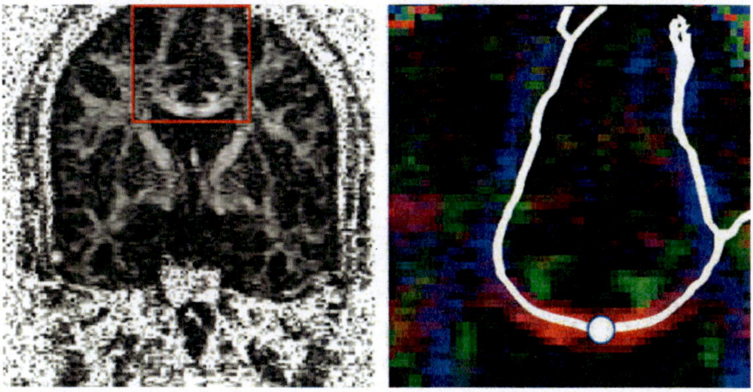

Fig. 8 Tracking performed for a DTI brain dataset. The red square in the left picture indicates the Corpus Callosum. The right picture shows a magnification of the red area. The white disc indicates the position where tracking was seeded. Our algorithm was able to detect the blossoming of the fibers of the Corpus Callosum below the cerebral cortex (leftmost branch) and the intersection between Corpus Callosum and Corona Radiata (rightmost branch)

resolve the crossing: the tracking stopped when it encountered a low FA voxel. Our algorithm (which is deterministic as well) recognised all the fiber segments and it was able to continue tracking in all the directions detected by the much more computationally demanding probabilistic method.

4.3 Brain Data

Results of the tracking for a DTI brain data set are shown in Fig. 8. A coronal slice of the brain is shown in the leftmost figure, colour coded according to the FA values (black represents FA=0 and white FA=1). Seeding the tracking in the lower part of the body of the Corpus Callosum, our method was able to detect the blossoming of the upper part of the Corpus Callosum where the fibers reach the cerebral cortex, and to resolve the low-FA area generated by the intersection of Corpus Callosum and Corona Radiata. Brain dataset courtesy of Gordon Kindlmann, Scientific Computing and Imaging Institute, University of Utah, and Andrew Alexander, W. M. Keck Laboratory for Functional Brain Imaging and Behavior, University of Wisconsin-Madison.

5 Conclusions

DTI tractography allows the inference of connectivity patterns within white matter of the brain. There are two main limitations of this technique. The first is that the tensor model does not always reflect the underlying white matter structure, as it is not able to deal with singularities such as crossing, branching, or kissing fibers. The second limitation is the inability of DTI tracking to reconstruct more than one trajectory per seed point.

In this paper we introduced an improved tracking technique that allows DTI streamlining to solve low-anisotropy regions and permits branching of trajectories. Our method performs interpolation for any low-anisotropy voxel met during tracking. Interpolation is computed in *Log-Euclidean space* [3] and collects directional information in a spherical neighbourhood of the voxel in order to reconstruct a tensor with a higher linear diffusion coefficient than the original. The weight of the contribution of a certain neighbouring voxel is proportional to its linear diffusion coefficient and inversely proportional to a power of the spatial Euclidean distance between the two voxels. This inverse power law provides our method with robustness against noise [21] . In order to resolve multiple fiber orientations, we divide the neighbourhood of the low-anisotropy voxel in 26 sectors, and compute an interpolated tensor in each sector according to the weighted tensor interpolation formula. The tracking then continues along the main eigenvector of the reconstructed tensor.

We tested our method on artificial, phantom and brain data, and compared with existing methods: (a) standard streamline tracking, (b) the Tensorlines method, (c) streamline tracking after an interpolation based on bilateral filtering, and (d) moving least square regularisation. We showed that in contrast to standard streamline tracking, our method is able to continue tracking in low-anisotropy areas, while Tensorlines tracking gets distorted already for low noise levels. Compared to streamline tracking after the interpolation based on bilateral filtering, our method is more effective in restoring high anisotropy values. Compared to the moving least square regularisation, our method generally performed better except for high noise levels. For phantom and real MRI data, we found that our method was able to detect the same tracts as probabilistic tracking.

Due to its ability to resolve cases of multiple fiber orientations in a single voxel our method gives the possibility to obtain, in comparison to standard deterministic tractography, results more similar to those provided by more powerful techniques that are intrinsically able to solve crossing fibers, such as high angular resolution diffusion imaging, which is not always available, or probabilistic tracking with multiple fiber orientations [39], which is computationally much more demanding: while probabilistic tractography requires hours of preprocessing, our method detects and resolves the crossing in a few seconds. Furthermore, in contrast to the probabilistic tractography, in our approach there is no need to perform heuristic thresholding for interpreting the results.

Future work will include a deeper study on the results achievable with this method in comparison with probabilistic tracking and other techniques that allow multiple fiber orientations per voxel. Validation of the results, e.g., by histological analysis, will certainly be a requirement for the practical application of this and other DTI techniques.

Acknowledgements We thank E. Fieremans, NYU Medical Center, and J. Sijbers, Univ. of Antwerp, for providing the phantom data set.

Appendix

Bilateral filtering of diffusion tensor data is achieved according to the following formula [19]:

$$T(x) = exp_m\left(\sum_{i=1}^{N} \frac{w_i(x)}{k(x)} log_m(T(\xi_i))\right) \tag{6}$$

$$w_i(x) = \alpha f_1(d_T(T(x), T(\xi_i)) + (1-\alpha) f_2(d_S(x, \xi_i)) \tag{7}$$

where $d_T(T(x), T(\xi_i))$ is the tensor dissimilarity between the two tensors $T(x)$ and $T(\xi_i)$. $d_S(x, \xi_i)$ is the spatial Euclidean distance between the voxels x and ξ_i. Here f_1 and f_2 are monotonically decreasing functions that map d_T and d_S in the interval [0, 1]. The value α weights the contribution of the two distances. We used a simplified version of bilateral filtering, by setting α to zero and thus weighting the contribution of the tensor $T(\xi_i)$ only according to the Euclidean distance between voxels x and ξ_i.

References

1. Alexander, A.: Analysis of partial volume effects in diffusion tensor MRI. Magn. Reson. Med. **45**(5), 770–780 (2001)
2. Alexander, D.C.: Multiple-fiber reconstruction algorithms for diffusion MRI. Annals N Y Acad Sci **1064**, 113–133 (2005)
3. Arsigny, V., Fillard, P., Pennec, X., Ayache, N.: Log-Euclidean metrics for fast and simple calculus on diffusion tensors. Magn. Reson. Med. **56**(2), 411–421 (2006)
4. Bammer, R., Acar, B., Moseley, M.E.: In vivo MR tractography using diffusion imaging. Eur. J. Radiol. **45**(3), 223–234 (2003)

5. Basser, P.J., Mattiello, J., Bihan, D.L.: Estimation of the effective self-diffusion tensor from the NMR spin-echo. J. Magn. Reson. **103**(3), 247–254 (1994)
6. Basser, P.J., Pajevic, S., Pierpaoli, C., Duda, J., Aldroubi, A.: In vivo fiber tractography using DT-MRI data. Magn. Reson. Med. **44**(4), 625–632 (2000)
7. Basser, P.J., Pierpaoli, C.: Microstructural and physiological features of tissues elucidated by quantitative-diffusion-tensor MRI. J. Magn. Reson. **111**(3), 209–219 (1996)
8. Beaulieu, C.: The basis of anisotropic water diffusion in the nervous system - a technical review. Nucl. Magn. Res. in Biomed. **15**(7-8), 435–455 (2002)
9. Behrens, T., Johansen-Berg, H., Jabdi, S., Rushworth, M.F., Woolrich, M.W.: Probabilistic diffusion tractography with multiple fibre orientations: What can we gain? Neuroimage **34**(1), 144–155 (2007)
10. Behrens, T.E.J., Woolrich, M.W., Jenkinson, M., Johansen-Berg, H., Nunes, R.G., Clare, S., Matthews, P.M., Brady, J.M., Smith, S.M.: Characterization and propagation of uncertainty in diffusion-weighted MR imaging. Magn. Reson. in Medicine **50**(5), 1077–1088 (2003)
11. Bergmann, O., Kindlmann, G., Peled, S., Westin, C.F.: Two-tensor fiber tractography. In: Proc. ISBI, 796–799 (2007)
12. Castano-Moraga, C., Rodrigues-Florido, M.A., Alvarez, L., Westin, C.F., Ruiz-Alzola, J.: Anisotropic interpolation of DT-MRI data. In: Proc. MICCAI, 343–350 (2004)
13. Chefd'hotel, C., Tschumperlé, D., Deriche, R., Faugeras, O.: Regularizing flows for constrained matrix-valued images. J. Math. Imaging Vision **20**(1-2), 147–162 (2004). DOI http://dx.doi.org/10.1023/B:JMIV.0000011324.14508.fb
14. Conturo, T.E., Lori, N.F., Cull, T.S., Akbudak, E., Snyder, A.Z., Shimony, J.S., McKinstry, R.C., Burton, H., Raichle, M.E.: Tracking neuronal fiber pathways in the living human brain. Proc. Natl. Acad. Sci. USA **96**(18), 10,422–10,427 (1999)
15. Corouge, I., Fletcher, P., Joshi, S., Gouttard, S., Gerig, G.: Fiber tract-oriented statistics for quantitative diffusion tensor MRI analysis. Med. Image Anal. **10**(5), 786–798 (2006)
16. Coulon, O., Alexander, D.C., Arridge, S.A.: A regularization scheme for diffusion tensor magnetic resonance images. In: Proc. of the 17th Int. Conf. Inf. Proc. Med. Imag., vol. 2082, 92–105 (2001)
17. Feddern, C., Weickert, J., Burgeth, B., Welk, M.: Curvature-driven PDE methods for matrix-valued images. Int. J. Comput. Vision **69**(1), 93–107 (2006). DOI http://dx.doi.org/10.1007/s11263-006-6854-8
18. Hahn, K., Pigarin, S., Putz, B.: Edge preserving regularization and tracking for diffusion tensor imaging. In: MICCAI, vol. 2208, 195–203 (2001)
19. Hamarneh, G., Hradsky, J.: Bilateral filtering of diffusion tensor magnetic resonance images. IEEE Trans Image Process. **16**(10), 2463–2475 (2007)
20. Hlawitschka, M., Scheuermann, G.: HOT- Lines: Tracking lines in higher order tensor fields. In: Proc. IEEE Visualization, 27–34 (2005)
21. Jalba, A., Roerdink, J.B.T.M.: Efficient surface reconstruction using generalized Coulomb potentials. IEEE Trans. Vis. Comput. Graph. **13**(6), 1512–1519 (2007)
22. Jones, D.K.: Determining and visualizing uncertainty in estimates of fiber orientation from diffusion tensor MRI. Magn. Reson. Med. **49**(1), 7–12 (2003)
23. Jones, D.K.: Studying connections in the living human brain with diffusion MRI. Cortex **44**(8), 936–952 (2008)
24. Jones, D.K., Simmons, A., Williams, S.C.R., Horsfield, M.A.: Non-invasive assessment of axonal fiber connectivity in the human brain via diffusion tensor MRI. Magn. Res. Med. **42**(1), 37–41 (1999)
25. Kanaan, R.A.A., Kim, J.S., Kaufmann, W.E., Pearlson, G.D., Barker, G.J., McGuire, P.K.: Diffusion tensor imaging in schizophrenia. Bio. Psychiatry **58**(12), 921–929 (2005)
26. Kingsley, P.B.: Introduction to diffusion tensor imaging mathematics: Part I. tensors, rotations, and eigenvectors. Concepts in Magn. Reson **28A**(2), 101–122 (2005)
27. Koch, M.A., Norris, D.G., Hund-Georgiadis, M.: An investigation of functional and anatomical connectivity using magnetic resonance imaging. NeuroImage **16**(1), 241–250 (2002)

28. Lazar, M., Weinstein, D.M., Tsuruda, J.S., Hasan, K.M., Arfanakis, K., Meyerand, M.E., Badie, B., Rowley, H.A., Haughton, V., Field, A., Alexander, L.A.: White matter tractography using diffusion tensor deflection. Human Brain Ma **18**(4), 306–321 (2003)
29. McGraw, T., Nadar, M.: Stochastic DT-MRI connectivity mapping on the GPU. IEEE Trans. Vis. Comp. Graphics **13**(6), 1504–1511 (2007)
30. Mori, S., Crain, B.J., Chacko, V.P., vanZijl, P.C.: Three dimensional tracking of axonal projections in the brain by magnetic resonance imaging. Ann. Neurol. **45**(2), 265–269 (1999)
31. Parker, G.J., Haroon, H.A., Wheeler-Kingshott, C.A.: A framework for a streamline-based probabilistic index of connectivity (PICo) using a structural interpretation of MRI diffusion measurements. J Magn Reson Imaging **18**, 242–254 (2003)
32. Parker, G.J.M., Stephan, K.E., Barker, G.J., Rowe, J.B., MacManus, D.G., Wheeler-Kingshott, C.A.M., Ciccarelli, O., Passingham, R.E., Spinks, R.L., Lemon, R.N., Turner, R.: Inizial demonstration of in vivo tracing of axonal projections in the macaque brain and comparison with the human brain using tensor imaging and fast marching tractography. NeuroImage **15**(4), 797–809 (2002)
33. Parker, G.J.M., Wheeler-Kingshott, C.A.M., Barker, G.J.: Estimating distributed anatomical connectivity using fast marching methods and diffusion tensor imaging. IEEE Trans. Med. Imaging **21**(5), 505–51 (2002)
34. Pennec, X., Fillard, P., Ayache, N.: A Riemannian framework for tensor computing. Int. J. Comput. Vis. **66**(1), 41–66 (2004)
35. Pierpaoli, C., Jezzard, P., Basser, P.J., Barnett, A., DiChiro, G.: Diffusion tensor MR imaging of the human brain. Radiology **201**(3), 637–648 (1996)
36. Poupon, C., Mangin, J., Frouin, V., Regis, J., Poupon, F., Pachot-Clouard, M., Bihan, D.L., Bloch, I.: Regularization of MR diffusion tensor maps for tracking brain white matter bundles. In: MICCAI, 489–498 (1998)
37. Poupon, C., Mangin, J.F., Clark, C.A., Frouin, V., Regis, J., Bihan, D.L., Bloch, I.: Towards inference of human brain connectivity from MR diffusion tensor data. Med. Image Anal. **5**(2), 1–15 (2001)
38. Sethian, A.: A fast marching level set method for monotonically advancing fronts. Prod. Natl. Acad. Sci. USA **93**(4), 1591–1955 (1996)
39. Smith, S.M., Jenkinson, M., Woolrich, M.W., Beckmann, C.F., Behrens, T.E., Johansen-Berg, H., Bannister, P.R., Luca, M.D., Drobnjak, I., Flitney, D.E., Niazy, R.K., Saunders, J., Vickers, J., Zhang, Y., Stefano, N.D., Brady, J.M., Matthews, P.M.: Advances in functional and structural MR image analysis and implementation as FSL. NeuroImage **23**(Supplement 1), 208 – 219 (2004). Mathematics in Brain Imaging
40. Sotiropoulos, S., Bai, L., Morgan, P.S., Auer, D.P., Constantinescu, C.S., Tench, C.R.: A regularized two-tensor model fit to low angular resolution diffusion images using basis directions. J. Magn. Reson. Imag. **28**(1), 199–209 (2008)
41. Staempfli, P., Jaermann, T., Crelier, G.R., Kollias, S., Valavanis, A., Boesiger, P.: Resolving fiber crossing using advanced fast marching tractography based on diffusion tensor imaging. NeuroImage **30**(1), 110–120 (2006)
42. Tschumperlé, D., Deriche, R., Deriche, R.: Diffusion tensor regularization with constraints preservation. In: IEEE Computer Society Conference on Computer Vision and Pattern Recognition, vol. 1, 948–953 (2001)
43. Tuch, D.S.: Q-ball imaging. Magn. Reson. Med. **52**(6), 1358–1372 (2004)
44. Tuch, D.S., Reese, T.G., Wiegell, M.R., Makris, N., Belliveau, J.W., Wedeen, V.J.: High angular resolution diffusion imaging reveals intra voxel white matter fiber heterogeneity. Magn. Reson. Med. **48**, 577–582 (2002)
45. Walker, J.C., Guccione, J.M., Jiang, Y., Zhang, P., Wallace, A.W., Hsu, E.W., Ratcliffe, M.B.: Helical myofiber orientation after myocardial infraction and left ventricular surgical restoration in sheep. J. Thoracic Cardiov. Surgery **129**(2), 382–390 (2005)
46. Watts, R., Liston, C., Niogi, S., Ulug, A.M.: Fiber tracking using magnetic resonance diffusion tensor imaging and its applications to human brain development. Ment. Retar. Dev. Disabil. Res. Rev. **9**(3), 168–177 (2003)

47. Wedeen, V.J., Hagmann, P., Tseng, W.Y., Reese, T.G., Weisskoff, R.M.: Mapping complex tissue architecture with diffusion spectrum magnetic resonance imaging. Magn. Res. Med. **54**(6), 1377–1386 (2005)
48. Weinstein, D.M., Kindlmann, G.L., Lundberg, E.C.: Tensorlines: Advection-diffusion based propagation through diffusion tensor fields. In: IEEE Visualization, 249–253 (1999)
49. Welk, M., Weickert, J., Becker, F., Schorr, C., Feddern, C., Burgeth, B.: Median and related local filters for tensor-valued images. Signal Process. **87**(2), 291–308 (2007)
50. Westin, C.F., Knutson, H.: Tensor field regularization using normalized convolution. In: Proc. 9th Int. Conf. Computer Aided Systems Theory, 564–572 (2003)
51. Wiegell, M.R., Larsson, H.B., Wedeen, V.J.: Fiber crossing in human brain depicted with diffusion tensor MR imaging. Radiology **217**, 897–903 (2000)
52. Wimberger, D.M., Roberts, T.P., Barkovich, A.J., Prayer, L.M., Moseley, M.E., , Kucharczyk, J.: Identification of premyelination by diffusion-weighted MRI. J. Comp. Assist. Tomogr. **19**(1), 28–33 (1995)
53. Zhukov, L., Barr, A.H.: Oriented tensor reconstruction: Tracing neural pathways from diffusion tensor MRI. In: Proc. IEEE Visualization, 378–394 (2002)
54. Zhukov, L., Barr, A.H.: Heart-muscle fiber reconstruction from diffusion tensor MRI. In: Proc. IEEE Visualization, 79–84 (2003)

Image-Space Tensor Field Visualization Using a LIC-like Method

Sebastian Eichelbaum, Mario Hlawitschka, Bernd Hamann, and Gerik Scheuermann

Abstract Tensors are of great interest to many applications in engineering and in medical imaging, but a proper analysis and visualization remains challenging. Physics-based visualization of tensor fields has proven to show the main features of symmetric second-order tensor fields, while still displaying the most important information of the data, namely the main directions in medical diffusion tensor data using texture and additional attributes using color-coding, in a continuous representation. Nevertheless, its application and usability remains limited due to its computational expensive and sensitive nature.

We introduce a novel approach to compute a fabric-like texture pattern from tensor fields motivated by image-space line integral convolution (LIC). Although, our approach can be applied to arbitrary, non-selfintersecting surfaces, we are focusing on special surfaces following neural fibers in the brain. We employ a multi-pass rendering approach whose main focus lies on regaining three-dimensionality of the data under user interaction as well as being able to have a seamless transition between local and global structures including a proper visualization of degenerated points.

S. Eichelbaum (✉) · G. Scheuermann
Abteilung für Bild- und Signalverarbeitung, Institut für Informatik, Universität Leipzig, Germany
e-mail: eichelbaum@informatik.uni-leipzig.de; scheuermann@informatik.uni-leipzig.de

M. Hlawitschka
Wissenschaftliche Visualisierung, Institut für Informatik, Universität Leipzig

B. Hamann
Institute for Data Analysis and Visualization (IDAV), Department of Computer Science, University of California, Davis, CA, USA
e-mail: hlawitschka@ucdavis.edu; hamann@ucdavis.edu

L. Linsen et al. (eds.), *Visualization in Medicine and Life Sciences II*, Mathematics and Visualization, DOI 10.1007/978-3-642-21608-4_11,

1 Motivation and Related Work

Since the introduction of tensor lines and hyperstreamlines [5], there have been many research efforts directed at the continuous representation of tensor fields, including research on tensor field topology [11, 23, 24]. Zheng and Pang introduced HyperLIC [31], which makes it possible to display a single eigendirection of a tensor field in a continuous manner by smoothing a noise texture along integral lines, while neglecting secondary directions. Recent approaches to visualize Lagrangian structures on tensor fields [12] provide information on one chosen tensor direction and are especially useful for diffusion tensor data, where the main tensor direction can be correlated to neural fibers or muscular structures, whereas the secondary direction only plays a minor role. More recently, Dick et al. [6] published an interactive approach to visualize volumetric tensor field for implant planning.

While glyph-based visualization techniques reveal a huge amount of the local information, even with advanced glyph placement techniques [16], an integration of this information into global structures and an interpolation of information in-between glyphs remains the responsibility of the user. On the other side, approaches focusing on global structures, like [17], which is used to calculate a skeleton of white matter tracts, are not able to provide local information or at least a smooth transition to local structures.

Hotz et al. [13] introduced Physically Based Methods (PBM) for tensor field visualization in 2004 as a means to visualize stress and strain tensors arising in geomechanics. A positive-definite metric that has the same topological structure as the tensor field is defined and visualized using a texture-based approach resembling LIC [3]. Besides other information, eigenvalues of the metric can be encoded by free parameters of the texture definition, such as the remaining color space. Whereas the method's implementation for parameterizable surfaces topologically equivalent to discs or spheres is straightforward, implementations for arbitrary surfaces remains computationally challenging. In 2009, Hotz et al. [14] enhanced their approach to isosurfaces in three-dimensional tensor fields. A three-dimensional noise texture is computed in the data set and a convolution is performed along integral lines tangential to the eigenvector field. LIC has been used in vector field visualization methods to imitate *Schlieren* patterns on surfaces experiments where a thin film of oil is applied to surfaces, which show patterns caused by the air flow. In vector field visualization, image-space LIC is a method to compute *Schlieren*-like textures in image space [9, 19, 20, 28, 29], intended for large and non-parameterized geometries. As these methods are based on vector fields, their application to a symmetric tensor field's major eigenvector is possible, with the limitation that the eigenvector field does not provide an orientation. For asymmetric tensor-fields, other approaches exist [30]. Besides the non-trivial application of image-space LIC to tensor data, image-space LIC has certain other drawbacks. Mainly because the noise pattern is defined in image space, it does not follow the movement of the surface and, therefore, during user interaction, the consistent surface impression is lost. A simple method proposed to circumvent this problem is animating the texture pattern by applying randomized trigonometric functions to the input noise. Weiskopf and

Ertl [27] solved this problem for vector field visualization by generating a three-dimensional texture that is scaled appropriately in physical space.

We implemented an algorithm similar to the original PBM but for arbitrary non-intersecting surfaces in image space. Our algorithm can perform at interactive frame rates for large data sets on current desktop PCs. We overcome the drawbacks present in image-space LIC implementations by defining a fixed parameterization on the surface. Thus, we do not require a three-dimensional noise texture representation defined at sub-voxel resolution on the data set. Our approach is capable of maintaining local coherence of the texture pattern between frames when (1) transforming, rotating, or scaling the visualization, and (2) changing the surface by, e.g., changing isovalues or sweeping the surface through space. In addition, we implement special application-dependent modes to ensure our method integrates well with existing techniques.

2 Method

We employ a multi-pass rendering technique that consists of four major rendering passes as outlined in Fig. 1. After generating the basic input textures once, the first pass projects all required data into image space. Pass two performs a silhouette detection that is used to guarantee integrity of the advection step computed by multiple iterations of pass three. Eventually, pass four composes the intermediate textures in a final rendering.

2.1 Projection into Image Space

First, we project the data into image space by rendering the surface using the default OpenGL rendering pipeline. Notably, the surface does not need to be

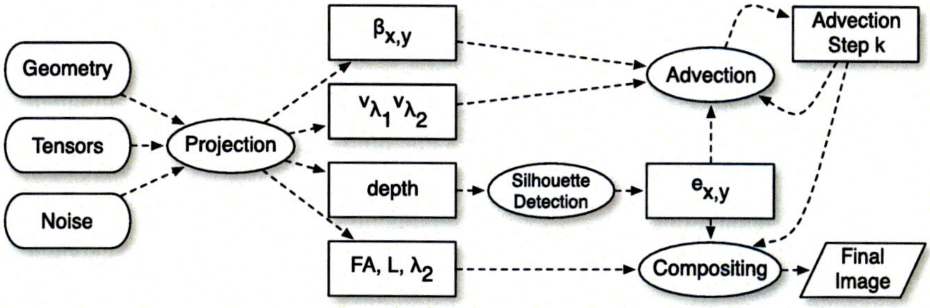

Fig. 1 Flowchart indicating the four major steps of the algorithm: *projection*, which transforms the data set in an image-space representation and produce the initial noise texture $\beta_{x,y}$ on the geometry; *silhouette detection*, required for the advection step and the final rendering; *advection*, which produces the two eigenvector textures; and the final *compositing*, which combines intermediate textures to the final visualization. Between consecutive steps, the data is transferred using textures

represented by a surface mesh. Any other representation that provides proper depth and surface normal information works just as well (e.g., ray-casting methods for implicit surfaces, cf. Knoll et al. [18]). In the same rendering step, the tensor field is transformed from world space to object space, i.e., each tensor T, which is interpolated at the point on the surface from the surrounding two- or three-dimensional tensor field is projected onto the surface by

$$T' = P \cdot T \cdot P^T, \tag{1}$$

with a matrix P defined using the surface normal n as

$$P = \begin{pmatrix} 1 - n_x^2 & -n_y n_x & -n_z n_x \\ -n_x n_y & 1 - n_y^2 & -n_z n_y \\ -n_x n_z & -n_y n_z & 1 - n_z^2 \end{pmatrix}. \tag{2}$$

The camera viewing system configuration and the available screen resolution imply a super- or sub-sampling of the data. We obtain an interpolated surface tensor in every pixel which is decomposed into the eigenvector/eigenvalue representation using a method derived from the one presented by Hasan et al. [10]. These eigenvectors, which are still defined in object space, are projected into image space using the same projection matrices M_M and M_P used for projecting the geometry to image space, usually the standard *modelview* and *projection* matrices OpenGL offers:

$$v'_{\lambda_i} = M_P \times M_M \times v_{\lambda_i}, \text{ with } (i \in 1, 2). \tag{3}$$

Even in the special case of symmetric second-order tensors in $\mathbb{R}^3$, which have three real-valued eigenvalues and three orthogonal eigenvectors in the non-degenerate case, in general, the projected eigenvectors are not orthogonal in two-dimensional space. To simplify further data handling, we scale the eigenvectors as follows:

$$\|v\|_\infty = max\{|v_x|, |v_y|\} \tag{4}$$

$$v''_{\lambda_i} = \frac{v'_{\lambda_i}}{\|v'_{\lambda_i}\|_\infty} \text{ with } i \in \{1, 2\}, \quad \text{and} \quad \|v'_{\lambda_i}\|_\infty \neq 0 \tag{5}$$

The special case $\|v'_{\lambda_i}\|_\infty = 0$ only appears when the surface normal is perpendicular to the view direction and, therefore, can be ignored. The maximum norm in Equation 4 ensures that one component is 1 or -1 and, therefore, one avoids numerical instabilities arising when limited storage precision is available, and can use memory-efficient eight-bit textures.

2.2 *Initial Noise Texture Generation*

In contrast to standard LIC approaches, to achieve a proper visual representation of the data, high-frequency noise textures, such as white noise, are not suitable for the compositing of multiple textures. Therefore, we compute the initial noise texture using the reaction diffusion scheme first introduced by Turing [25] to simulate the mixture of two reacting chemicals, which leads to larger but smooth "spots" that are randomly and almost uniquely distributed (cf. Fig. 2, right). For the discrete case, the governing equations are:

$$\begin{aligned}\Delta a_{i,j} &= F(i,j) + D_a \cdot (a_{i+1,j} + a_{i-1,j} + a_{i,j+1} + a_{i,j-1} - 4 \cdot a_{i,j}),\\ \Delta b_{i,j} &= G(i,j) + D_b \cdot (b_{i+1,j} + b_{i-1,j} + b_{i,j+1} + b_{i,j-1} - 4 \cdot b_{i,j}), \text{ where}\\ F(i,j) &= s(16 - a_{i,j} \cdot b_{i,j}) \text{ and } G(i,j) = s(a_{i,j} \cdot b_{i,j} - b_{i,j} - \beta_{i,j}).\end{aligned} \tag{6}$$

Here, we assume continuous boundary conditions to obtain a seamless texture in both directions. The scalar s allows one to control the size of the spots where a smaller value of s leads to larger spots. The constants D_a and D_b are the diffusion constants of each chemical. We use $D_a = 0.125$ and $D_b = 0.031$ to create the input textures.

2.3 *Noise Texture Transformation*

Mapping the initial texture to the geometry is a difficult and application-dependent task. Even though there exist methods to parameterize a surface, they employ restrictions to the surface (such as being isomorphic to discs or spheres), require additional storage for texture atlases (cf. [15,21]) and, in general, require additional and often time-consuming pre-processing.

Another solution, proposed by Turk et al. [26], calculates the reaction diffusion texture directly on the surface. A major disadvantage of this method is the computational complexity. Even though these approaches provide almost distortion-free texture representations, isosurfaces, for example, may consist of a large amount of unstructured primitives, which increases the pre-processing time tremendously.

Whereas previous approaches for image space LIC either use parameterized surfaces to apply the initial noise pattern to the surface or use locally or globally defined three-dimensional textures [27], we define an implicit parameterization of the surface that provides an appropriate mapping of the noise texture to the surface.

We start by implicitly splitting world space in voxels of equal size, filling the geometry's bounding box, i.e., we define a regular grid. Each voxel i is described by its base coordinate b_i and a constant edge length l. The seamless reaction diffusion texture is mapped to the surface of each of these voxels. To assign a texture coordinate to each vertex, the object space coordinate is transformed to the voxel space that is described by a minimum and maximum coordinate whose connecting line is the bounding box' diagonal. Points v_g on the geometry are transformed to v_{voxel} using

$$v_{voxel} = v_g \cdot \begin{pmatrix} l & 0 & 0 & -b_{min_x} \\ 0 & l & 0 & -b_{min_y} \\ 0 & 0 & l & -b_{min_z} \\ 0 & 0 & 0 & 1 \end{pmatrix} \tag{7}$$

and transformed into the local coordinate system by

$$v_{hit} = v_{voxel} - \lfloor v_{voxel} \rfloor. \tag{8}$$

The two texture coordinates are chosen to be those two components of v_{hit} that form the plane that is closest to the tangential plane of the surface in this point.

$$t = (v_{hit_i}, v_{hit_j}), \text{ with } i \neq j \neq k \wedge (n_k = max\{n_i, n_j, n_k\}). \tag{9}$$

In other words, this method transforms the world coordinate system to a system defined by one voxel, assuring that every component of every point v_{hit} is in $[0, 1]$. The texture coordinate is determined by the surface's normal and, in particular, by the voxel side-plane whose normal is most similar to the surface's one (in terms of angle between them). The use of the term "voxel" is for illustration purpose only; those voxels are never created explicitly. As a result, the texture β contains the image space mapped input noise texture as shown in Fig. 2. This texture is used as the initial pattern that is advected in the advection step.

Regardless of its simplicity, this method allows a continuous parameterization of the surface space that only introduces irrelevant distortions for mapping the noise texture (cf. Fig. 2). The mapping is continuous but not C^1-continuous, which is not required for mapping the noise texture as discontinuities in the first derivatives automatically vanish in the advection step.

Another positive aspect of this mapping is the possibility of a change of scale that is not available in the approaches of, e.g., Turk et al. [26]. By changing the size of voxels during the calculation, different frequencies of patterns can easily be produced and projected to the geometry. This capability allows one to change the

Fig. 2 Illustration of the reaction diffusion texture used (left) and the noise texture mapped to geometry $\beta_{x,y}$ (right)

Fig. 3 Comparison of two different values of l to demonstrate the possibility for dynamic refinement of the input noise to achieve different levels of detail

resolution of the texture as required for automatic texture refinement when zooming. A comparison of two different levels of detail is shown in Fig. 3.

2.4 *Silhouette Detection*

To avoid advection over geometric boundaries, a silhouette of the object is required to stop advection in these areas [19]. Otherwise, tensor advection would lead to a constant flow of "particles" across surface boundaries which makes the surface's geometry and topology unrecognizable.

A standard three-by-three Laplacian filter, defined by the convolution mask

$$\begin{bmatrix} 0 & 1 & 0 \\ 1 & -4 & 1 \\ 0 & 1 & 0 \end{bmatrix} \tag{10}$$

applied to the depth values followed by thresholding has proven to be suitable for our purposes. The silhouette image $e_{x,y}$ for each pixel (x, y) is then in the red color channel of its output texture.

2.5 *Advection*

We have discussed how to project the geometry and the corresponding tensor field to image space. With the prepared image space eigenvectors and the input noise texture, mapped to geometry, advection can be done. We use a simple Euler integration applied to both vector fields. With Euler's method some particle can be followed along a stream. In our case, we do not calculate streamlines at each position of both vector fields, as normally done in LIC. We directly advect the noise input texture with the given vector fields, which has the same result as locally filtering the data along pre-computed streamlines. This decision was based on the fact that massively parallel architectures like modern GPUs are able to perform this task in parallel for each pixel a hundred times per second. Formally, the advection step can be described as follows: First, we assume an input field P to be a continuous function, defined in a two-dimensional domain:

$$f_P : (x, y) \rightarrow p, \text{ with } x, y, p \in [0, 1], \tag{11}$$

i.e., it is a function returning the input value of the field P at a given position. Continuity is ensured with interpolating values in between. With this in mind, the iterative advection on each point (x, y) on the image plane can now be described by

$$\forall x, y \in [0, 1] : \forall \lambda \in \{\lambda_1, \lambda_2\} :$$

$$\begin{aligned} p_0^\lambda &= \beta_{x,y}, \\ p_{i+1}^\lambda &= k \cdot \beta_{x,y} + (1 - k) \cdot \frac{f_{p_i^\lambda}((x, y) + v'_\lambda) + f_{p_i^\lambda}((x, y) - v'_\lambda)}{2}. \end{aligned} \tag{12}$$

The iterative advection process has to be done for each eigenvector field separately with separate input and output fields p_i as can be seen in Fig. 4. The value at a given point is a mix of the input noise and the iteratively advected input noise from the prior steps. Since the eigenvectors v'_{λ_j} do not have an orientation, the advection has to be done in both directions. The iteration can be stopped when the value change exceeds a threshold, i.e., if $|p_{i+1}^\lambda - p_i^\lambda| < \epsilon$. We have chosen $\epsilon = 0.05$, which is very conservative but ensures a proper rendering.

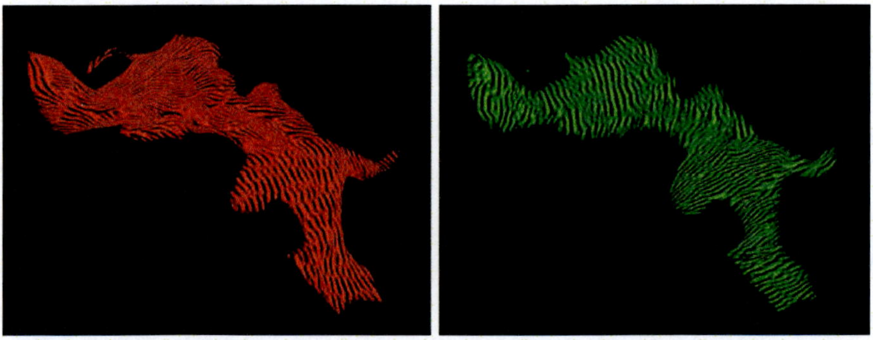

Fig. 4 Advection texture after ten iterations. Left: red channel containing $p_9^{\lambda_1}$, the advection image of eigenvector field 1; right: green channel containing $p_9^{\lambda_2}$, the advection image of eigenvector field 2

2.6 Compositing

The final rendering pass composes the temporary textures for final visualization. Whereas pixels that are not part of the original rendering of the geometry are discarded using the information from the depth buffer, for all other pixels the color values at a point (x, y) in image space after k iterations is defined by:

$$\begin{aligned} R &= \frac{r \cdot f_{p_k^{\lambda_2}}(x,y)}{8 \cdot f^2_{p_k^{\lambda_1}}(x,y)} + e_{x,y} + light(\mathcal{L}_{x,y}), \\ G &= \frac{(1-r) \cdot f_{p_k^{\lambda_1}}(x,y)}{8 \cdot f^2_{p_k^{\lambda_2}}(x,y)} + e_{x,y} + light(\mathcal{L}_{x,y}), \\ B &= e_{x,y} + light(\mathcal{L}_{x,y}), \end{aligned} \tag{13}$$

where $p_k^{\lambda_1}$ and $p_k^{\lambda_2}$ are the fields generated from the eigenvector advection and e is the silhouette image. The scalar factor r is used to blend between the two chosen tensor directions. This approach creates a mesh resembling the tensor field's structure. To reduce the effect of light sources on the color coding, we use a separate lighting function $light$ that, while manipulating the intensity, does not affect the base color of the mesh structure. Even though Blinn-Phong shading [2] has proven to provide the required depth cues, additional emphasis of the third dimension using depth-enhancing color coding has proven to provide a better overall understanding of the data [4]. Finally, further filters can be applied on the composed image, like contrast enhancement or sharpening filters common to vector field LIC [9, 27]. Figure 5 shows the result of Equation 13 combined with Blinn-Phong shading and an applied sharpening filter.

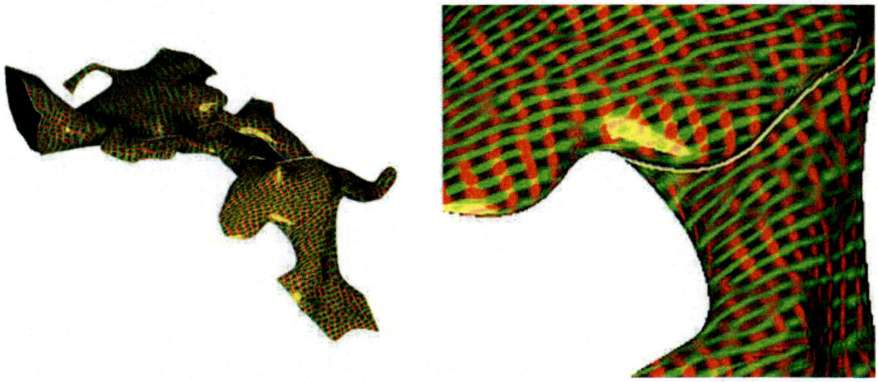

Fig. 5 The final image produced by the output processing shader with lighting. Left: the whole geometry. Right: a zoomed part of the geometry to show the fabric structure on the surface

2.7 Implementation

Our implementation is not limited to a special kind of geometry. It is able to handle almost every tensor field defined on a surface. It is, for example, possible to calculate an isosurface on a derived scalar metric, like fractional anisotropy or on a second data set to generate a surface in a three-dimensional data domain. Other methods include hyper-stream surfaces [5], wrapped streamlines [8], or domain-dependent methods like dissection-like surfaces presented in [1]. The only requirement for the surface is that it is non-selfintersecting and smooth normals are provided as they are required for the projection step and for proper lighting. The noise texture can be pre-calculated in a pre-processing step or stored in a file as it is independent of the data.

Our implementation is basing on an multipath rendering approach, where each of the four processing steps in Fig. 1 is implemented as shader program and rendered one after the other utilizing frame-buffer-objects (FBO) and textures for data transfer.

The first step projects the geometry into image space, simply by rendering the geometry and pre-calculating the Phong light intensity $\mathscr{L}_{x,y} \in [0,1]$ at every rendered fragment with the coordinates x and y. In the same step, the tensors are projected as well using (1)–(5). Tensor projection is done fragment-wise as interpolation of vertex-wise calculated and projected eigensystems causes problems. Please note, that the texture with the projected eigenvectors needs to be initialized in black, as the RGBA-quadruple $(0,0,0,0)$ denotes both eigenvectors to be of length 0 and can be used as abort criterion during advection. The precalculated two-dimensional noise texture (c.f. Sect. 2.2) is also mapped to each fragment during fragment processing using (7)–(9). As the projection step is done using an FBO, the resulting values can be written to multiple textures, which then can be used as input textures for the consecutive steps.

Followed by the iterative advection of the mapped noise in image-space, the silhouette detection step applies a Laplacian convolution kernel fragment-wise to the depth buffer to provide the needed border information to the advection step (c.f. Sect. 2.4). The input of the advection step is the noise, mapped on the geometry. Figure 2, right shows this. As Equation 12 shows, the advection uses the previously advected texture which, in the first iteration, is the geometry-mapped noise and advects it in direction of the eigenvector field. The resulting texture is then combined with the original geometry-mapped noise. During each render loop of the complete scene, we apply advection three times. The resulting texture is then used as input for the advection step during the next render pass. Figure 4 shows these advected images, separate for each of the eigenvector fields.

The advected images are finally composed to the final image shown on the screen, showing the fabric-like structure in combination with color-mapping. Our compositing is mainly the combination of both advected noise textures according to (13).

3 Results

We have introduced a method to create a fabric-like surface tensor LIC in image space, similar to the one introduced in [13]. We used ideas from [19] to transform the algorithm into image space. Our implementation, using this method, is able to reach frame rates high enough for real-time user interaction. The only bottleneck is the hardware's ability in rendering large and triangle-rich geometry. All further steps can be done in constant time, see Table 1.

3.1 *Artificial Test Data Sets*

We first applied our method to artificial test data sets with different complex topologies. The Bretzel5 data set shown here is defined implicitly:

$$((x^2 + .25 * y^2 - 1) * (.25 * x^2 + y^2 - 1))^2 + z^2 - 0.1 = 0. \tag{14}$$

We used the Hessian matrix of the corresponding scalar fields on the surfaces as tensor fields which, in fact, describe the curvature. The results displayed in Fig. 6 show that neither the topology nor our artificial parameterization of the input noise texture influences the quality of the final rendering. In the center of the sphere in Fig. 6, a degenerate point can be seen. On this point, the first and second eigenvalue are both zero. Our method can handle these points properly.

3.2 *Modification for Medical Data Processing*

Even though many higher-order methods have been proposed, due to scanner, time, and cost limitations, second-order tensor data is still dominant in clinical application. Medical second-order diffusion tensor data sets differ from engineering data sets because they indicate one major direction whereas the secondary and

Table 1 Frames per second (fps) for different data sets with given number of triangles and numbers of tensors. The frame rates are compared to simple rendering of the geometry using Phong shading. The frame rates were obtained for an AMD Athlon(tm) 64 X2 Dual Core Processor 3800+ (512K L2 Cache) with an NVIDIA G80 GPU (GeForce 8800 GTS) and 640MB of graphics memory at a resolution of 1024 × 768 pixels. The geometry share relates the time used by the GPU to rasterize the geometry to the overall rendering time, which contains all steps of the pipeline. The time used to render the geometry clearly dominates the rendering times and reaches up to 90% of the overall rendering time even for medium-sized geometries

Figures	Nb Triangles	Nb Tensors	fps	fps (Phong only)	Ø Geometry Share
Figure 8	41472	63075	32	61	72%
Figure 5	58624	88803	30	60	69%
Figure 7	571776	861981	14	16	90%

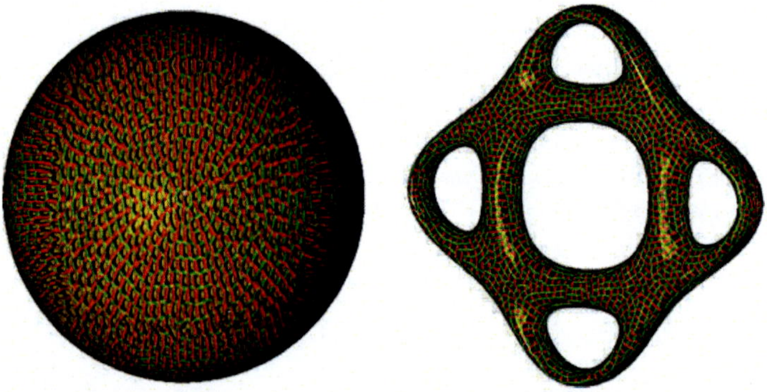

Fig. 6 Analytic test data sets. We applied our method to isosurfaces and the scalar field's Hessian matrix, showing the curvature on the surface, to demonstrate the suitability for topologically more complex surfaces. Shown here are the final images using our method for a sphere and Bretzel5 data set (14). The image from the sphere data set has been improved by applying a further bump-mapping filter as introduced in [7]. The eigenvalues and eigenvectors of the Hessian matrix denote the change of the normal on the surface, which corresponds to the gradient in the scalar field

ternary directions only provide information in areas where the major direction is not well-defined, i.e., the fractional anisotropy—a measure for the tensor shape—is low. Almost spherical tensors, which indicate isotropic diffusion, occur in areas where multiple fiber bundles traverse a single voxel of the measurement or when no directional structures are present. Therefore, we modulate the color coding using additional information: In areas where one fiber direction dominates, we only display this major direction using the standard color coding for medical data sets, where x, y, and z alignment are displayed in red, green, and blue, respectively. In areas where a secondary direction in the plane exists, we display this information as well but omit the secondary color coding and display the secondary direction in gray-scale rendering mode and always below the primary direction (cf. Fig. 8). We use the method of Anwander et al. [1] to extract surfaces that are, where possible, tangential to the fiber directions. Hence, we can guarantee that the projection error introduced by using our method in the surface's domain remains sufficiently small, which can be seen in Fig. 9. Even in areas where the fractional anisotropy is low and the color coding does no longer provide directional information, such as in some parts of the pyramidal tract in Fig. 8, the texture pattern still provides this information.

The applicability of our method, especially to medical second-order tensor data, is mainly due to its real-time ability and its ability to provide a smooth transition between local and global structures. Neuroscientist researcher can explore tensor data interactively, by using slices inside the volume or by calculating surfaces in interesting regions.

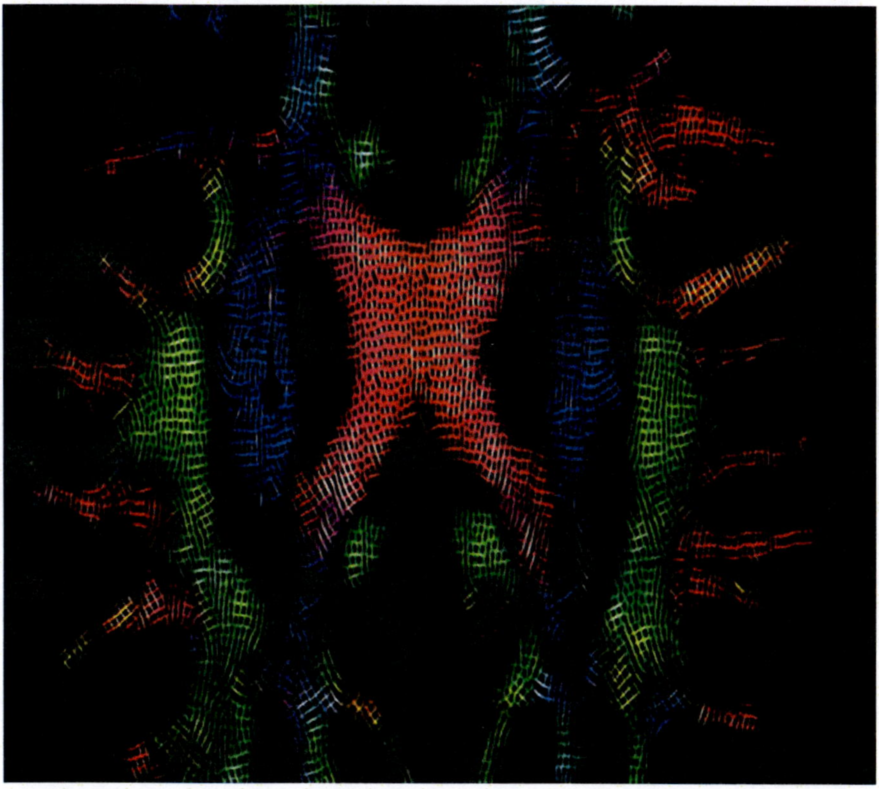

Fig. 7 An axial slice through a human brain: Corpus callosum (CC) (red), pyramidal tract (blue), and parts of the cinguli (green in front and behind the CC) are visible. The main direction in three-dimensional space is indicated by the RGB color map, where red indicates the lateral (left–right), green anterior–posterior, and blue superior–inferior direction. The left–right structure of the CC can clearly be seen in its center, whereas color and pattern indicate uncertainty towards the outer parts. The same is true for the cinguli's anterior–posterior structure. As seen from the blue color, the pyramidal tract is almost perpendicular to the chosen plane and, therefore, secondary and ternary eigenvectors dominate the visualization. Alternatively, we could easily fade out those out-of-plane structures in cases where they distract the user

3.3 Performance

As indicated before, the only "bottleneck" in the visualization pipeline that is strongly geometry-dependent is the projection step. Since the surface needs to be rendered repeatedly in case of user interaction, the performance measures of our method consider repeated rendering of the geometry. The frame rate with geometry not being moved and, therefore, making the projection step and the edge detection step unnecessary, is considerably higher. Our implementation requires only few advection iterations per frame, which ensures high frame rates and smooth

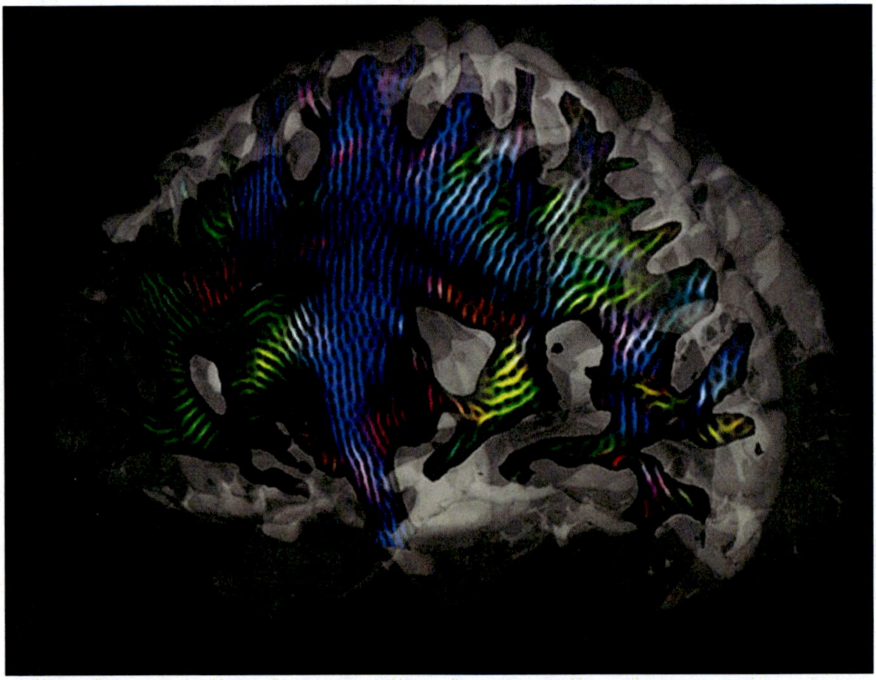

Fig. 8 Diffusion tensor data set of a human brain. We employed the method by Anwander et al. [1] to extract a surface following neural fibers and applied our method with an alternative color coding that is more suitable and can be incorporated more easily into medical visualization tools

interaction. To make the frame rates comparable, in the following tables, user interaction is assumed and, therefore, rendering a single frame always consists of:

- One projection step, including geometry rendering.
- One edge detection pass.
- Three advection iterations.
- One output processing pass.

As seen in the previous sections, fragments not belonging to the geometry are discarded as soon as possible without using deferred shading. This also generates performance gain in advection and output processing. In Table 1, a selection of data sets with their corresponding number of triangles and tensors are listed. These measurements where taken on a 1024×768 viewport with about $\frac{3}{4}$ of the pixels covered by the surface.

The assumption that geometry rendering with projection is the weakest component in this pipeline and that edge detection, advection, and output processing perform at a data-independent frame rate is confirmed by the frame rates shown in Table 1. It confirms that for large geometries, rendering the geometry alone is the dominating component. Since the vertex-wise calculations during projection are limited to tensor projection (1) and vertex projection (7), the most expensive

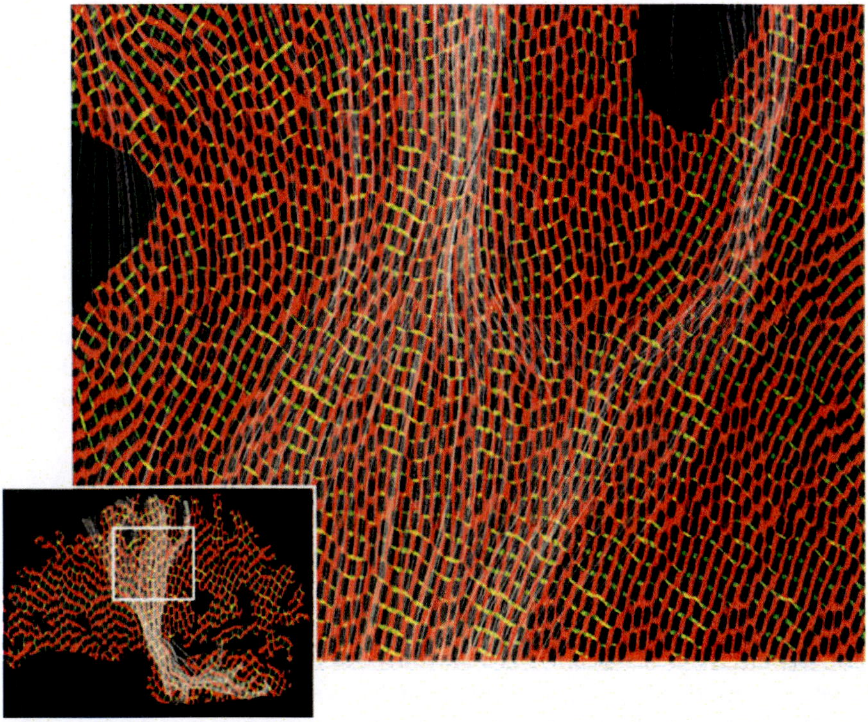

Fig. 9 For validation, we calculated the fiber-tracts near the used surface. The directions can now be compared. It shows that the major eigenvector direction correlates with the direction of the fiber-tracts. As the used slice is near to the pyramidal tract, there is a strong diffusion in axial direction. But not all the shown fiber-tracts match the direction of the shown surface tensor, which is caused by not having all fiber-tracts directly on the surface. Most of them are slightly in front of it and therefore can have other local directions

calculations during projection are executed per fragment. This means that the expensive eigenvalue decomposition and eigenvector calculations are only required for fragments (pixels) actually visible on the screen. Independent of the geometry's complexity, the number of fragments that require the costly processing steps reaches a saturation value and therefore does not require further processing time even if the geometry complexity increases. Table 1 states that this saturation point has been reached in Fig. 7. The computation for per-fragment Phong lighting consumes much more time as it is calculated before the invisible fragments have been discarded, which explains the nearly equal framerates. To further decouple the calculation effort from the geometry's size, the depth test should be performed before performing the eigendecomposition. This goal can be achieved by first rendering the projected tensors to a texture, and computing the decomposition on visible fragments only. Nevertheless, this is not necessary for our current data set and screen sizes where the time required to render the geometry itself clearly dominates the time required to compute the texture pattern in image space. This can be seen in the increasing values in Table 1 with increasing size of vertices rendered.

4 Conclusions and Possible Directions for Future Research

We have presented a novel method for rendering fabric-like structures to visualize tensor fields on almost arbitrary surfaces without generating three-dimensional textures that span the whole data set at sub-voxel resolution. Therefore, our method can be applied to complex data sets without introducing texture memory problems common to methods relying on tree-dimensional noise textures. As major parts of the calculation are performed in image space, the performance of our algorithm is almost independent of data set size, provided that surfaces can be drawn efficiently, e.g., by using acceleration structures to draw only those parts of the geometry that intersect the view frustum or using ray tracing methods.

Whether the surface itself is the domain of the data, a surface defined on the tensor information (e.g., hyper stream surfaces), or a surface defined by other unrelated quantities (e.g., given by material boundaries in engineering data or anatomical structures in medical data) is independent from our approach. Nevertheless, the surface has to be chosen appropriately because only in-plane information is visualized. To circumvent this limitation, information perpendicular to the plane could be incorporated in the color coding, but due to a proper selection of the plane that is aligned with our features of interest, this has not been necessary for our purposes.

Especially in medical visualization, higher-order tensor information is becoming increasingly important and different methods exist to visualize these tensors, including local color coding, glyphs, and integral lines. Nevertheless, an extension of our approach is one of our major aims. In brain imaging, experts agree that the maximum number of possible fiber directions is limited. Typically, a maximum of three or four directions in a single voxel are assumed (cf., [22]). Whereas the number of output textures can easily be adapted, the major remaining problem is a lack of suitable decomposition algorithms on the GPU. Image-space techniques, by their very nature, resample the data and, therefore, require one to use such proper interpolation schemes. In addition, maintaining orientations and assigning same fibers in higher-order data to the same texture globally is not possible today and, therefore, is a potential topic for further investigation.

Acknowledgements We thank Alfred Anwander and Thomas R. Knösche from Max Planck Institute for Human Cognitive and Brain Sciences, Leipzig, Germany, for providing the human brain image data sets, and for fruitful discussions and comments. We thank the members of the Visualization and Computer Graphics Research Group of the Institute for Data Analysis and Visualization, Department of Computer Science, UC Davis, and the members of the Abteilung für Bild- und Signalverarbeitung des Instituts für Informatik der Universität Leipzig, Germany.
Mario Hlawitschka was supported by NSF grant CCF-0702817.

References

1. Anwander, A., Schurade, R., Hlawitschka, M., Scheuermann, G., Knsche, T.: White matter imaging with virtual klingler dissection. NeuroImage **47**(Supplement 1), S105 – S105 (2009). DOI 10.1016/S1053-8119(09)70916-4. Organization for Human Brain Mapping 2009 Annual Meeting
2. Blinn, J.F.: Models of light reflection for computer synthesized pictures. In: SIGGRAPH '77: Proceedings of the 4th annual conference on Computer graphics and interactive techniques, 192–198. ACM, New York, NY, USA (1977). DOI http://doi.acm.org/10.1145/563858.563893
3. Cabral, B., Leedom, L.C.: Imaging vector fields using line integral convolution. In: SIGGRAPH '93: Proceedings of the 20th annual conference on Computer graphics and interactive techniques, 263–270. ACM, New York, NY, USA (1993). DOI http://doi.acm.org/10.1145/166117.166151
4. Chu, A., Chan, W.Y., Guo, J., Pang, W.M., Heng, P.A.: Perception-aware depth cueing for illustrative vascular visualization. In: BMEI '08: Proceedings of the 2008 International Conference on BioMedical Engineering and Informatics, 341–346. IEEE Computer Society, Washington, DC, USA (2008). DOI http://dx.doi.org/10.1109/BMEI.2008.347
5. Delmarcelle, T., Hesselink, L.: Visualization of second order tensor fields and matrix data. In: VIS '92: Proceedings of the 3rd conference on Visualization '92, 316–323. IEEE Computer Society Press, Los Alamitos, CA, USA (1992)
6. Dick, C., Georgii, J., Burgkart, R., Westermann, R.: Stress tensor field visualization for implant planning in orthopedics. IEEE Transactions on Visualization and Computer Graphics **15**(6), 1399–1406 (2009). DOI http://doi.ieeecomputersociety.org/10.1109/TVCG.2009.184
7. Eichelbaum, S., Hlawitschka, M., Hamann, B., Scheuermann, G.: Fabric-like visualization of tensor field data on arbitrary surfaces in image space. Submitted to Dagstuhl Seminar 09302
8. Enders, F., Sauber, N., Merhof, D., Hastreiter, P., Nimsky, C., Stamminger, M.: Visualization of white matter tracts with wrapped streamlines. In: C.T. Silva, E. Gröller, H. Rushmeier (eds.) Proceedings of IEEE Visualization 2005, 51–58. IEEE Computer Society, IEEE Computer Society Press, Los Alamitos, CA, USA (2005)
9. Grabner, M., Laramee, R.S.: Image space advection on graphics hardware. In: SCCG '05: Proceedings of the 21st spring conference on Computer graphics, 77–84. ACM, New York, NY, USA (2005). DOI http://doi.acm.org/10.1145/1090122.1090136
10. Hasan, K.M., Basser, P.J., Parker, D.L., Alexander, A.L.: Analytical computation of the eigenvalues and eigenvectors in DT-MRI. Journal of Magnetic Resonance **152**(1), 41 – 47 (2001). DOI 10.1006/jmre.2001.2400
11. Hesselink, L., Levy, Y., Lavin, Y.: The topology of symmetric, second-order 3d tensor fields. IEEE Transactions on Visualization and Computer Graphics **3**(1), 1–11 (1997). DOI http://dx.doi.org/10.1109/2945.582332
12. Hlawitschka, M., Garth, C., Tricoche, X., Kindlmann, G., Scheuermann, G., Joy, K.I., Hamann, B.: Direct visualization of fiber information by coherence. International Journal of Computer Assisted Radiology and Surgery, CARS, CUARC.08 Special Issue (2009)
13. Hotz, I., Feng, L., Hagen, H., Hamann, B., Joy, K., Jeremic, B.: Physically based methods for tensor field visualization. In: VIS '04: Proceedings of the conference on Visualization '04, 123–130. IEEE Computer Society, Washington, DC, USA (2004). DOI http://dx.doi.org/10.1109/VIS.2004.80
14. Hotz, I., Feng, Z.X., Hamann, B., Joy, K.I.: Tensor field visualization using a fabric-like texture on arbitrary two-dimensional surfaces. In: T. Möller, B. Hamann, R.D. Russel (eds.) Mathematical Foundations of Scientific Visualization, Computer Graphics, and Massive Data Exploration. Springer-Verlag Heidelberg, Germany (2009)
15. Iwakiri, Y., Omori, Y., Kanko, T.: Practical texture mapping on free-form surfaces. In: PG '00: Proceedings of the 8th Pacific Conference on Computer Graphics and Applications, 97. IEEE Computer Society, Washington, DC, USA (2000)

16. Kindlmann, G.: Superquadric tensor glyphs. In: Proceedings of IEEE TCVG/EG Symposium on Visualization 2004, 147–154 (2004)
17. Kindlmann, G., Tricoche, X., Westin, C.F.: Anisotropy creases delineate white matter structure in diffusion tensor MRI. In: Ninth International Conference on Medical Image Computing and Computer-Assisted Intervention (MICCAI'06), Lecture Notes in Computer Science 4190, 126–133. Copenhagen, Denmark (2006)
18. Knoll, A., Hijazi, Y., Hansen, C., Wald, I., Hagen, H.: Interactive ray tracing of arbitrary implicits with simd interval arithmetic. In: RT '07: Proceedings of the 2007 IEEE Symposium on Interactive Ray Tracing, 11–18. IEEE Computer Society, Washington, DC, USA (2007). DOI http://dx.doi.org/10.1109/RT.2007.4342585
19. Laramee, R.S., Jobard, B., Hauser, H.: Image space based visualization of unsteady flow on surfaces. In: VIS '03: Proceedings of the 14th IEEE Visualization 2003 (VIS'03), 18. IEEE Computer Society, Washington, DC, USA (2003). DOI http://dx.doi.org/10.1109/VISUAL.2003.1250364
20. Laramee, R.S., van Wijk, J.J., Jobard, B., Hauser, H.: Isa and ibfvs: Image space-based visualization of flow on surfaces. IEEE Transactions on Visualization and Computer Graphics **10**, 637–648 (2004). DOI http://doi.ieeecomputersociety.org/10.1109/TVCG.2004.47
21. Purnomo, B., Cohen, J.D., Kumar, S.: Seamless texture atlases. In: SGP '04: Proceedings of the 2004 Eurographics/ACM SIGGRAPH symposium on Geometry processing, 65–74. ACM, New York, NY, USA (2004). DOI http://doi.acm.org/10.1145/1057432.1057441
22. Schultz, T., Seidel, H.P.: Estimating crossing fibers: A tensor decomposition approach. IEEE Transactions on Visualization and Computer Graphics **14**(6), 1635–1642 (2008). DOI http://doi.ieeecomputersociety.org/10.1109/TVCG.2008.128
23. Tricoche, X.: Vector and tensor field topology simplification, tracking, and visualization. Ph.D. thesis, University of Kaiserslautern, Germany (2002)
24. Tricoche, X., Scheuermann, G., Hagen, H.: Tensor topology tracking: A visualization method for time-dependent 2D symmetric tensor fields. In: Eurographics 2001 Proceedings, Computer Graphics Forum 20(3), 461–470. The Eurographics Association, Saarbrücken, Germany (2001). DOI http://dx.doi.org/10.1111/1467-8659.00539
25. Turing, A.: The chemical basis of morphogenesis. Philosophical Transactions of the Royal Society of London **237**(641), 37 – 72 (1952)
26. Turk, G.: Generating textures on arbitrary surfaces using reaction-diffusion. In: SIGGRAPH '91: Proceedings of the 18th annual conference on Computer graphics and interactive techniques, 289–298. ACM, New York, NY, USA (1991). DOI http://doi.acm.org/10.1145/122718.122749
27. Weiskopf, D., Ertl, T.: A hybrid physical/device-space approach for spatio-temporally coherent interactive texture advection on curved surfaces. In: GI '04: Proceedings of Graphics Interface 2004, 263–270. Canadian Human-Computer Communications Society, School of Computer Science, University of Waterloo, Waterloo, Ontario, Canada (2004)
28. van Wijk, J.J.: Image based flow visualization. In: SIGGRAPH '02: Proceedings of the 29th annual conference on Computer graphics and interactive techniques, 745–754. ACM, New York, NY, USA (2002). DOI http://doi.acm.org/10.1145/566570.566646
29. van Wijk, J.J.: Image based flow visualization for curved surfaces. In: VIS '03: Proceedings of the 14th IEEE Visualization 2003 (VIS'03), 17. IEEE Computer Society, Washington, DC, USA (2003). DOI http://dx.doi.org/10.1109/VISUAL.2003.1250363
30. Zhang, E., Yeh, H., Lin, Z., Laramee, R.S.: Asymmetric tensor analysis for flow visualization. IEEE Transactions on Visualization and Computer Graphics **15**, 106–122 (2009). DOI http://doi.ieeecomputersociety.org/10.1109/TVCG.2008.68
31. Zheng, X., Pang, A.: Hyperlic. In: VIS '03: Proceedings of the 14th IEEE Visualization 2003 (VIS'03), 33. IEEE Computer Society, Washington, DC, USA (2003). DOI http://dx.doi.org/10.1109/VISUAL.2003.1250379

Towards a High-quality Visualization of Higher-order Reynold's Glyphs for Diffusion Tensor Imaging

Mario Hlawitschka, Younis Hijazi, Aaron Knoll, and Bernd Hamann

Abstract Recent developments in magnetic resonance imaging (MRI) have shown that displaying second-order tensor information reconstructed from diffusion-weighted MRI does not display the full structure information acquired by the scanner. Therefore, higher-order methods have been developed. Besides the visualization of derived structures such as fiber tracts or tractography (directly related to stream lines in fluid flow data sets), an extension of Reynold's glyph for second-order tensor fields is widely used to display local information. At the same time, fourth-order data becomes increasingly important in engineering as novel models focus on the change in materials under repeated application of stresses. Due to the complex structure of the glyph, a proper discrete geometrical approximation, e.g., a tessellation using triangles or quadrilaterals, requires the generation of many such primitives and, therefore, is not suitable for interactive exploration. It has previously been shown that those glyphs defined in spherical harmonic coordinates can be rendered using hardware acceleration. We show how tensor data can be rendered efficiently using a similar algorithm and demonstrate and discuss the use of alternative high-accuracy rendering algorithms.

M. Hlawitschka (✉) · B. Hamann
Scientific Visualization Group, University of Leipzig, Germany
e-mail: hlawitschka@informatik.uni-leipzig.de; bhamann@ucdavis.edu

Y. Hijazi
University of Strasbourg, France

Frauenhofer ITWM and University of Kaiserslautern, Germany
e-mail: younis.hijazi@lsiit-cnrs.unistra.fr; younis.hijazi@itwm.frauenhofer.de

A. Knoll
University of Kaiserslautern, Germany
e-mail: aknoll@uni-kl.de

L. Linsen et al. (eds.), *Visualization in Medicine and Life Sciences II*, Mathematics and Visualization, DOI 10.1007/978-3-642-21608-4_12,

1 Introduction

When implementing and testing novel visualization techniques, basic methods for displaying data are important to verify their correctness. One of the best-known techniques is the visualization of glyphs, i.e., small icons representing the local data values. Whereas in vector visualization a single direction indicated by an arrow can be used to display the local information, second-order information can be represented by displaying scaled eigenvectors derived from the tensor's matrix representation. Even though this representation displays all information, surface glyphs are often preferred as they show the continuous behavior and, in general, reduce visual clutter: Spheres spanned by the scaled eigenvectors are the most general representation for positive-definite symmetric second-order tensor fields, but a generalization to higher-order tensors is hard to derive. Therefore, the Reynold's glyph for second-order tensor fields has been extended to higher-order data.

For second-order tensors, the representation of the glyph is straightforward and can be implemented by sampling a sphere and scaling the radius according to the function

$$f(\mathbf{x}) = \mathbf{x}^T D \mathbf{x}, \tag{1}$$

where $\mathbf{x}$ is a unit vector and D the tensor's matrix representation. Rewriting the equation using Einstein's sum convention, the extension to higher-order tensors becomes obvious [6] and is given by

$$f(\mathbf{x}) = T_{i_1 i_2 i_3 \ldots i_n} \mathbf{x}_{i_1} \mathbf{x}_{i_3} \mathbf{x}_{i_3} \cdots \mathbf{x}_{i_n}. \tag{2}$$

where the sum is implicitly given over same indices. The standard way of rendering those glyphs is by sampling a tessellation of a sphere. The two most common methods used to display higher-order glyphs are sampling the glyph along the azimuthal and longitudinal coordinates of a sphere, which leads to an unbalanced distribution of sampling points close to the poles, and sampling the glyph using a subdivision of basic shapes, usually triangulated platonic solids (tetrahedra, octahedra, and icosahedra). Applying those subdivision schemes produces several hundred triangles per glyph and, when displaying slices of the data set with several hundreds of glyphs, the increasing memory consumption negatively influences the performance of the whole visualization system. Even though the described method introduces an almost uniform sampling on the sphere, it does not provide a uniform sampling on the surface, which should be sampled depending on the curvature of the glyph, i.e., a refined sampling where large curvatures occur and a coarse sampling in flat areas. While increasing the smoothness of the glyph's representation, this method leads to an increased computational complexity. Given the fact that the function f relates to the spherical harmonic representation, which can be seen as a Fourier transform on the sphere, it can be shown that higher-order tensors introduce more high-frequency components on the surface that require finer tessellation. The increasing angular resolution of diffusion-weighted magnetic resonance scans and the increasing angular precision provided by post-processing tools require the data to be represented at order eight to twelve (cf. Tournier et al. [20]) which exceeds the point where the generation of geometry is no longer reasonable.

We review recent work on higher-order glyph visualization and show how to apply these techniques efficiently to tensor data. Several issues arise with this technique when moving to higher-order representations that we resolve. We propose alternative rendering schemes for high-quality rendering.

2 Related Work

Glyph rendering has a long history in visualization. With the raise of modern graphics boards, hardware acceleration has become a major topic for efficient rendering of large amounts of glyphs for high-resolution displays. Starting from the ray tracing of spheres and ellipsoids [4] where an analytical projection is possible, sphere tracing and ray tracing became important for superquadrics (e.g., Sigg et al. [7, 18]) where no analytical intersection can be calculated. Hlawitschka et al. [5] presented a method of rendering superquadrics on the GPU-based evaluation of the glyph's gradient function along the ray of sight and using a gradient descent method to approach the surface. In both cases, heuristics have been used to discard unused fragments early on to speed up calculations. Both methods fail in terms of performance for more complex surface functions.

Only recently, Peeters et al. [15] were the first to publish a method to display fourth-order glyphs in spherical harmonic representation using hardware-accelerated ray tracing for spherical harmonic functions given by

$$\Phi(\theta, \varphi) = \sum_{l=0}^{\infty} \sum_{m=-l}^{l} a_l^m Y_l^m(\theta, \varphi),$$

where θ and ϕ are the polar and the azimuthal angle, respectively, Y_l^m is the spherical harmonic function of degree l and order m, and a_l^m are the factors defining the function. After a bounding-sphere test for early ray termination, the ray is sampled at a constant step size using a sign test on an implicit function derived from (2) to check whether the surface is hit. If a possible intersection is found, a binary search refines the intersection up to a visually reasonable level.

Several publications describe rendering implicit surfaces on the GPU [8, 9], but most of them are not suitable because the simplicity of Peeter et al.'s approach simply outperforms the "optimizations" suitable for more complex settings.

Our method presented here differs from the method by Peeters et al. in various ways. First, we compute all values using the Cartesian tensor representation to avoid the use of trigonometric functions. Second, our method is not limited to symmetric glyphs of order four, but can be used for a wider range of glyphs, especially glyphs of higher order. Third, we present a method that automatically optimizes the code to the given tensor representation and, therefore, the method is optimal in memory requirement and necessary computations for lower-order glyphs as well as for higher-order glyphs. Finally, we tested our method on higher-order glyphs to ensure its suitability for data representations such as those required for the spherical deconvolution method by Tournier et al. [20].

3 Method

We first derive a function representation suitable for rendering higher-order tensor data. We then show how to optimize the rendering and, finally, introduce a different approach using high-quality ray casting.

3.1 Implicit Function Representation

Spherical harmonics basis representations have proven to be a standard method of computing and storing derived data from medical images [3, 6, 16, 17]. In general, the explicit, parameterized representation of the surface described in spherical coordinates is

$$f : S^2 \to \mathbb{R}^3$$

$$f(\theta, \varphi) = v(\theta, \varphi) \sum_{l,m} a_l^m Y_l^m(\theta, \varphi),$$

where $v(\theta, \phi)$ denotes a normalized vector pointing in direction (θ, ϕ), and Y_l^m is called the spherical homogeneous polynomial of degree l and order m. Let θ and ϕ be the latitudinal and longitudinal angles indicating a point $\mathbf{p} \in \mathbb{R}^3$. The function can be written as an implicit function with the variable $\mathbf{p}$

$$v(\theta, \varphi) \sum_{l,m} a_l^m Y_l^m(\theta, \varphi) - p = \mathbf{0}$$

or simply

$$\sum_{l,m} a_l^m Y_l^m(\theta, \varphi) - \|\mathbf{p}\| = 0.$$

Nevertheless, a transform from spherical coordinates to Cartesian coordinates seems appropriate to avoid trigonometric functions and the evaluation of Legendre polynomials, which both are numerically unstable at the poles and tend to be computationally challenging. Given data in spherical harmonic coordinates described by the linearized version of the weighting factors $\mathbf{w}_i$, the spherical harmonic basis VY_i, and the tensor $\mathbf{T}_i$, the matrix given by

$$M : m_{ij} = \frac{\langle Y_i(S^2), T_j(S^2)\rangle_{S^2}}{\|T_j\|_{S^2}}$$

defines the transform from spherical harmonic space to tensor space [2, 14]. The expression $\langle\cdot,\cdot\rangle_{S^2}$ denotes the scalar product on the sphere and the linearized tensor after the transform is

$$\mathbf{t} = M\mathbf{w}.$$

A change of coordinate system leads to a representation in harmonic polynomials that can be written using an n-th order tensor $T^{(n)}$ as

$$f_n(\mathbf{x}) = \mathbf{x}T_{i_1 i_2 i_3 \ldots i_n}\mathbf{x}_{i_1}\mathbf{x}_{i_3}\mathbf{x}_{i_3}\cdots\mathbf{x}_{i_n}. \tag{3}$$

When evaluating f at an arbitrary point $\mathbf{p}$, an implicit function representation for $\|\mathbf{p}\| \neq 0$ is derived from

$$\|f_n(\mathbf{p})\| - \|\mathbf{p}\|^{n+1} = 0,$$

which takes into account that $f(\mathbf{p})$ is a polynomial of order $n+1$ regarding the radial directions (i.e., regarding $r = \|\mathbf{p}\|$). [1]

3.2 Surface Normal

Whereas the normal in glyph-based techniques is usually estimated using finite differences to determine the tangent space and calculate this tangent space's normal,

[1] We can rewrite f_n to a function f'_n equivalent to the spherical harmonics case by scaling the function by its distance from the center

$$f'_n(\mathbf{p}) = \frac{\mathbf{p}}{\|\mathbf{p}\|}\frac{f_n(\mathbf{p})}{\|\mathbf{p}\|^n}$$

and using the implicit function

$$f'_n(\mathbf{p}) - \|\mathbf{p}\| = 0$$

we get

$$\frac{\mathbf{p}}{\|\mathbf{p}\|}\frac{f_n(\mathbf{p})}{\|\mathbf{p}\|^n} - \|\mathbf{p}\| = 0$$

$$\frac{1}{\|\mathbf{p}\|}\frac{f_n(\mathbf{p})}{\|\mathbf{p}\|^n} - 1 = 0$$

$$f_n(p) - \|\mathbf{p}\|^n = 0$$

using the previous definition of the spatial function, we can compute the glyph's normal implicitly. The normal is given by the gradient at points on the isosurface (contour), i.e.,

$$\begin{aligned}\frac{\partial f(\mathbf{p}) - \|\mathbf{p}\|}{\partial p_i} &= \frac{\partial f(\mathbf{p})}{\partial p_i} - \frac{\partial \|\mathbf{p}\|}{\partial p_i} \\ &= \frac{\partial f(\mathbf{p})}{\partial p_i} - 2 p_i \|\mathbf{p}\|\end{aligned}$$

4 Implementation

We use hardware-accelerated ray tracing for rendering the implicit function given in (3). As the approach of Peeters et al. [15] targets the same class of functions, we closely relate their approach that consists of 1) copying the data to the vertex shader; 2) utilizing early ray termination using a bounding sphere; 3) approximation of intersections of the ray with the implicit surface by sampling the path using a fixed step size and doing a sign check of the implicit function; 4) refining the point by a bisection algorithm; and, finally 5) computing the surface normal for lighting. In the following sections, we only point out the major differences of the two approaches and refer the reader to the original paper for technical details.

4.1 Setup

We transform the ray into a local coordinate system so that the glyph lies in $[-1, 1]^3$ and its center lies at the origin. This simplifies the ray-glyph intersection computation as all calculations are performed relative to the glyph's center.

4.2 Copying Data to the GPU

As we are dealing with higher-order tensors, the number of scalars representing a single tensor value is large and, starting with the 28 values of a symmetric sixth-order tensor, reaches the limit current desktop hardware allows us to pass to the shaders and between the shaders (typically 24 floating-point variables). To bypass this issue, we have to store all data in texture memory and access the data independently for each pixel *fragment shader*. An example of the memory layout we use is shown in Fig. 3 using a symmetric fourth-order tensor. The exact location of the data is given by the tensor's index, which can be stored in any free vertex attribute. Obviously, using texture memory and the need to extract the tensor values

Table 1 Comparison of the performance for different rendering modes using frames per seconds. The same data set is rendered repeatedly under the same viewing angle for about two seconds and the average frame rate is shown here

Step size	Bisect 0	Bisect 1	Bisect 2	Bisect 3	Interpolation
0.01	5.0	5.0	5.0	5.0	4.8
0.02	8.6	9.3	9.1	9.0	8.5
0.05	18.6	19.0	18.7	18.0	18.0

per fragment affects the speed of the algorithm, but we keep the number of texture look-ups small (four look-ups for order four, seven look-ups per pixel for order six, $\lceil (n+1)(n+1)/8 \rceil$ look-ups for order n). In addition, the texture look-ups are not required for the early ray termination step and, therefore, all texture look-ups are performed after this step.

4.3 *Ray–Surface Intersection*

The preliminary ray–glyph intersection step is the most important step in the algorithm. A failure to detect an intersection here will discard the fragment and this error cannot be corrected later on. Therefore, the sampling step size has to be sufficiently small to ensure that all rays that hit the glyph are correctly detected and, in addition, that they intersect with the right "lobe" of the glyph. If they accidentally miss a part of the glyph, sampling artifacts occur (Table 1).

4.4 *Refinement Step*

Given a proper pair of points, one lying outside the glyph towards the eye point and one lying inside the glyph, bisection is in fact an efficient method to increase the quality as it requires a single function evaluation to reduce the interval by a factor two, giving a binary digit in precision. In our experiments we found that a final linear interpolation of the given interval improves the quality of the rendering tremendously *without* requiring an additional interpolation because we store the previous pair of function values and the size of the search interval. Let f^a, f^b be the function values for parameter a and b, respectively, we compute the ray parameter t

$$t = \frac{-f^a}{f^b - f^a}(b-a) + a.$$

This reduces stair-stepping artifacts when using lower sampling rates along the ray as seen in Fig. 1.

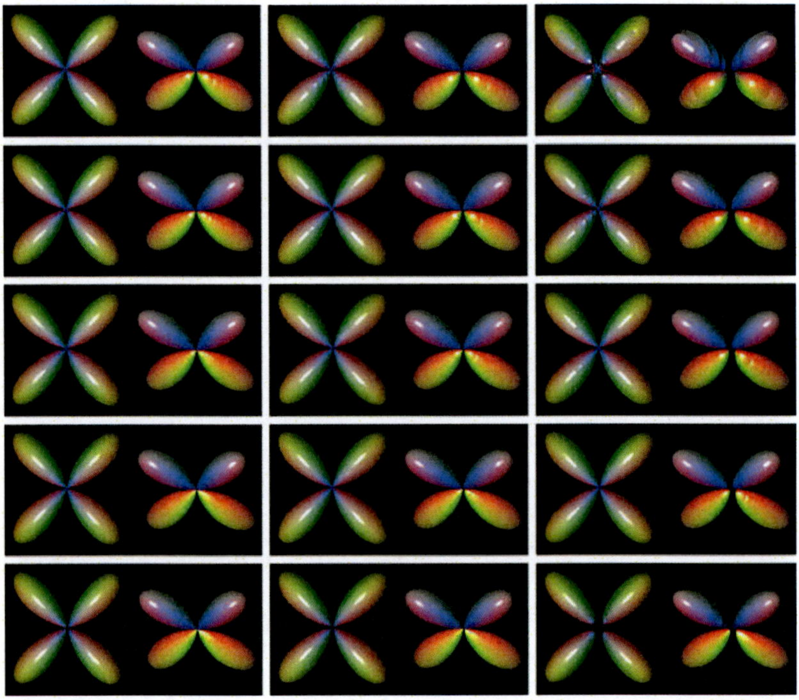

Fig. 1 Rows from top to bottom: comparison of different rendering modes using fixed step size; fixed step size combined with one, two, and three bisection steps; and fixed step size combined with interpolation, respectively. Columns from left to right: two glyphs rendered using a step size of 0.01, 0.02, and 0.05, respectively. None of the methods can improve the quality of the shape in those parts that are not captured by the initial sampling. Those artifacts become especially visible in the center of the right glyph, which has the same shape as the left one but is slightly tilted which makes it numerically challenging. When the glyph is correctly sampled, a linear interpolation performs better than two bisection steps. Depending on the application and the size of the glyphs, a step size of 0.2 and linear interpolation as shown in the center bottom panel seems to provide the best tradeoff: A coarser initial sampling is crucial especially for rays that do not hit the glyph, as those account for most function evaluations and the linear interpolation does not need any additional function evaluations but provides visually better results than three bisection steps

4.5 Overcoming the Singularity at the Origin

Similarly to the spherical harmonics definition, the implicit function defined here has high-frequency components close to the glyph's center, i.e., the origin of our local coordinate system. Whereas moving the coordinate system to the glyph's center in general ensures higher numerical precision for some operations, it does not solve the problem of under-sampling by the simple ray casting. Even though the glyphs shown in Fig. 1 are rare in standard imaging techniques like q-Ball imaging [21, 22], in the post-processing step and to ensure better visibility, a sharpening

Fig. 2 The problem of under-sampling the glyph in the center (first picture) can be solved in different ways. The introduction of a basic isotropic component to every glyph avoids high-frequency components in the implicit function and, therefore, produces a picture without gaps while only slightly changing the glyph (second picture). A better solution is an approach based on adaptive sampling towards the center of the glyph. The two pictures are rendered with a step size of 0.2 and 0.5 (third and fourth picture, respectively), and increase the sampling step size to a third of their original step size towards the center. While slowing down rendering speed slightly (from 18 fps to 15 fps in our test case), we are able to produce more precise visualizations

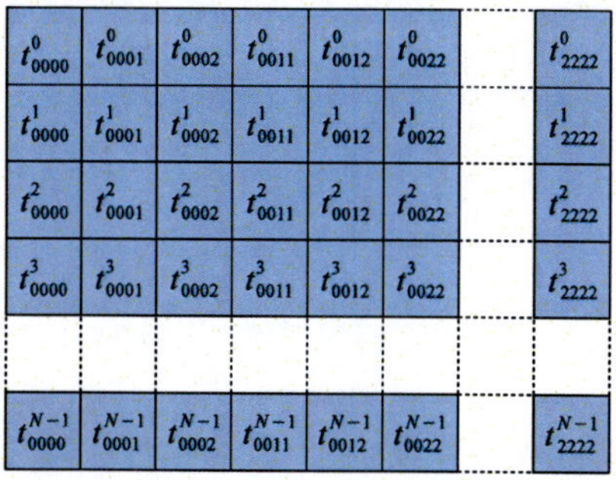

Fig. 3 Layout of the tensor in texture memory using the example of a symmetric fourth-order tensor: Depending on the graphics board used, the use of 1D or 2D textures may result in better performance. Usually, the tensor data is not a power of two and, therefore, leaving memory elements empty may be necessary to achieve better frame rates. Those texture elements can be used as additional information for example for color coding or scaling

filter is applied to the surface function that introduces high-frequency components and, in general, removes the major part of the isotropic behavior. Therefore, it is not uncommon to have glyphs that touch their center point. There are two ways to overcome this problem: First, one can use a redefinition of the glyph that avoids those situations and, second, one could use a finer sampling of the data. Even though the first approach changes the glyph, it can be used to visualize the data since the user is aware of this fact.

As a finer sampling of the data reduces the speed of the algorithm tremendously (which is mainly due to the fact that a large number of rays never hit the surface and, therefore, account for the majority of function evaluations), we adapt the sampling to the expected frequency pattern of the surface by changing the step size to a finer step size in closer vicinity to the center of the glyph while maintaining the coarser step size at the outer parts. A result of this approach is shown in Fig. 2.

4.6 Function Implementation

Depending on the type of input data, we automatically generate the suitable rendering code. This is useful as the tensor function simplifies for lower-order tensors as well as for symmetric tensors. Especially the reduced amount of texture look-ups during rendering leads to an increase in performance.

We compute the function depending on the data using

$$f(x, y, z) = \sum_i \left[x^{h_{i,x}} y^{h_{i,y}} z^{h_{i,z}}\right] - \sqrt{x^2 + y^2 + z^2}^{n+1} \tag{4}$$

for non-symmetric tensors. The optimized version for symmetric tensors is

$$f(x, y, z) = \sum_i \left[\pi_i x^{h_{i,x}} y^{h_{i,y}} z^{h_{i,z}}\right] - \sqrt{x^2 + y^2 + z^2}^{n+1} \tag{5}$$

where

$$\pi_i = \frac{3^n}{h_{i,x}!h_{i,y}!h_{i,z}!}$$

is the number of occurrences of each term and $h_{i,x}$ is the number of occurrences of the digit 0 (1, 2) in the number i represented to the base 3, i.e., the power of x (y, z), respectively.

For higher shading quality and faster evaluation of the gradient, we also provide the partial derivatives of the function used as surface normal, which can be derived analytically from (5), e.g., for the symmetric case

$$\frac{\partial f(x, y, z)}{\partial x} = \sum_i \left[\nu\pi_i h_{i,x} x^{h'_{i,x}} y^{h_{i,y}} z^{h_{i,z}}\right] - 2x(n+1)\sqrt{x^2 + y^2 + z^2}^{n},$$

$$\frac{\partial f(x, y, z)}{\partial y} = \sum_i \left[\nu\pi_i h_{i,y} x^{h_{i,x}} y^{h'_{i,y}} z^{h_{i,z}}\right] - 2y(n+1)\sqrt{x^2 + y^2 + z^2}^{n}, \text{ and}$$

$$\frac{\partial f(x, y, z)}{\partial z} = \sum_i \left[\nu\pi_i h_{i,z} x^{h_{i,x}} y^{h_{i,y}} z^{h'_{i,z}}\right] - 2z(n+1)\sqrt{x^2 + y^2 + z^2}^{n},$$

with

$$\begin{cases} h'_{i,a} = h_{i,a} - 1 & \text{and } \nu = 1 \quad \forall h_{i,a} > 0, \\ h'_{i,a} = 0 & \text{and } \nu = 0 \quad \text{otherwise.} \end{cases}$$

Given the flexibility of automatic glyph function generation, an implementation of the spherical harmonics case described by Peeters et al. [15] is possible with only minor changes in our existing code. We used the real-valued definition as defined by Descoteaux et al. [3] using sine and cosine of θ and φ, and the radius r as input parameters of a modified spherical harmonic function. Again, we store the parameter vector a linearly in a texture similarly to the tensor parameter texture discussed before.

4.7 Optimization

The obvious drawback of this automatic code generation is the missing manual optimization of code. There are many calculations of powers that are used in several places across the functions that could be reused. This is even partially true across functions; the function evaluation itself shares major parts with the gradient evaluation. For three reasons, we did not explore this additional potential: When looking at the generated compiled code, many of these optimizations already were performed by the compiler and we doubt there is much room for optimization at this point. Second, most of these optimizations require additional storage which is limited on the graphics board and, therefore, may limit the number of threads running on the GPU in parallel. Third, besides the additional memory requirement, cross-function storage between the function evaluation and the gradient vector calculation is not possible when interpolating the parameter to increase precision (as done in the bisection and in our linear interpolation approach).

4.8 Rendering

We implemented our approach using NVIDIA's Cg and OpenGL in our existing visualization system. We trigger the fragment shader by rendering the front-facing quadrilaterals of a bounding box around the glyph, which is projected to image space in the vertex shader and parameterized using local coordinates that help with the calculation the ray parameters. Even though we could add another early ray termination step here by ignoring all pixels that do not lie within the projection of the bounding sphere, we currently skip this step for simplicity. The sphere intersection test is responsible for discarding about $\frac{\pi}{6}$ of the rays hitting the bounding box [12] and, therefore, is a mandatory optimization. Based on the ray–sphere intersection, the ray is parameterized and the main algorithm starts as described in the preceding paragraphs.

4.9 Color Coding

Whereas there are many different ways to color-code second-order tensor information, to the best of our knowledge, only two major color-coding schemes are used for higher-order tensors: The frequently used color coding by scalar value and the less frequently used color coding by direction. Both schemes fit seamlessly in our approach as the information required can be obtained from the surface position itself. Given a point **p** on the surface and the normalized vector $\bar{\mathbf{p}}$ pointing from the origin towards **p**,

$$c = \mathsf{colormap}(\|\mathbf{p}\|)$$

represents the first and

$$c = \mathsf{RGB}(\mathrm{abs}(\bar{p}_x), \mathrm{abs}(\bar{p}_y), \mathrm{abs}(\bar{p}_z))$$

the latter color coding scheme. Both can be implemented efficiently in the fragment shader, as the only required information, the hit point **p**, is already computed. Examples of different color codings are shown in Fig. 5.

5 Robust Intersection Using Interval and Affine Arithmetic

Even though the constant step size approach has proven to be efficient and we have shown how to improve the rendering quality even further, we introduce another glyph rendering approach, relying on interval analysis, for even more accurate results. This approach is based on Knoll et al.'s implicit surface rendering [8, 9], which can be used to render almost any implicit surface up to a user-defined precision. The algorithm is based on interval arithmetic (IA) and reduced affine arithmetic (RAA) which are sketched below. We integrated Knoll's algorithm in our software and may use it as an option for the glyph rendering.

5.1 Interval Arithmetic and Reduced Affine Arithmetic

We first implemented an IA and an RAA library in Cg and re-implemented the glyphs functions using these two new types. Interval arithmetic was introduced by Moore [10] as an approach to bounding numerical rounding errors in floating point computation. IA is particularly well-known for its robust rejection test, especially for ray tracing, but it can suffer from overestimation problems. To address this issue, a few decades later, affine arithmetic (AA) was developed by Comba & Stolfi [1]. In practice we might want to truncate the number of elements composing an affine form due to memory consumption, in which case it is referred to as reduced affine arithmetic. Intuitively, if IA approximates the convex hull of a function f with a bounding box, AA employs a piecewise first-order bounding polygon, such as the parallelogram in Fig. 4. For our class of glyph functions, RAA is up to five times faster than IA.

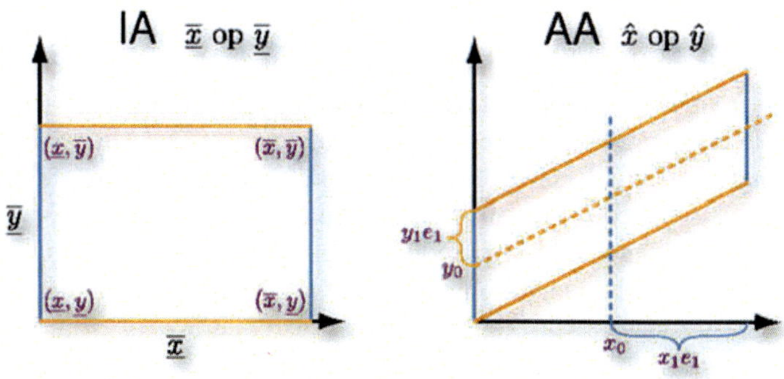

Fig. 4 Bounding forms resulting from the combination of two interval (left) and affine (right) quantities

5.2 *Rejection Test*

Moore's fundamental theorem of interval arithmetic [10] states that for any function f defined by an arithmetical expression, the corresponding interval evaluation function F is an *inclusion function* of f: $F(\overline{x}) \supseteq f(\overline{x}) = \{f(x) \mid x \in \overline{x}\}$ where $\overline{x}$ is an interval. Given an implicit function f and a n-dimensional bounding box B defined as a product of n intervals, we have a very simple and reliable rejection test for the box B not intersecting the image of the function f (in our case, the glyph surface), $0 \notin F(B) \Rightarrow 0 \notin f(B)$. This property can be used in ray tracing or mesh extraction for identifying and skipping empty regions of space.

5.3 *Comparison with Our Previous Glyph Rendering Algorithm*

IA/AA numerical approaches are more robust than point-sampling methods (e.g. using the rule of signs) but they require more computations and therefore are slower; they work well when a function is generally non-Lipschitz (Knoll et al. [9]), as is the case at singularities in our functions. However, our glyph functions are generally Lipschitz outside these singularities, causing point-sampling approaches to work well in practice. Note that brute-force methods like point-sampling work best on the GPU [19]. When combined with a bisection scheme (with enough iterations) on the GPU, IA-based approaches provide high-quality glyphs at interactive rates [9] regardless singularities such as the one at the center; in this way IA helps solving the singularity problems discussed previously.

6 Results

We applied our algorithm to several real-world data sets. The first data set shown in Fig. 5 is a $3 \times 3 \times 3 \times 3$ stiffness tensor of order four generated by Alisa Neeman [11].

The second data set is a human brain image data set of a healthy volunteer and was provided by the Max Planck Institute for Human Cognitive and Brain Sciences, Leipzig, Germany. It had been acquired using 60 gradient directions, using three-times averaging and 19 b0 images on a three Tesla Siemens Trio scanner. The input was mapped to symmetric tensors using least-squares fitting [6]. It turns out that

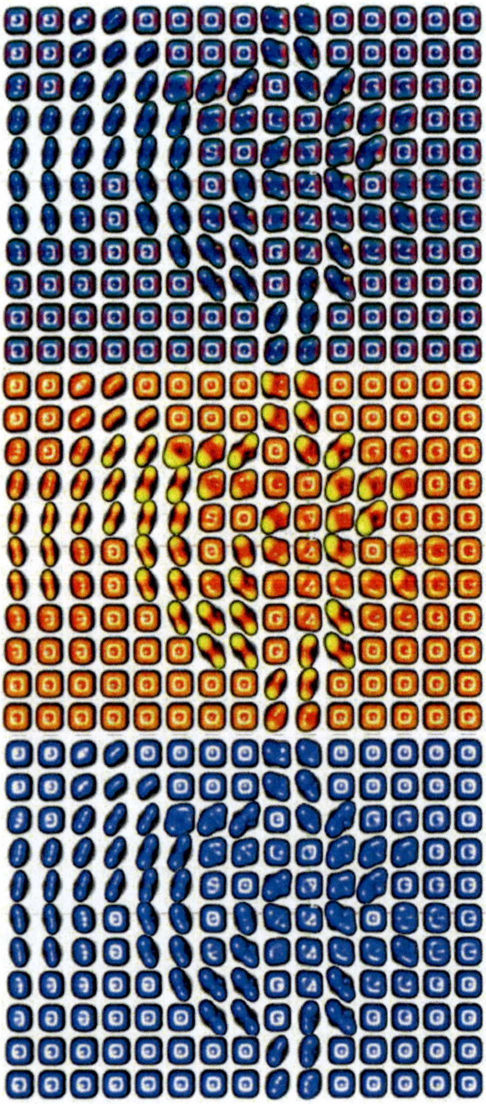

Fig. 5 A fourth-order stiffness tensor data set from material sciences shown using RGB color coding, a magnitude color coding ranging from black to red to yellow, and a uniform blue color coding. The data shows a simulated force applied to a brick of complex material

Table 2 Comparison of our method to a direct rendering of the gyph geometry

System	#Glyphs	Tensor	#Triangles	Time generation	fps rendering
L	258	4n	(res 7)	1s	$60s^{-1}$
L	258	4n	(res 10)	3s	$59s^{-1}$
L	258	4n	(res 20)	9s	$50s^{-1}$
L	258	4n	ray tracing	<1s	$18s^{-1}$

the standard resolution glyphs are shown is quite small and, therefore, a lower-quality rendering can be employed without notable difference in quality of the final visualization.

6.0.1 Performance

We tested our algorithm on a late 2008 Apple MacBook Pro 15" laptop computer with the build-in Nvidia GeForce 9600M GT graphics board and 512MB of VRAM. There, using an OpenGL window size of 800x600 pixel, we achieve a rendering typical speed of 15 to 20 frames per second. Even though this is lower than the speed of standard tessellation-based approaches, there is almost no overhead when pre-processing the geometry. Therefore, changing the glyph's location, e.g., by changing the plane in the data set or modifying a region of interest, can be done at almost the same frame rates (Table 2).

6.0.2 Memory Requirements

The memory use highly depends on the amount of glyphs displayed, whether display lists are generated or not, and the order of the tensor information. The texture memory used has the same size as the data in main memory. Only when an additional padding is used, e.g., when power-of-two textures lead to increased performance, slightly more memory is used. Additional memory on the GPU may be required to store the bounding box information, which could be generated on the fly using geometry shaders.

7 Discussion

Due to the flexibility of the implicit function ray tracing, the presented approach can be used to display different kinds of glyphs: Using symmetric, positive-definite second-order tensors as input, the output is the Reynold's glyph [6, 11]. Using the function

$$f(\mathbf{p}) = \frac{1}{p_x^2} T_{xx} + \frac{2}{p_x p_y} T_{xy} + \frac{2}{p_x p_z} T_{xz} + \frac{1}{p_y^2} T_{yy} + \frac{2}{p_y p_z} T_{yz} + \frac{1}{p_z^2} T_{zz} = c$$

leads to a representation of the ellipsoidal glyph, which can be selected as an option in our implementation. Using the functions provided by Özarslan and Mareci [13] we could even render this glyph directly from higher-order data. However, as there is an explicit ray-ellipsoid intersection algorithm, this is not recommended. Even though the superquadric tensor glyph [7] can be represented by implicit functions, currently, it does not fit in this scheme as the function is based on the relation of two parameters (cf. [7] for details.) Even though this could be implemented as well, we advice to use more efficient implementations as presented by Hlawitschka et al. [5], for example.

In contrast to displaying geometry, which is usually copied to the GPU in smaller packets, our method relies on most data residing in GPU memory. This implies that the GPU's main memory and the order of the glyph limits the number of glyphs that can be displayed in one rendering pass. This can be circumvented by splitting the data set into smaller subsets that are rendered successively.

Future research will target optimizations to reduce the number of glyphs that are rendered even though they are occluded by other glyphs. Deferred shading is not suitable in our case because most of the time is used for the actual ray-glyph intersection, which has do be done anyway. In addition, deferred shading requires additional data lookups and, when a small number of glyphs is occluded, this would slow down the system tremendously.

Acknowledgements We thank Alisa Neemann for providing the fourth-order material tensor data set and Alfred Anwander, Max Planck Institute for Human Cognitive and Brain Sciences, Leipzig, Germany, for the human brain data set. We thank all members of the Institute for Data Analysis and Visualization (IDAV), Department of Computer Science, UC Davis, CA, USA, and the members of the Image and Signal Processing Group, University of Leipzig, Germany, for their comments and support. This research was funded in part by NSF grant CCF-0702817.

References

1. Comba, J.L.D., Stolfi, J.: Affine arithmetic and its applications to computer graphics. In: Proc. VI Brazilian Symposium on Computer Graphics and Image Processing (SIBGRAPI'93), 9–18 (1993). URL citeseer.ist.psu.edu/dihlcomba93affine.html
2. Descoteaux, M., Angelino, E., Fitzgibbons, S., Deriche, R.: Apparent diffusion coefficients from high angular resolution diffusion imaging: Estimation and applications. Magnetic Resonance in Medicine **56**, 395–410 (2006)
3. Descoteaux, M., Angelino, E., Fitzgibbons, S., Deriche, R.: Regularized, fast, and robust analytical q-ball imaging. Magnetic Resonance in Medicine **58**, 497–510 (2007)
4. Gumhold, S.: Splatting illuminated ellipsoids with depth correction. In: VMV, 245–252 (2003)
5. Hlawitschka, M., Eichelbaum, S., Scheuermann, G.: Fast and memory efficient GPU-based rendering of tensor data. In: Proceedings of the IADIS International Conference on Computer Graphics and Visualization 2008 (2008)

6. Hlawitschka, M., Scheuermann, G.: HOT-lines — tracking lines in higher order tensor fields. In: C.T. Silva, E. Gröller, H. Rushmeier (eds.) Proceedings of IEEE Visualization 2005, 27–34 (2005)
7. Kindlmann, G.: Superquadric tensor glyph. Joint Eurographics – IEEE TCVG Symposium on Visualization (2004)
8. Knoll, A., Hijazi, Y., Hansen, C., Wald, I., Hagen, H.: Interactive ray tracing of arbitrary implicits with simd interval arithmetic. Proceedings of the 2nd IEEE/EG Symposium on Interactive Ray Tracing 11–18 (2007)
9. Knoll, A., Hijazi, Y., Kensler, A., Schott, M., Hansen, C., Hagen, H.: Fast ray tracing of arbitrary implicit surfaces with interval and affine arithmetic. Computer Graphics Forum 2008 **28**(1), 26–40 (2008)
10. Moore, R.E.: Interval Analysis. Prentice Hall, Englewood Cliffs, NJ (1966)
11. Neeman, A.: Visualization techniques for computational mechanics. Dissertation, University of California, Santa Cruz (2009). URL http://users.soe.ucsc.edu/~aneeman/
12. Nienhuys, H.W.: Ray tracing news (1997)
13. Özarslan, E., Mareci, T.H.: Generalized diffusion tensor imaging and analytical relationships between diffusion tensor imaging and high angular resolution diffusion imaging. Magnetic Resonance in Medicine **50**, 955–965 (2003)
14. Özarslan, E., Vemuri, B.C., Mareci, T.H.: Higher rank tensors in diffusion MRI. In: J. Weickert, H. Hagen (eds.) Visualization and Processing of Tensor Fields, 177–187. Springer–Verlag Berlin Heidelberg (2006)
15. Peeters, T., Prčkovska, V., van Almsick, M., Vilanova, A., ter Haar, R.: Fast and sleek glyph rendering for interactive HARDI data exploration. In: Pacific Visualization (2009)
16. Schultz, T.: Feature extraction for visual analysis of DW-MRI data. Ph.D. thesis, Universität des Saarlandes (2009)
17. Schultz, T., Seidel, H.P.: Estimating crossing fibers: A tensor decomposition approach. IEEE Transactions on Visualization and Computer Graphics (Proc. IEEE Visualization) **14**(6), 1635–1642 (2008)
18. Sigg, C., Weyrich, T., Botsch, M., Gross, M.: GPU–based ray–casting of quadric surfaces. Eurographics Symposium on Point–Based Graphics (2006)
19. Singh, J.M., Narayanan, P.: Real-time ray tracing of implicit surfaces on the GPU. IEEE Transactions on Visualization and Computer Graphics **99**(2). DOI http://doi.ieeecomputersociety.org/10.1109/TVCG.2009.41
20. Tournier, J.D., Calamante, F., Connelly, A.: Robust determination of the fibre orientation distribution in diffusion MRI: Non-negativity constrained super-resolved spherical deconvolution. NeuroImage **35**, 1459–1472 (2007)
21. Tuch, D.S.: Q-ball imaging. Magnetic Resonance in Medicine, 1358–1372 (2004). DOI 10.1002/mrm.20279
22. Tuch, D.S., Wisco, J.J., Khachaturian, M.H., Ekstrom, L.B., Kötter, R., Vanduffel, W.: Q-ball imaging of macaque white matter architecture. Philosophical Transactions of The Royal Society B (2005)

Part V
Visualizing Genes, Proteins, and Molecules

VENLO: Interactive Visual Exploration of Aligned Biological Networks and Their Evolution

Steffen Brasch, Georg Fuellen, and Lars Linsen

Abstract To understand life, it is fundamental to elucidate the evolution and function of biological networks in multiple species. Recently it has become possible to reconstruct the evolution of specific biological networks for several species. The data resulting from these reconstructions consists of ancestral networks and gene trees. To analyze such data, interactive visual methods are needed. We present a system that is able to visualize the evolution of biological networks in many species. We start with providing a comprehensible overview of the entire data set and provide details of the data upon demand via interaction mechanisms to select interesting subsets of the data. The selected subsets can be visualized using two main visualization types: (a) as network alignments in 2.5D (or other known) layouts or (b) as an animation of evolving networks. We developed a graph layout algorithm supporting the comparison of networks across both species and time steps without changing the graph layout while switching between the overview, the animated view, and the alignment view. We evaluate our system by applying it to real-world data.

S. Brasch (✉)
Department of Mathematics and Computer Science, Ernst-Moritz-Arndt Universität, Greifswald, Germany
e-mail: steffen.rudnick@gmx.de

G. Fuellen
Institute for Biostatistics and Informatics in Medicine and Aging Research, University Rostock, Germany
e-mail: fuellen@alum.mit.edu

L. Linsen
School of Engineering and Science, Jacobs University, Bremen, Germany
e-mail: l.linsen@jacobs-university.de

L. Linsen et al. (eds.), *Visualization in Medicine and Life Sciences II*, Mathematics and Visualization, DOI 10.1007/978-3-642-21608-4_13,

1 Introduction

Analyzing biological networks plays a key role in understanding life and health. Many diseases are related to a malfunction of a biological network. The cause of the malfunction can be multifactorial resulting from the influence of various components of a network and its interactions. Diseases frequently have different characteristics in related species. Understanding these differences and their origin can lead to insights into the contribution of networks and their components to biological function. Hence, one is interested in comparing networks of different species and their evolution. In general, a better understanding of evolution is a major goal in biology.

We present a system to visually explore the evolution of biological networks of a set of species. Biological networks exist in different flavors including interaction networks, regulatory networks, and metabolic networks. We show how our visualization system can be applied to the analysis of the evolution of protein interaction networks. Such networks can be aligned using automatic alignment software. The input to our system are networks together with the respective gene trees, i.e., one gene tree for each set of proteins, that evolved from a single ancestor. Given this input, we know how proteins and protein interactions of different networks relate to each other.

In the course of evolution, proteins or sets of proteins can duplicate (forming paralogs) and interactions between proteins can pop up or vanish. Moreover, entire networks can be duplicated by speciation (forming orthologs) and evolve independently of each other. In Sect. 2, we describe the biological background on protein interaction networks, describe the data, and formulate the driving biological questions and their implications on specifying visualization problems. For analysis of the data, one is interested in observing intra-network connections in form of protein interactions and paralogy relationships as well as inter-network connections in form of orthology relationships. Obviously, the different types of connections should be visually distinguishable.

A canonical representation of networks is the use of graphs as a data structure. In Sect. 3, we discuss existing approaches to graph drawing, in general, and to evolving graphs and aligned network visualization, in particular. In Sect. 4, we present the design goals for our visualization task.

Our layout for alignments is based on the one described by Brasch et al. [9] for network alignment visualization. In this paper we describe the visualization of their evolution. Section 4 gives all the detail on the layout algorithm.

For the exploration of the data set, we provide two main visualization strategies. The first strategy provides a static view of selected networks, i.e., the selected subset of the complete data set, using the alignment visualization [9], accompanied by various interaction mechanisms to support a comparative analysis. The second strategy provides a dynamic view by displaying the evolution of the selected networks in an animation. One can select which networks are to be included in the animation. If several networks of species existing at the same time in the course of evolution are selected, they are displayed utilizing the alignment visualization. The initial interaction to select species and the visualization strategy is supported by a 2D layout of the species tree including all networks. It provides a good overview over the data, yet it already indicates complexity and large-scale changes of the networks within the species tree. In Sect. 5 our interactive visual exploration system and its application within a scenario is described.

2 Biological Background

Proteins are linear polymers built from 20 or more building blocks called amino acids. Parts of a protein that are found conserved in many other proteins are called domains; these domains are usually responsible for the protein interactions.

2.1 Protein Interaction Networks

Protein interactions are transient or permanent connections between proteins, and they are important for many biological phenomena such as signaling, transcriptional regulation, and multi-enzyme complexes. They are explained by molecular adhesive forces between parts of the proteins (domains) which in turn can be tracked down to the atomic level.

Interaction networks evolve by the loss and gain of nodes (proteins) and links (interactions). Biologists assume that the complex networks interconnecting the components of an organism such as a human being are, like all of life, the result of a more or less gradual evolutionary process of descent with modification. Emergence of biological complexity is nevertheless poorly understood, and a deeper understanding is of utmost importance.

The most simple kind of evolutionary scenario yielding a complex interaction network is depicted in Fig. 1. A single protein without any interaction partners except itself is the most simple network (Fig. 1a). This protein is duplicated (by duplication of the underlying genomic material, i.e. the gene), and we observe two identical proteins that interact with each other. Evolution goes on, and the proteins become gradually different in shape and function, and we observe a small network of two distinct proteins related by duplication (Fig. 1b). These two are called *paralogs*, and inherited interaction is called paralogous. Further duplications give rise to a more complex network of paralogs (Fig. 1c); note that some interactions were lost after duplication. Assume that we have followed the evolution of our proteins and their network up to the common ancestor of two species, for example, human and fly. After *speciation*, entities evolve differently in the two lineages of descent that yield the two different species (Fig. 1d and d'). The proteins related by speciation are called *orthologs* and the inherited interactions and the networks themselves are called orthologous. Conservation of protein interactions across species has been

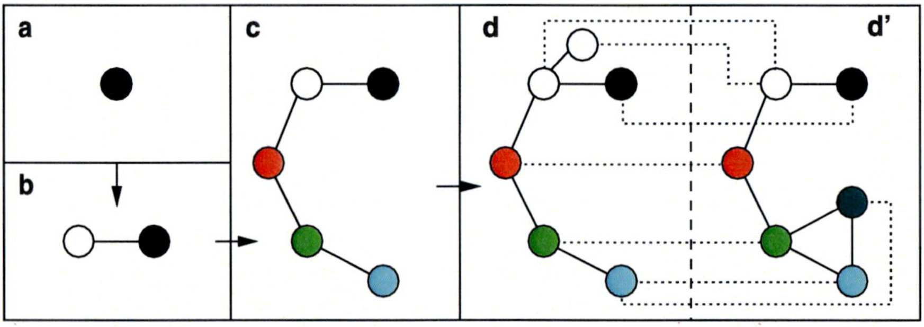

Fig. 1 Evolution of protein interaction networks. A single protein (**a**) duplicates and a small interaction network is built up (**b**). For some time the network grows more complex as depicted in (**c**). A speciation event splits the line of descent into two lineages (**d**) and (**d'**), that are evolving independently thereafter. Any pair of orthologous proteins in the resulting species in (**d**) and (**d'**) are connected by a dotted line. Thus the dotted lines represent the alignment of (**d**) and (**d'**)

observed in many cases [3,30,35], for example for the transcriptional network which regulates the development of the heart. It is regulated by interactions which have been conserved at least since the last common ancestor of human and fly [10].

We do not know how the protein interaction networks of life evolved. However, due to a lot of experimental work we know more and more protein interaction networks, or at least parts of them, for different species living today. Additionally, we know for many proteins in different species that they are orthologs, and for many proteins in the same species that they are paralogs. Therefore we are able to align protein interaction networks. A *network alignment* for a number of networks from different species specifies which nodes (representing the proteins) in one network correspond to (i.e. are orthologous to) which nodes in the other networks. This correspondence may be one to one, or it may relate a set of paralogs in one species to an orthologous set of paralogs in another species. More precisely, we view proteins from one species to be paralogous if they evolved by duplication after the speciation event splitting the last common ancestor. Two proteins in one species that evolved from the same protein are *not* understood as paralogous *if* they were already distinct proteins in the protein interaction network of the last common ancestor. In other words, we take the *last* speciation event as the reference point. For the two networks in Fig. 1d and d' the alignment is shown using the dotted lines there.

In biology, scientists are not only faced with protein interaction networks but with many other kinds of biological networks. In regulatory networks, the type of nodes stays the same (these are genes/proteins as in the case of protein interaction networks). The edges are directed, however, so it is theoretically possible that the direction of an arrow changes in evolution. Such cases are extremely rare, but as a consequence directionality of edges must be highlighted in the visualization of network evolution, and edges may need to be labeled to point out any possible change of directionality. Metabolic networks feature more types of nodes, including enzymes and metabolites. Enzymes (proteins) can be treated the same as

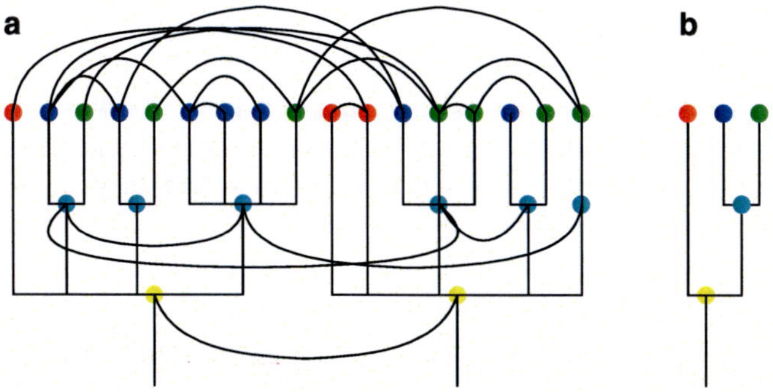

Fig. 2 One very small toy example showing how the data looks like. The example consists of two gene trees (left) in which the species the single proteins belong to are color coded. The curved lines indicate the interactions between the proteins. On the right hand side the species tree for the example is shown

genes/proteins; they also follow an evolutionary history. Metabolites are expected to stay constant; there is no straightforward way to estimate a phylogenetic history of metabolites. Such a history is not necessary either, assuming that constancy is likely true for very long periods of time.

2.2 Data Description and Problem Specification

The data we are dealing with describes the evolution of networks of several species. It therefore contains the *species tree*, the phylogenetic tree for the evolution of these species, see Fig. 2b. For the evolutionary information of the single proteins the data include *gene trees*, the phylogenetic trees for the proteins, see the two trees consisting of straight lines in Fig. 2a. These gene trees differ from a species tree in the fact that a branching in the species tree means a speciation event in the evolution and a branching in the gene tree can mean a speciation event or a duplication of a protein within a species yielding two paralogous proteins. Obviously these trees have to fit together (the biologists speak of reconciled trees) in the sense that the speciation events in each gene tree should be the same as the ones in the species tree [32]. The *interactions* of the single proteins only exist between proteins of the same species, forming a protein interaction network for each species, in Fig. 2a the networks consisting of equal colored nodes connected by curved lines. Note that, although the example in Fig. 2 is only consisting of two gene trees and less than seven interactions per network it is hard to identify any pattern here.

In summary, we have inter-network connections that link proteins of different species and intra-network connections that link proteins within one species. A visualization method for such data should integrate all these links in one system, but should clearly distinguish them visually without overloading the visual representation. In particular, views that highlight just one of these aspects should be naturally included.

2.3 Key Questions

Evolution in general as well as the evolution of biological networks is very complex and not fully understood, even though biologists today know which basic concepts of evolution exist (duplication, speciation, etc.). Therefore, they are interested in finding common patterns in the evolution of networks. With this background some important questions are:

- Which are the specific features of the network evolution that is investigated? For example, are there interactions between paralogous groups of proteins that only evolve (or disappear) in certain species?
- Which rules are predominant in the evolution of interaction networks? For example, can we identify frequent cases of *preferential attachment* [14] (also known as *the rich get richer*).
- Are there unexpected events in the data that indicate mistakes in the reconstruction software or even in the model used for the reconstruction?

In general, the biologists want to learn how complex systems of interacting entities evolve, and we wish to support this goal with our visualization system.

2.4 Visualization Requirements

The user is interested in having a good overview of the complete data set with the possibility to identify interesting parts and to explore them. Within the evolution of the networks it should be possible to trace single proteins or interactions through time and across networks. It is also important that the single evolution events, such as speciation, duplication and loss of proteins, and gain and loss of interactions, should be visible and the changes should become clear. Additionally the visualization system should provide the comparative visualization of the different networks of one time step or of the different evolution stages of one species over time.

Within these tasks we are dealing with the visualization of graphs as node link diagrams and also with the visualization of alignments, which are both essential parts of data sets concerning the evolution of biological networks. For the layout of the single networks some general requirements for graph layout, such as:

- All nodes should be clearly separated.
- Long edges are to be avoided.
- The number of edge crossings should be minimized.
- Available space should be used in an optimal way.

should be met by our visualization system. Brasch et al. [8, 9] defined some requirements for the layout of alignments, which are also valuable for us. These are:

- The structure of individual networks should be easily identifiable.

- Individual networks should be clearly separated.
- Alignment relations should be visible.

3 Related Work

Network evolution is an interesting and vivid field in biology. Some basic mechanisms of evolution are known, but nevertheless it is not completely understood. Lots of research is done to change this [17, 31] and these approaches also indicate a need for data visualization in this field.

3.1 Graph Drawing

It is intuitive to represent biological networks such as protein interaction networks as graphs. In a protein interaction network the proteins can be represented as vertices of a graph and the protein interactions as edges of the graph. Therefore, visualizing biological networks is a special subject of graph drawing which is a well-studied field in information visualization [19].

The layout of a graph is most important because it determines the human perception of the graph [2]. In graph drawing one is generally interested in optimizing the layout of the graph with respect to some properties and constraints. Many different approaches exist, depending on the properties of the graph or on the information one wants to visually extract or highlight. Graphs are commonly drawn using a 2D layout where vertices are drawn as nodes and edges represented by lines. Plenty of algorithms exist for automated graph drawing [12].

Force-based layout algorithms like the ones by Fruchterman and Reingold [16] or by Kamada and Kawai [26] are commonly used. Another algorithm iteratively approving the layout is to define an energy function penalizing bad layout properties and using an optimization algorithm like simulated annealing [11]. The advantage of these algorithms is their flexibility, i. e. defining the forces according to a special design goal, or the energy function which makes these algorithms suitable for many different graphs in diverse applications. Even within the field of biology, a wide range of graph layout algorithms are being used as can be seen in the numerous visualization tools for biological networks such as Cytoscape [37], ProViz [24], VisANT [23], or VANTED [25].

3.2 Visualizing Aligned Networks and Evolving Graphs

In each time step of evolution there is a set of aligned networks. Obviously, the evolution of aligned networks cannot be globally visualized in a comprehensive way, if a single time step cannot be understood easily. Therefore, it is essential to have an effective and intuitive layout for network alignments. Aligned networks can be regarded as a set of graphs, where the alignment establishes connections between the graphs or, more precisely, between entities of the graphs (e.g., some of the nodes are aligned across the networks). For visualizing an alignment of protein interaction networks different approaches have been considered and are being used today. For a detailed survey on the state of the art in visualizing aligned biological networks we refer to Brasch et al. [8]. There the approaches are divided into two main classes, namely the classes called "side by side" and "all in one".

The "side by side" approach, cf. Fig. 1, follows the idea to draw the individual aligned networks next to each other in a 2D layout and to highlight the aligned nodes by the same relative position and/or additional edges connecting them [28, 29, 38]. The "all in one" approach draws the aligned networks in just one node-link diagram where one node represents the orthologous proteins of all networks [1, 20]. Some disadvantages of this approach can be alleviated to some degree by using the idea of metagraphs [22].

An appropriate solution that combines the advantages of both classes is given by using 2.5D layouts [5], where the individual networks are laid out in 2D and the relationship of the entities is implied by drawing all 2D layouts simultaneously using the third dimension and by placing corresponding entities on top of each other. Schreiber [36] used such an approach for the comparison of different biological networks in the context of metabolic pathways. However, his approach does not support the visualization of paralogous entities (proteins). Moreover, he did not provide any interactive exploration methods and his approach is specialized for metabolic pathways and a KEGG [27] like layout.

Within our task we are not only dealing with aligned networks, but with the evolution of aligned networks. This is a more complex version of the visualization of evolving graphs. Several approaches for so-called dynamic graph drawing exist [6, 7, 13, 18]. Additionally the ideas of evolving graphs can also be related to the layout of aligned graphs. The split representation of evolving graphs, i.e., each time step is shown in a separate drawing window, corresponds to the "side by side" layout and the merged representation, i.e., all time steps integrated into one drawing window, corresponds to the "all in one" layout. Some dynamic graph drawing approaches also consider a 2.5D approach with each time step drawn in a separate layer where the layers are placed on top of each other [4, 15].

4 The Layout

Our visualization system presents a set of graphs (the protein interaction networks) with trees (the gene trees) interconnecting these graphs. We have to lay out the graphs to arrange them according to the specified requirements and to allow for a comprehensible display of the inter-graph relations. Therefore we have to decide on how to visualize the inter-graph relations before implementing the overall layout algorithm.

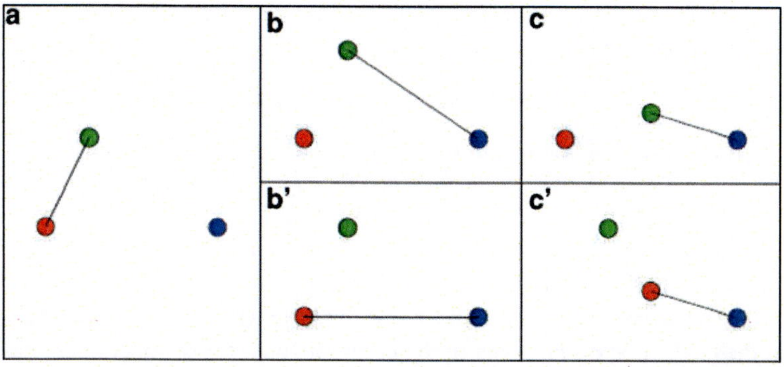

Fig. 3 In (**a**) the network of the common ancestor is laid out. After a speciation event this network evolves into the networks shown in (**b**) and (**b'**). The newly computed layout for these two networks is shown in (**c**) and (**c'**)

Evolution of networks is a process in time and it can be shown as a sequence of graphs. Hence it can be visualized using methods from dynamic graph drawing such as animated evolving graphs or graphs in a 2.5D layered set-up on top of each other. The latter is mainly useful for linear sequences taken from the complete species tree. As the user follows the nodes due to their position, it is necessary to preserve the mental map [34], this means for one time step, that only small changes are allowed. In Fig. 3 an example is shown for the speciation of a network and the computation of new layouts for the two networks, preserving the mental maps. Although following the changes in the single networks is easy, comparing the two networks in c) and c') is hard, as two proteins have the same position, which are not orthologs. This layout would be misleading.

Thus we decided to not allow changes of position of orthologous nodes over time. All the descendants of a protein of the common ancestor have to be placed around the position of the node of the common ancestor in the course of evolution. Also, nodes that come into being as a result of duplication at some point in the evolution will not change their position anymore once they have been placed.

4.1 Layout Algorithm

From the considerations made above we can deduce that the layout of all the graphs is governed by the graph layout of the common ancestor. Once, the graph of the common ancestor has been laid out, the position of the nodes of its descendants is given with respect to the design choice of a mental map. The descendants of each node of the ancestor are required to maintain the position of the parent node in the descendants' graphs. This procedure is iterated over the entire evolutionary path. What remains to be settled is what happens in the case of duplication. If a node gets duplicated, a parent node has multiple children in one descendant. Obviously, they cannot maintain the parent node's position without coinciding. Hence, they have to be laid out "around" the parent node's position.

4.1.1 Computation of the Layout

The main task of the layout algorithm is to position the nodes of the protein interaction network of the common ancestor. The edges in the graph represent the interactions of the proteins within the common ancestor's network. During evolution usually this network gains more and more complexity, cf. Sect. 5.3. Taking into account the edges that do not exist in the common ancestor but occur at some point during evolution, we enhance the graph that represents the common ancestor. An additional edge between two nodes in the graph is established, if any two of their descendant nodes exhibit an interaction in any of the descendant networks. The layout algorithm, applied to the enhanced graph of the common ancestor's network, positions nodes close to each other that have links at a later point in evolution. Hence, not only the layout of the common ancestor's graph is optimized, but also those of all its descendants.

Having set up all the links of the graph considering the interaction in the common ancestor's network as well as in all its descendants' networks, the actual computation of the graph layout can be obtained using any standard general-purpose graph layout algorithm. We include different known algorithms in our software system. Two force-directed algorithms, namely the one by Fruchterman and Reingold [16] and the one by Kamada and Kawai [26], have been implemented. In addition, we provide an implementation of an iterative algorithm using simulated annealing [11], punishing long edges, node overlap, edge crossings, and small angles between edges. The latter can also be used to subsequently improve layouts computed by the other algorithms; only small changes are allowed then. The user is free to choose his/her preferred algorithm or just test the different ones in order to find a pleasing layout. Changes can also be done manually by drag and drop.

The user can save pleasing layouts once computed and reload them in later sessions.

The best results are reached using our simulated annealing approach, but this might last up to 30s for larger evolutions with three recent networks each with more than 200 nodes on a Intel Xeon 3.06GHz, but less than one second for evolution with three recent networks each with about 40 proteins.

4.1.2 Laying Out the Paralogs

One protein may duplicate several times in the course of evolution and all these descendants in the gene tree should be placed around the position of their common ancestor. This should happen in a way that proteins getting assigned a position once do not change their position anymore. Furthermore, all their descendants (due to ongoing duplication) are also placed nearby each other and their common ancestor,

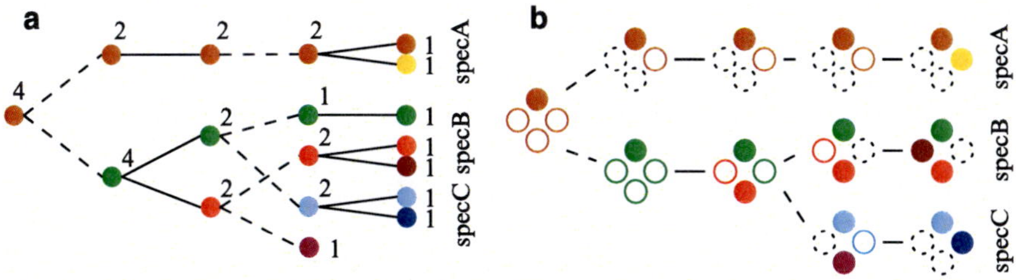

Fig. 4 Computation of the layout for paralogs across species. The detection of positions needed is shown in (**a**) using the gene tree. The mapping of the positions to the paralogs in each network is shown in (**b**). There the species tree is shown accompanied with the proteins in each species. The gene tree in (**a**) is drawn with one edge crossing to have the proteins of each species grouped together. The line pattern for the edges in both trees indicates whether we have a speciation event (dashed line) or a duplication event (solid line). In each time step the proteins have unique colors, allowing to align (**a**) and (**b**). In Figure (**b**) the circles represent positions, which are assigned to the node with the same color. Positions that are not assigned to any node are represented by black dotted circles. Usually the single events (speciation, duplication) are not separated like this in a gene tree, we use this depiction here for better comprehensiblity. We also added some events although nothing happened there

i.e. all descendant nodes stay near to the common ancestor node not just near to their direct ancestor node.

We decided to place all descending nodes for each node in the most common ancestor network around this ancestor at the corners of a regular polygon. Computing the positions for all descendants of one common ancestor protein works in two steps. In the first step for each node the number of positions needed is computed and in a second step the positions are assigned to the nodes using the knowledge from step one.

Number of positions needed: The number of positions needed for a node is the maximum number of descendants of this node, which should be placed near to it. The numbers are shown for an example in Fig. 4a. For computing these numbers we start in the gene tree at the leaves. Each leaf obviously only needs one position for itself. Going towards the root the numbers are computed iteratively. For each node the number of positions needed is the sum of the numbers of its descendants if duplications happened (solid lines in Fig. 4) or the maximum of the number of the descendants if speciation happened (dashed lines in Fig. 4).

Assigning the positions: The number of positions needed for a node in the most common ancestor network equals the number of corners of the polygon on which the descendants are placed. This is shown in Fig. 4b. Assigning the position starts from the root of the tree. The single protein at the root gets all the assigned positions and itself it is rendered on the first of them, where counting starts at the top and goes clockwise. Then it assigns the positions to its descendants, each one gets the number of positions it needs. The first child gets the first positions and the next child the following ones. This ensures that the positions the proteins get assigned are disjoint and all positions of one proteins are adjacent corners of the polygon. This process is applied recursively till the leaves are reached.

4.1.3 Generating the 2.5D Layout

From the graph layout the 2.5D representation of the aligned networks is obtained by assigning each network an individual layer displayed in Cartesian coordinates at equidistant heights z. For each node, a three-dimensional primitive is rendered at (x, y, z) where (x, y) are the coordinates computed by the algorithm and z is the assigned height for the network. The edges are connecting the nodes inside each individual network. Hence, these intra-network connections automatically lie in one layer. The orthologous groups of paralogous proteins are rendered on top of each other, i.e., the polygons, on which edges they are rendered, have the same (x, y)-coordinates and just defer by their height z. Therefore, orthologs are easy to identify just by position. No edges between different layers are necessary.

The 2.5D layout can be applied to networks of one time step, i.e., the aligned networks of different species are stacked on top of each other (order given by gene tree), and it can be applied to the evolution of one network, i.e., the aligned networks of one species and all its ancestors are stacked on top of each other (where order is given by evolution).

4.2 Edge Bundling

We experienced that there are often many interactions between two sets of paralogous proteins. As each of these interactions is represented by a single line, there are many, nearly parallel, edges drawn between these two sets of paralogous proteins. This causes a cluttered view with lots of edge crossings. We alleviate this problem by bundling these edges [21]. This method was invented to bundle edges of adjacency relations in hierarchical data, but it can be adapted for bundling edges between sets of paralogous proteins. In hierarchical data, edges are bundled, if the connected leaves are related via the same inner nodes of the hierarchy. Analogously, we are bundling edges, connecting proteins of the same sets of paralogs. Therefore, two sets of paralogous nodes are connected by one bundle of edges, which are connecting the single nodes.

Edge bundling is achieved by drawing the edges not as straight lines, but as a cubic NURBS (non uniform rational b-spline [33]) using suitable control points for each edge. In our case the vertices representing the paralogous proteins all lie on a circle and therefore, we use the two center points of the respective circles as additional control points (in addition to the two endpoints of the curve), see Fig. 5.

As seen in Fig. 5(b), it is sometimes impossible to distinguish between the different edges, after bundling. Therefore, providing a possibility for changing the bundling strength is recommended, as it is done in the classical edge bundling [21].

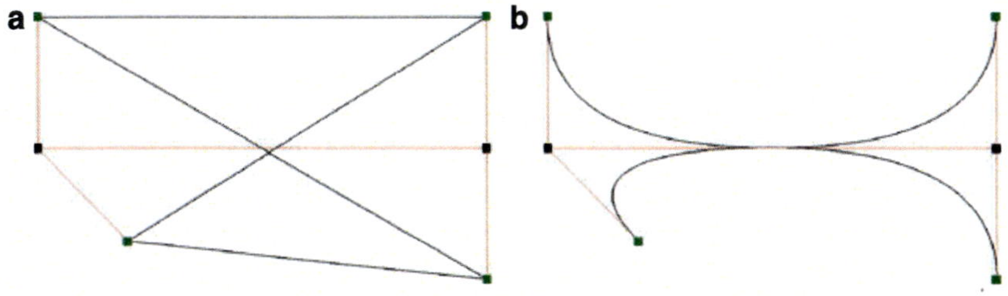

Fig. 5 An example network (vertices are green and edges black) for explaining the edge bundling. In (**a**) the network is shown using straight lines for the edges. In (**b**) the edges are drawn as NURBS and applying edge bundling. The control polygons are shown in red and the additional control points (center points of circles, the paralogs are drawn on) are shown as black points

Controlling the bundling strength is achieved by changing the control points, where the inner control points are linearly interpolated between the center point of the circle for the paralogs and the curve's endpoint. Thus, for bundling strength 1 the control points are the centers of the circles and for 0 they are identical to the end points, which causes no bundling at all. The result for bundling strength 1 is the strong bundling shown in Fig. 5(b) and for bundling strength 0 the straight lines as shown in Fig. 5(a). An example for a weaker bundling with strength of 0.75 is shown in Fig. 7.

5 Interactive Visual Exploration of Evolution

Using our ideas of a visual exploration system for network evolution data, we implemented an interactive system called VENLO. VENLO is an acronym for "Visualization of Evolving Networks using Layout Optimization" . In this Section we describe this system and show an exploration scenario.

5.1 Interacting with Overview and 2.5D Layout

When the data are loaded to the system, the display is initialized with an overview of the entire data. The system displays the species tree. However, instead of representing each species at one time step by just rendering one node, it is being displayed by rendering its network, see Fig. 6. Different colors are being used for different species, which allows the user to identify the species during the interactive data exploration in each filtered view. A default color assignment being used, but the colors can, of course, be easily modified by the user to accommodate for his/her preferred coloring scheme.

The user can select networks by simply clicking at them with the mouse in order to display them in the described 2.5D layout for direct comparison. Different views

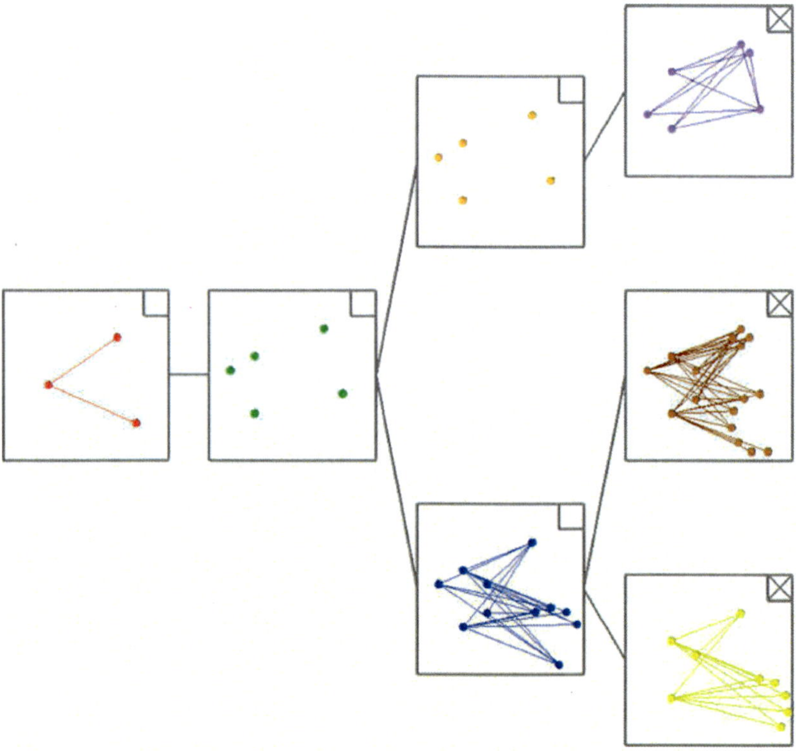

Fig. 6 Overview of the example evolution data set for the eIF/LARP network. The root network, from a pre-fungi/metazoa species, is displayed as the root in the very left. The evolution goes from left to right and a branching in the tree is a speciation event in the evolution. The three current species on the right are yeast (purple), fly (brown), and worm (yellow)

on the data are supported. Examples are given in the exploration scenarios described below.

When some of the networks have been selected the user decides whether he/she wants to compare the networks by displaying them in the 2.5D layout or whether the evolution of the networks should be shown in an animation. If several networks of one time step are chosen for the animated view, the aligned networks of one time step are displayed using the 2.5D layout. In the static 2.5D layout as well as in the dynamic animated 2D or 2.5D layout, the user is provided with interactive means for rotation of the scene, zooming, or panning. These interactions are supported by intuitive use of mouse and keyboard.

5.2 *The Animated Evolution View*

Animating the evolution of protein interaction networks differs from the animation known from dynamic graph drawing, because additionally duplication of nodes and even duplication of networks (speciation) happens. Speciation occurs, when between two time steps the number of networks changes. We visualize speciation by duplicating the respective network and adding an additional layer in the 2.5D layout. The duplicated network transitions smoothly to the new layer by using a linear interpolation between old and new layer height. When a new layer is inserted, other layers are moved respectively to make space for this new network. After the speciation, the course of evolution of both networks is animated.

Duplication of individual nodes (rather than entire networks) can be handled in a local fashion. Hence, the node is duplicated and the new node is animated moving linearly to its new position. The loss of nodes and the gain or loss of edges is visually represented by linearly decreasing or increasing the size of the respective geometric primitives. In summary, in each step things visually change, if and only if they change due to evolution. For the transformation from one network to the next in time, the color for nodes and edges is linearly interpolated between those of the two networks. Thus, the coloring is consistent and shows what transition, i.e., from which network to which network, is currently being visualized. The colors had been specified in the overview layout.

Using a data set with only one gene tree, our system provides a new way of visualizing gene trees. With our understanding of evolution data as gene trees combined with interaction data, a single gene tree is a special case of network evolution data. The gene tree can be visualized using the animated view, such that the differentiation of speciation events and duplication events becomes very clear. In addition to usual gene tree visualization also the interactions between the respective proteins are included in the visualization.

The animation runs at interactive frame rates for smaller and medium networks. Using three recent networks with together around 800 nodes and its evolution data, the frame rates might go down to 10 fps using the bundled edges, or 20 fps using simple lines on a Intel Xeon 3.06GHz with a GeForce6800 Ultra graphics card.

5.3 Exploration Scenario

To give the reader an impression on how to use our visual exploration tool, we give a short example with real world data. The example shows the evolution of the eIF/LARP network from a pre-fungi/metazoa species up to the current networks of fly (*Drosophila melanogaster*), yeast (*Saccharomyces cerevisiae*), and worm (*Ceanorhabditis elegans*). As mentioned, the exploration of the data starts with the overview given in the species tree style, see Fig. 6.

In this overview the growing complexity of the networks can be seen and also the differences in complexity between the networks of current species is obvious. The user is free to chose a set of networks he wants to investigate in more detail. In Fig. 6, only the three networks of the three current species are selected by clicking at the respective networks. Selected networks are indicated by the cross in the upper right of the selected box. These can now be viewed as an alignment; the user is free to choose the style, 2.5D, side by side, or all in one. In Fig. 7 the chosen networks

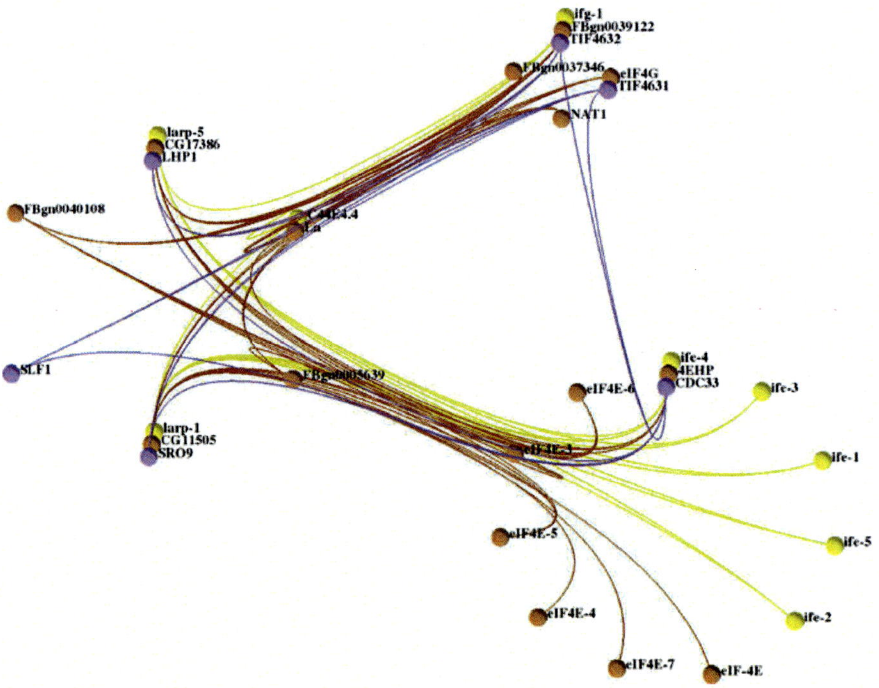

Fig. 7 The networks of the three current species from the example in Fig. 6 are shown as an alignment in 2.5D, using edge bundling

are shown as an alignment using the 2.5D layout. In this Figure we used the edge bundling with a bundling strength of 0.75. The advantage of this bundling is, that one gets a quick overview of the connections between the set of paralogs, which is a huge advantage in larger networks and which, in the example, draws the user's attention immediately to the two purple edges from top right to the bottom, where no other edges exist. This means that only in yeast interactions between pairs of proteins evolved, while this is not the case in worm and fly.

Choosing the two networks after the first speciation event shown in Fig. 6 and the two upper networks from the current species (yeast and fly), allows the user to explore the evolution of these networks as changing network alignments in an animated style. Screenshots from this animation are given in Fig. 8. The animation evolves from left to right. In the very left screenshot there are just the two networks the animation starts with. In the animation it can be seen, how two extra edges in yeast evolve from the first screenshot to the second. Furthermore, one can observe that the single protein shown in the bottom in both species duplicated independently in yeast and fly, this duplication did not happen in their common ancestor, although one could suggest this only seeing the orthologous sets of paralogs in the networks. This insight cannot be obtained by inspecting the alignment of yeast and fly

Fig. 8 Screenshots from an animation view of the above evolution data example. On the left, there are the two networks that we start with, the next two screenshots are taken at time points between begin and end of the evolution process, and the last screen shot shows the final networks of the current species, yeast and fly

individually. Also the duplication of the top protein into four, respective two proteins in the top and the lost of two proteins on the right can be seen well.

6 Conclusion and Future Work

We presented the first visualization system that is able to visualize evolutionary data for protein interaction networks. With the help of our system researchers are able to visually explore their data and gain a deeper insight. The system includes an overview of the complete data set using a 2D layout of the species tree, where the user can select several networks for a more detailed and comparative visualization. These selected networks are visualized in a 2.5D set-up or in an animated set-up. Within both set-ups the user might zoom, rotate, and pan to have the desired view on the scene and examine interesting details. The system is designed to help the user to answer some of the key questions on evolution and to give them a better understanding of their data and evolution, in general. Additionally, our system can be used to visualize and explore gene trees in a novel way supporting a better differentiation of duplication and speciation. We presented an exploration example using real world data to show how the visualization works.

For future work we would like to include the evolution of transcription factor binding sites (TFBS). Nowadays, their evolution can also be reconstructed and scientists are interested in exploring these data with the help of visual means. Moreover, we can envision that our system will be included in tools like Cytoscape, which will be able to handle 3D graphics in its next version.

Acknowledgements The authors wish to thank Janusz Dutkowski et al. for providing us with evolution data of protein interaction networks they have reconstructed.

References

1. Sourav Bandyopadhyay, Roded Sharan, and Trey Ideker. Systematic identification of functional orthologs based on protein network comparison. *Genome Res.*, 16(3):428–435, March 2006.

2. Jim Blythe, Cathleen McGrath, and David Krackhardt. The effect of graph layout on inference from social network data. In Franz J. Brandenburg, editor, *Graph Drawing, Passau, Germany, September 20-22, 1995*, 40–51. Springer, 1996.
3. P. Bork, L.J. Jensen, C. von Mering, A.K. Ramani, I. Lee, and E.M. Marcotte. Protein interaction networks from yeast to human. *Curr Opin Struct Biol*, 14(3):292–299, 2004.
4. Ulrik Brandes and Steven R. Corman. Visual unrolling of network evolution and the analysis of dynamic discourse. *Information Visualization*, 2(1):40–50, 2003.
5. Ulrik Brandes, Tim Dwyer, and Falk Schreiber. Visual understanding of metabolic pathways across organisms using layout in two and a half dimensions. *Journal of Integrative Bioinformatics*, 1(1):119–132, 2004.
6. Ulrik Brandes and Dorothea Wagner. A bayesian paradigm for dynamic graph layout. In *GD '97: Proceedings of the 5th International Symposium on Graph Drawing*, 236–247, London, UK, 1997. Springer-Verlag.
7. Jrgen Branke. Dynamic graph drawing. In M Kaufmann and D Wagner, editors, *Graph Drawing - Models and Algorithms*, 228–246. Springer, Berlin, 2001.
8. Steffen Brasch, Lars Linsen, and Georg Fuellen. Visualization of aligned biological networks: A survey. In Franz-Erich Wolter and Alexei Sourin, editors, *Proc. 2007 International Conference on Cyberworlds*, 49–53. IEEE Computer Society, 10 2007.
9. Steffen Brasch, Lars Linsen, and Georg Fuellen. Vanlo interactive visual exploration of aligned biological networks. *BMC Bioinformatics*, 10(327), 2009.
10. R.M. Cripps and E.N. Olson. Control of cardiac development by an evolutionarily conserved transcriptional network. *Dev Biol*, 246(1):14–28, 2002.
11. Ron Davidson and David Harel. Drawing graphs nicely using simulated annealing. *ACM Transactions on Graphics*, 15(4):301–331, 1996.
12. G. Di Battista, P. Eades, R. Tamassia, and I. G. Tollis. Algorithms for drawing graphs: An annotated bibliography. *Comput. Geometry: Theory Appl.*, 4:235–282, 1994.
13. Stephan Diehl and Carsten Görg. Graphs, they are changing. In *GD '02: Revised Papers from the 10th International Symposium on Graph Drawing*, 23–30, London, UK, 2002. Springer-Verlag.
14. Eli Eisenberg and Erez Y. Levanon. Preferential attachment in the protein network evolution. *Phys. Rev. Lett.*, 91(13):138701, Sep 2003.
15. Cesim Erten, Stephen G. Kobourov, Vu Le, and Armand Navabi. Simultaneous graph drawing: Layout algorithms and visualization schemes. *J. Graph Algorithms Appl.*, 9(1):165–182, 2005.
16. Thomas M. J. Fruchterman and Edward M. Reingold. Graph drawing by force-directed placement. *Software - Practice and Experience*, 21(11):1129–1164, 1991.
17. Todd A. Gibson and Debra S. Goldberg. Reverse Engineering the Evolution of Protein Interaction Networks, 2009.
18. Carsten Görg, Peter Birke, Mathias Pohl, and Stephan Diehl. Dynamic graph drawing of sequences of orthogonal and hierarchical graphs. In *Graph Drawing*, 228–238. Springer Berlin, Heidelberg, 2004.
19. Ivan Herman, Guy Melançon, and M. Scott Marshall. Graph visualization and navigation in information visualization: A survey. *IEEE Transactions on Visualization and Computer Graphics*, 6(1):24–43, /2000.
20. E. Hirsh and R. Sharan. Identification of conserved protein complexes based on a model of protein network evolution. *Bioinformatics*, 23(2), January 2007.
21. Danny Holten. Hierarchical edge bundles: Visualization of adjacency relations in hierarchical data. *IEEE Transactions on Visualization and Computer Graphics*, 12(5):741–748, 2006.
22. Zhenjun Hu, Joe Mellor, Jie Wu, Minoru Kanehisa, Joshua M. Stuart, and Charles Delisi. Towards zoomable multidimensional maps of the cell. *Nature Biotechnology*, 25(5):547–554, May 2007.
23. Zhenjun Hu, Joseph Mellor, Jie Wu, and Charles Delisi. Visant: an online visualization and analysis tool for biological interaction data. *BMC Bioinformatics*, 5:17, 2004.
24. F. Iragne, M. Nikolski, B. Mathieu, D. Auber, and D.J. Sherman. Proviz: protein interaction visualization and exploration. *Bioinformatics*, 21(2):272–274, 2005. Received on August 4,

2003; revised on June 15, 2004; accepted on August 18, 2004. Advance Access Publication September 3, 2004.
25. Bjorn H. Junker, Christian Klukas, and Falk Schreiber. Vanted: A system for advanced data analysis and visualization in the context of biological networks. *BMC Bioinformatics*, 7:109, 2006.
26. T. Kamada and S. Kawai. An algorithm for drawing general undirected graphs. *Inf. Process. Lett.*, 31(1):7–15, 1989.
27. M. Kanehisa and S. Goto. KEGG: Kyoto Encyclopedia of Genes and Genomes. *Nucleic Acids Res.*, 28:27–30, 2000.
28. B. P. Kelley, R. Sharan, R. M. Karp, T. Sittler, D. E. Root, B. R. Stockwell, and T. Ideker. Conserved pathways within bacteria and yeast as revealed by global protein network alignment. *Proc Natl Acad Sci U S A*, 100(20):11394–11399, September 2003.
29. M. Koyutürk, Y. Kim, S. Subramaniam, W. Szpankowski, and A. Grama. Detecting conserved interaction patterns in biological networks. *J Comput Biol*, 13(7):1299–1322, September 2006.
30. L.R. Matthews, P. Vaglio, J. Reboul, H. Ge, B.P. Davis, J. Garrels, S. Vincent, and M. Vidal. Identification of potential interaction networks using sequence-based searches for conserved protein-protein interactions or "interologs". *Genome Res*, 11(12):2120–2126, 2001.
31. Ratmann O, Wiuf C, and Pinney JW. From evidence to inference: probing the evolution of protein interaction networks. *HFSP Journal*, 3(5):290–306, December 2009.
32. Roderic D. Page and Michael A. Charleston. From gene to organismal phylogeny: Reconciled trees and the gene tree/species tree problem. *Molecular Phylogenetics and Evolution*, 7(2):231–240, April 1997.
33. Les Piegl and Wayne Tiller. *The Nurbs Book.* Springer Berlin/ Heidelberg/ New York, 1997.
34. Helen C. Purchase, Eve Hoggan, and Carsten Goerg. How important is the mental map? - an empirical investigation of a dynamic graph layout algorithm. In Michael Kaufmann and Dorothea Wagner, editors, *LNCS, Graph Drawing*, 184–195. Springer, Berlin / Heidelberg, 2007.
35. A.K. Ramani and E.M. Marcotte. Exploiting the co-evolution of interacting proteins to discover interaction specificity. *J Mol Biol*, 327(1):273–284, 2003.
36. Falk Schreiber. Visual comparison of metabolic pathways. *J. Vis. Lang. Comput.*, 14(4):327–340, 2003.
37. P. Shannon, A. Markiel, O. Ozier, N. S. Baliga, J. T. Wang, D. Ramage, N. Amin, B. Schwikowski, and T. Ideker. Cytoscape: a software environment for integrated models of biomolecular interaction networks. *Genome Res*, 13(11):2498–2504, November 2003.
38. Roded Sharan and Trey Ideker. Modeling cellular machinery through biological network comparison. *Nature Biotechnology*, 24(4):427–433, April 2006.

Embedding Biomolecular Information in a Scene Graph System

Andreas Halm, Eva Eggeling, and Dieter W. Fellner

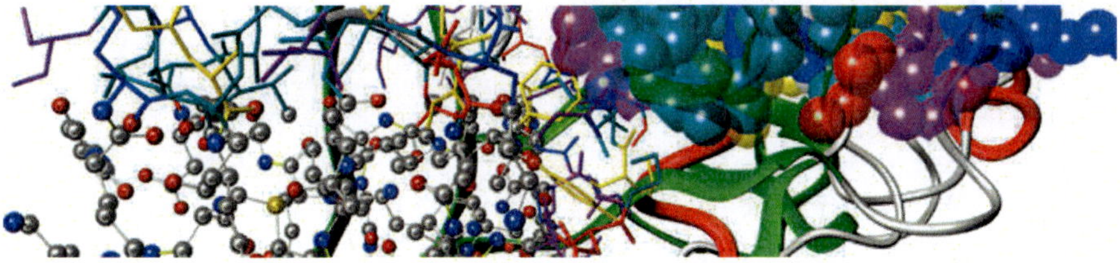

Abstract We present the Bio Scene Graph (BioSG) for visualization of biomolecular structures based on the scene graph system OpenSG. The hierarchical model of primary, secondary and tertiary structures of molecules used in the organic chemistry is mapped to a graph of nodes when loading molecular files.

We show that using BioSG, displaying molecules can be integrated in other applications, for example in medical applications. Additionally, existing algorithms and programs can be easily adapted to display the results with BioSG.

1 Introduction

Data about biomolecular structures is widely available today, as for example through the protein data bank (PDB [22]). The PDB contains 57944 biomolecular structures as of June 02, 2009. Molecular visualization however, is still way below

A. Halm (✉) · E. Eggeling
Fraunhofer Austria Research GmbH, Austria
e-mail: andreas.halm@fraunhofer.at; eva.eggeling@fraunhofer.at

D.W. Fellner
Institute of Computer Graphics and Knowledge Visualization, TU Graz, Austria and GRIS, TU Darmstadt, Germany
e-mail: d.fellner@tugraz.at

L. Linsen et al. (eds.), *Visualization in Medicine and Life Sciences II*, Mathematics and Visualization, DOI 10.1007/978-3-642-21608-4_14,

its best. Most visualization programs rely on third-party software such as raytracing programs to generate nice results. Integration with other systems is non-existent.

The Bio Scene Graph (BioSG) is a visualization system based on the BioBrowser [7, 8]. It integrates biomolecular visualization into the scene graph system OpenSG. This allows for two-way interchange of technology. On the one hand, the BioSG can make use of anything developed for the OpenSG scene graph system with no additional work. This includes, for example, annotation systems already programmed for different purposes, or medical, chemical or biological visualizations. On the other hand, the BioSG itself can be easily integrated into any OpenSG-based applications, for embedding biomolecular information in medical applications as well as other research areas.

2 Data Layout

2.1 Basics of a Scene Graph

A scene graph system can be described as some kind of management software that processes objects and abstract links between them. These objects are often three dimensional objects which are organized hierarchically. The items stored in the hierarchy are called "nodes" of the scene graph and usually represent displayable parts. They may also be used to logically group nodes together.

A simple 3D model of a car, for example, could consist of the car body and the four wheels. The car body would be implemented as one node in the scene graph and the four wheels being child nodes. Moving the car body does also move the wheels, but when driving in a curve, the wheels need to move independently for steering. Internally, the scene graph system would only store one 3D model of a wheel and use that model four times. This allows storing complex models very memory efficient.

A scene graph system does not only implement the graph itself, it also contains a lot of management tools. Usually a functionality is available not only to render the graph nodes and import geometry, but also to run arbitrary algorithms on the graph. There must be nodes available for defining cameras and lighting conditions. Scene graph systems designed to work in clustered environments also need network features as well as the functionality to duplicate a given graph on a different computer, which may also have different memory layout (MSB vs LSB). Last but not least, interfaces to the most used GUI toolkits are needed.

The main reason to implement molecular visualizations using a scene graph system is because molecules and scene graphs have a similar structure. Molecules are hierarchically structured, especially in organic chemistry. A small number of atoms build an amino acid. A number of amino acids whose three dimensional organization follows a special rule build a so called secondary structure, for example a helix or a sheet. Some of these secondary structures form a chain, and a molecule is composed of one or more (usually similar) chains. Such a strictly hierarchically structure is perfectly suited for processing with a scene graph system.

Usually a scene graph is only used to hold the data of the graphic representation of the scene, as for example with the tool PDB2VRML [6], which can be used to generate a set of VRML files, each containing the static triangular data of a specific visualization of the molecule in a scene graph. Another example is FPV [4], which generates a Java3D scene graph also containing static meshes, additionally flattening the scene graph after loading to a smaller scene graph containing much fewer nodes (6 for spacefill display) because of performance issues. BioSG on the other hand, is a scene graph specifically designed to hold all the semantic information available in the PDB file. No triangular data is generated when loading the file, but the structure of the molecule is resembled in the scene graph. The data needed for visualization (like triangles) are dynamically generated at runtime, which allows for a level-of-detail approach. Additionally, it means no information is lost – the PDB could theoretically be recreated from the scene graph.

Most existing molecular visualization packages implement a fully-customized scene graph, or generate rendering commands by iterating over in-memory molecule data structures. The approach chosen with BioSG however is to stay as generic and as compatible (with other OpenSG programs) as possible. Combination with other programs or modules are possible with minimal adaption only. Embedding BioSG into other programs that already use OpenSG, but have no or only rudimentary visualization components yet, would be very easy.

2.2 The Choice of the Scene Graph System

Several such scene graph systems exist, as for example OpenSG [18], OpenSceneGraph [17] and the Nvidia Scene Graph SceniX [21]. Additionally, game engine packages like OGRE [16] contain structures similar to a scene graph, but they are obviously not developed for scientific research. Game engines usually concentrate on fast rendering (or fast rejection of invisible parts), and are very limited pertaining running arbitrary algorithms on the scene graph data. Additionally, game engines usually presume that large parts of the geometry are static, which is counterproductive when animating or dynamically changing molecules.

The scene graph system that we selected as a basis for BioSG is OpenSG (Version 2), because of the reasons mentioned below.

Performance is one of the core requirements of any interactive 3D graphics system. Although it is a very powerful scene graph system, the aspect of rendering performance has always been the main goal during OpenSG development. OpenSG uses a wide variety of optimization techniques to efficiently use the available horsepower [20]. In the case of BioSG, this means that scene graph traversal and rendering are very fast, and optimization techniques that would remove the semantic information embedded (as described in [4]) need not be applied.

OpenSG supports a very general and flexible multi-threading model that gives totally independent threads write access to the scene graph without interfering with each other. Therefore, multithreaded algorithms can work on the scene graph data (the molecular data in this case), and even multiple algorithms can run their computations at once.

One of the strengths of OpenSG is its capability to render simultaneously on multiple networked computers [25]. This allows operation in CAVE-like environments without adaption of the code. Programs could also utilize that functionality and implement distributed computation on PC clusters. This would be very helpful for complex algorithms.

OpenSG supports a lot of different platforms, from common desktop operating system (Windows, Linux, MacOSX, Solaris, etc.) to handheld devices (using OpenGL ES). This opens a large variety of application fields for BioSG.

2.3 BioSG Implementation

From the biological point of view, an organic molecule can be seen as a hierarchical structure. A molecule may contain several chains, which consist of secondary structures like helices or sheets. These secondary structures are made of amino acids concatenated together. Amino acids in turn consist of atoms and bonds (Fig. 1). All

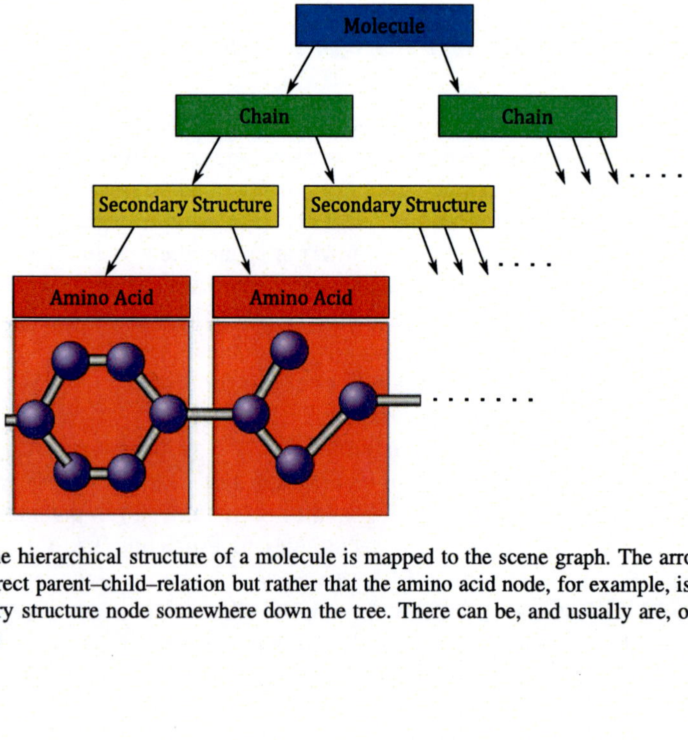

Fig. 1 The hierarchical structure of a molecule is mapped to the scene graph. The arrows do not mean a direct parent–child–relation but rather that the amino acid node, for example, is a child of a secondary structure node somewhere down the tree. There can be, and usually are, other nodes inbetween

hierarchy elements, from whole molecules down to amino acids are implemented as nodes in BioSG. They are created and linked together when loading a PDB file.

The implementation for these four hierarchy levels is straightforward. The classes for residues, secondary structures, chains and the molecule itself contain all the data which is read from the PDB file, like secondary structure type for example. Additionally, temporary data is created for visualization purposes. The residue class is special, however, as it contains additional arrays for atoms and bonds. Atoms and bonds are not child nodes of the amino acids. This reduces memory consumption and increases the performance of traversing the scene graph. On the other hand, having both atoms and bonds implemented as nodes would be of no benefit at all.

OpenSG itself does not expose any kind of ordering of the child nodes of one parent node. In the case of BioSG, it is crucial to have the child nodes ordered as there are bonds between the amino acids, and there are also bonds connecting secondary structures (from the last amino acid of one secondary structure to the first amino acid of the next secondary structure). For every amino acid, it is unambiguous which amino acid is prior to it and which one is next. Therefore the BioSG nodes contain additional links to their previous and next neighbors to allow iteration along a molecular chain. These links are built automatically when loading a PDB file. However, these additions to the scene graph do not break any existing OpenSG functionality, as these links are currently only used to render the bonds connecting neighbouring amino acids. External algorithms that modify the scene graph (like deleting or inserting nodes) should make sure that these links are kept intact.

OpenSG contains, apart from the hierarchical structure, a mechanism to store the actual data contained in the nodes. This concept is called "NodeCore" in OpenSG. In this case the nodes do not contain their associated data but rather link to "core" structures which then contain the data. This allows to save memory for scenes that have a lot of repeating elements, because the data is only stored once but can be instantiated as often as needed. This feature is currently not used in the BioSG but could strip down the memory usage even more. Since big molecules can contain thousands of amino acids, but there is only a limited number of existing amino acids (currently there are 22 known natural amino acids), it makes sense to use the data splitting mechanism here. We are currently investigating if this feature comes at the expense of rendering performance.

2.4 Advantages of the Approach

As the scene graph system OpenSG itself is not modified for this implementation, BioSG is compatible with all programs using OpenSG. The links to previous and next amino acids are implemented as extensions and usually do not interfere with OpenSG functionality, however insertion and deletion of nodes should be implemented to handle these links correctly or display errors may occur (like missing bonds between residues).

An interesting feature of OpenSG, especially in this context, is that applications using OpenSG to render a given scene graph do not need to know about the types of the nodes. OpenSG applications can dynamically load extension modules which enables them to load new file types and visualize new node types. BioSG is such an extension module, which allows any OpenSG application to load PDB files and visualize molecules. However, if BioSG is used in such a way, the OpenSG application has no information about the molecular structure like the fact that a specific node resembles a secondary structure and which type of structure that is, for example. It is still able to visualize a molecule and after selecting one node of the scene graph, which is always possible using built-in functionality, attach other nodes for adding textual information or hyperlinks, or modify transformations and thereby the threedimensional structure of the molecule. The BioSG has to be integrated into the main application for complete two-way-interaction like information exchange.

The fact that BioSG does not depend on the implemented nodes being direct parents or direct childs of each other allows for molecular animations very easily. The scene graph is constructed with extra transformation nodes between the special node types. These transformation nodes are standard node types in OpenSG and can be modified by other modules. Using these nodes algorithms can animate or modify the position of the residues in the molecule on any hierarchical level, like for example rotate one chain in relation to the others with just one call.

3 Selected Visualization Aspects

The BioSG currently supports four standard visualization styles as shown in Fig. 2. These are implemented in a number of different ways to make sure to get the best possible display quality on a variety of computers.

Three of the rendering methods (see Fig. 2a, c and d) are implemented at residue level. This means that a single residue node in a scene graph is able to display itself using one of these three visualizations. It is possible to have these nodes in a scene graph without the other node types presented in this paper.

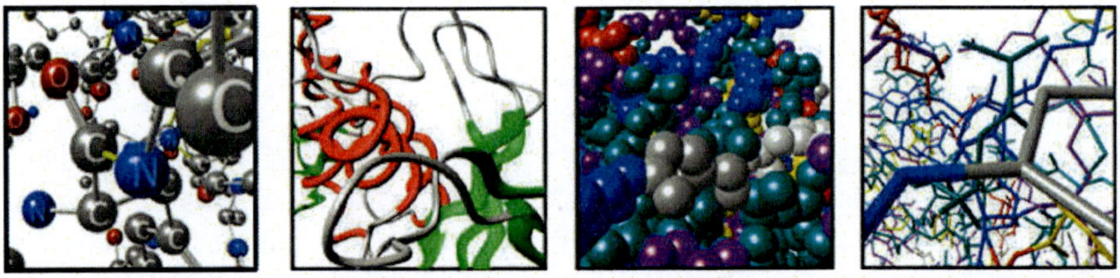

(a) Ball-and-Stick (b) Ribbon (c) Spacefill (d) Sticks

Fig. 2 Visualization styles supported by the BioSG

The ribbon display style (Fig. 2b) is implemented on the secondary structure level, whose child nodes need to be residue nodes to generate the base mesh. Apart from that, it is also independent of the other node types.

3.1 Classic Visualization Techniques

The visualization styles Spacefill and Sticks use level-of-detail (LOD) meshes (Fig. 3) or depth sprites (Fig. 4) on low end hardware, subject to pixel shader support. The depth sprite shader is implemented using the assembler-like shading language on *Shader Model 1.0*, so it does even run on GeForce3-class hardware. The visualizations are based on the algorithms described in [8].

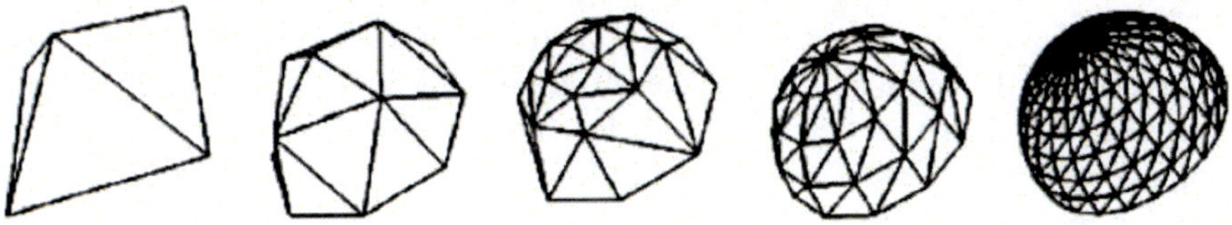

(a) Level 0 (b) Level 1 (c) Level 2 (d) Level 3 (e) Level 4

Fig. 3 LOD-Meshes used for visualizing a half sphere. The meshes are precalculated and loaded on startup

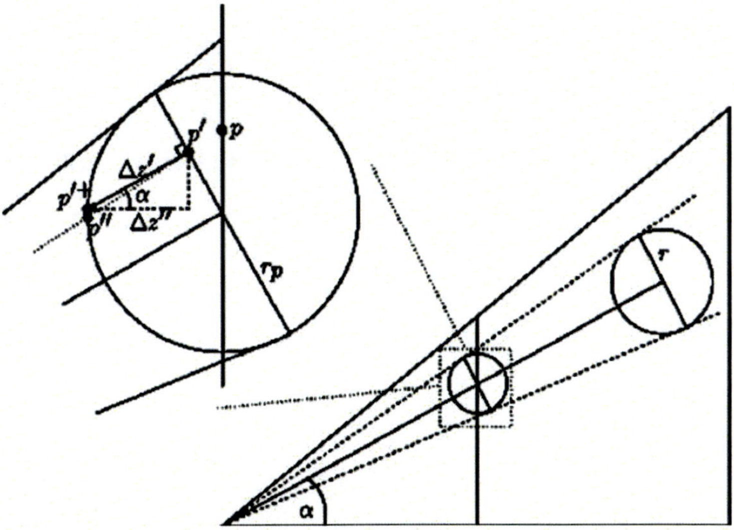

Fig. 4 Depth Sprites. The depth offset $\Delta z = \sqrt{1 - x^2 - y^2}$ is saved in a texture, and is added to the z coordinate of the billboard via pixel shader

The Ball-and-Stick visualization style uses simple billboards to render both the atomic sphere as well as the bonds between them on low end hardware.

The Ribbon visualization style uses a dynamic level-of-detail approach which generates as detailed triangle meshes as needed to retain smooth outlines. When moving around, the algorithm reevaluates visible parts and adds or removes triangles as needed. When run on low end hardware, the algorithm generates less detailed meshes in order to retain a refresh frequency of 30 frames per second. The algorithm is explained in [7].

3.2 *Raycasting*

If the hardware supports programmable shaders and is fast enough, most elements are rendered using raycasting. This is done using fragment shaders which compute a ray-sphere intersection. The result of the intersection computation is a Boolean telling if the object has been hit or not, the depth the hit has taken place at which is then written to the depth buffer, plus the surface normal at that point which is used for lighting. As the correct depth is computed and used for depth buffer writes and comparisons, such spheres intersect correctly such that they are good for rendering styles as the Spacefill style. The Ball-and-Stick style and the Spacefill style use ray-sphere-intersection shaders for rendering the atoms, and with the Sticks style round edges between the cylinders are rendered using these shaders. The Ball-and-Stick and the Sticks visualization additionally use ray-cylinder intersection shaders to render the atomic bonds [3].

3.3 *Comparison to Widely Used Molecular Viewers*

The visualization algorithms explained in section *Classic visualization techniques* are quite common, although not commonly used in molecular visualization programs. Most programs allow setting a static level of detail and generate triangle meshes for spheres and cylinders. TexMol [1] uses dynamic LOD and a Cg implementation of raycasting for spheres and cylinders. PyMol [5] uses OpenGL shaders for depth sprites. VMD [26] is able to use GLSL for sphere raycasting. Molekel [14] uses GLSL for per-pixel lighting only.

3.4 *Special Shaders*

The bonds between residues are rendered using a special shading technique to draw attention to these spots. As a bond is actually a logical abstraction of two atomic electron shells interfering each other, the atomic bond is more a smooth transition between atoms than a sharp edge. The shader that renders the bonds between two residues uses colored stripes to resemble this context (Fig. 5).

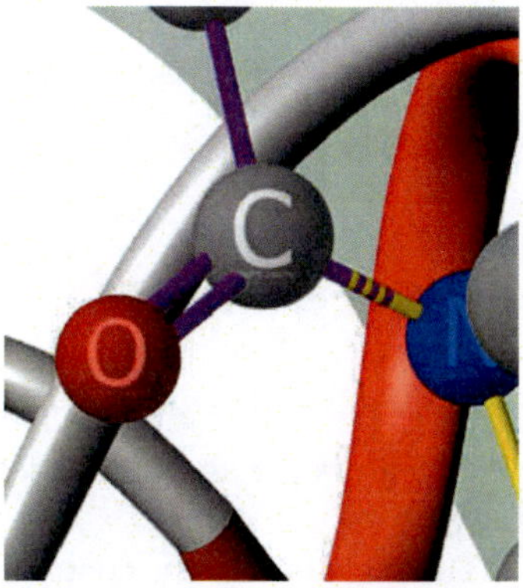

Fig. 5 Bonds between residues use a special two colored shader

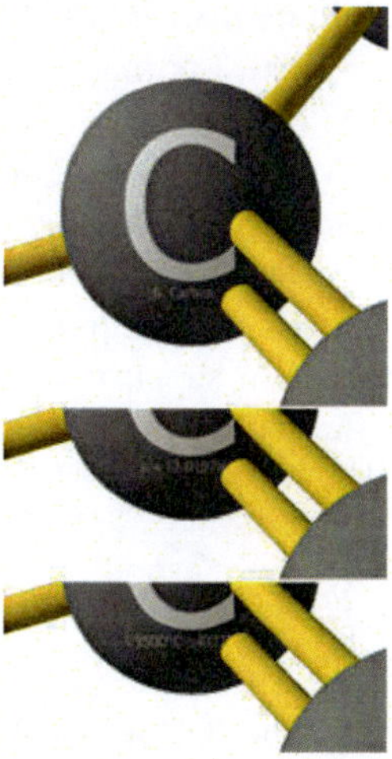

Fig. 6 The shader used for the atoms displays additional information (atom name and number, atom weight, melting and boiling point)

This feature is especially useful in the context of the hover mode (see below), where single secondary structures can be displayed using the Ball-and-Stick visualization or the Sticks visualization while the remaining parts of the molecule are displayed using the ribbon style. In this case, it would otherwise be difficult to identify which atom of which amino acid is connected to the remainder of the molecule. These parts are very easy to spot when the striped connection shader is used.

The Ball-and-Stick visualization renders the atoms using a shader that allows to display additional detail information directly on the atoms (Fig. 6). This is done by rendering up to three pieces of textual information directly into the atom texture - one text line in red, one in green and one in blue. The shader selects the information to display depending on an index which currently changes once a second, but could also be selected manually or automatically depending on the program context. This shows a new way to embed additional information without the need to clutter the display.

Of course this way of displaying information is not really suitable when rendering large molecules, however it may be useful when displaying small parts like single amino acids.

4 GUI Features

The current GUI to the BioSG kernel is based on Microsoft Fluent UI controls, which allows for an intuitive interface to the visualization features [2, 13] (Figs. 7 and 8).

The main button bar contains fast access to all visualization features. The bar is highly customizable so an unwanted feature can be hidden, or a feature often used could be mapped to keyboard shortcuts. The button bar can also be minimized to work like a normal menu. Additionally a fullscreen mode is available which hides all GUI elements.

All output messages which the program generates are sorted into the following four message types: general messages, file information, OpenGL-related messages and OpenSG-related warnings and errors. The messages are printed into the corresponding output pane and can be accessed via the tabbed interface. The output pane also supports an autohide feature which only displays the pane when new information is available.

4.1 Hover Mode

When the "hover mode" is enabled, every time the mouse is moved a ray intersection is calculated with the mesh used for the Ribbon visualization (Fig. 2b). The result

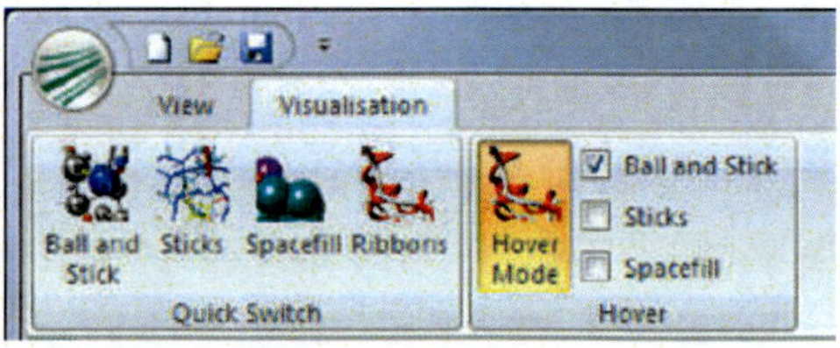

Fig. 7 Fluent UI button bar

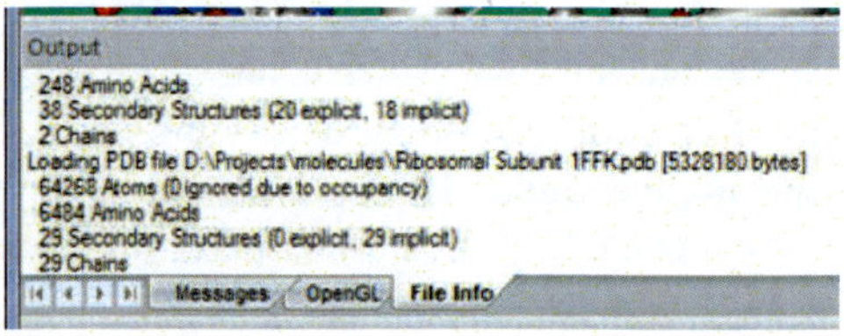

Fig. 8 Output pane

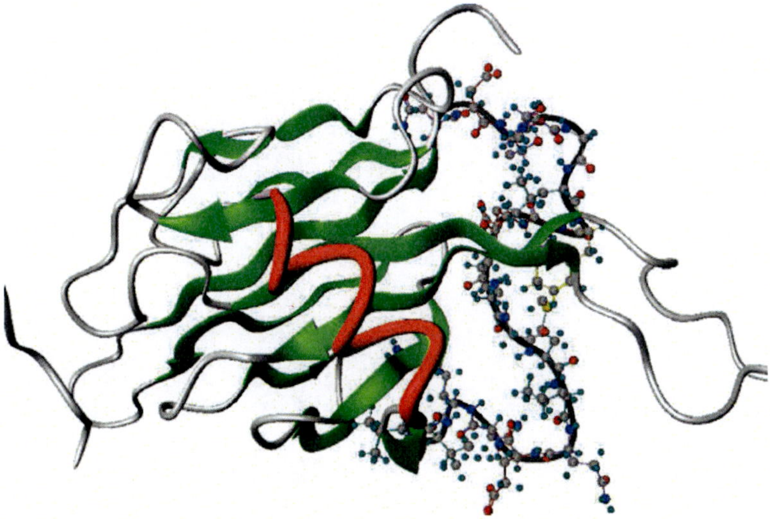

Fig. 9 The hover mode reveals the amino acids under the selected Ribbon structure

of such a query is the scene graph node under the mouse pointer. This node is set to being partially transparent, and all its child nodes (the amino acids contained in that secondary structure) are displayed in one of the more detailed Ball-and-Stick, Sticks or Spacefill visualization styles (Fig. 9). As soon as the mouse leaves the Ribbon, the visualization is switched back.

This method enables the researcher to view big molecules as Ribbons, and only display more details when needed, thus helping to resolve the problem of cluttered display.

4.2 Screenshots

The screenshot utility (Fig. 10) allows to save an image of the current visualization independent of the current active display resolution. This is accomplished using Frame Buffer Objects (FBO). The rendering is temporarily rerouted from the standard OpenGL viewport to a FBO which is created at the selected resolution. This way not only standard sizes can be rendered, but also much bigger sizes than available on a desktop. The maximum possible resolution for an FBO on a Nvidia Geforce 8800 card is 8192x8192 pixels.

Resolution settings are predefined for saving images in extreme quality (the highest resolution being 8192x6144 with a 4:3 aspect ratio) as well as rendering images in HD (1280x720, 16:9 aspect ratio) and FullHD (1920x1080, 16:9 aspect ratio) formats, which are getting more and more common today. The HD and FullHD formats are especially well suited for generating content used in videos.

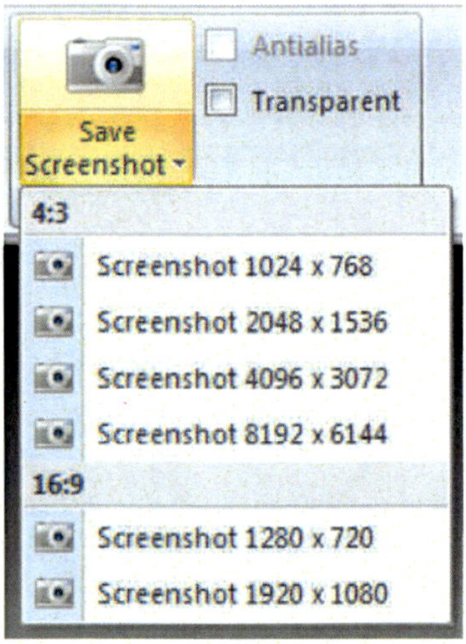

Fig. 10 Screenshots can be saved as PNG or JPG files in high resolution

4.3 *Camera Flight*

Most molecular visualization tools provide some way to generate high-quality videos or at least still images. This is usually accomplished using scripting languages to generate a camera flight path, and using external raytracing packages for rendering. Chimera [19] uses a scripting language to generate camera flight paths and encode videos, additionally data can be exported to the rendering and modelling package Maya to generate high-quality videos. PyMol [5] uses a scripting language to generate video sequences whereby simple animations can also be initiated using the menu. VMD [26] can be scripted using Tcl or Python and recommends using an external raytracer to generate high quality images. Jmol [10] as well as RasMol [24] also define its own scripting languages (which are very similar) with which image sequences can be generated, which have to be reencoded into a video file in a separate step.

This application however does not use external programs, in fact the visualization algorithms already generate extremely detailed high quality output so that using a raytracer instead is not needed. Using the simple camera replay controls this program is well suited for demonstration purposes, without the need to generate videos beforehand.

We implemented a camera flight tool which is very simple to setup and use. After loading and positioning a molecule, pressing the '+' button adds a waypoint to the waypoint list (Fig. 11). A miniature screenshot is generated and displayed in the waypoint list for users guidance. More waypoints can be added by rotating, zooming and panning the molecule and pressing the *add* button repeatedly. After a minimum of three waypoints have been set, the camera flight spline is displayed, which is implicitly defined by the waypoints (Fig. 12) and generated automatically. The waypoints can be reviewed and edited by double-clicking them in the list.

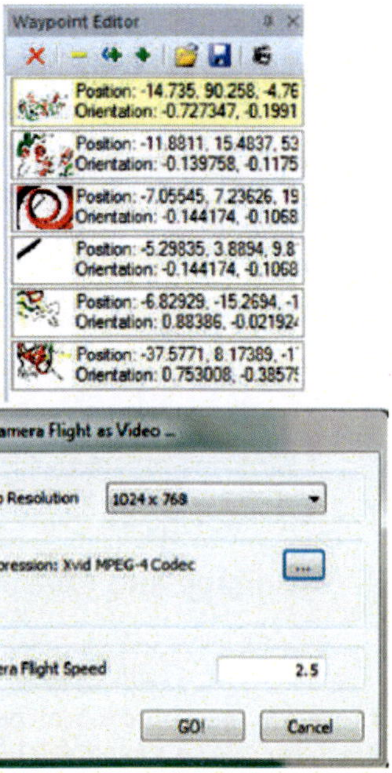

Fig. 11 The waypoint editor and the video export dialog

Fig. 12 The yellow striped spline marks the calculated camera path

The integrated camera flight utility serves several purposes. For example, it is the perfect tool for demonstration purposes. Camera flights can be loaded and started from the commandline as well as from the GUI (Fig. 13), and the program can optionally switch to fullscreen mode during playback.

Additionally, it is possible to render a camera flight into an AVI video file by simply pressing the *export to AVI* button in the waypoint editor. Some options can be set via the export dialog as the resolution and the compression settings. This utility makes generation of high quality videos very simple.

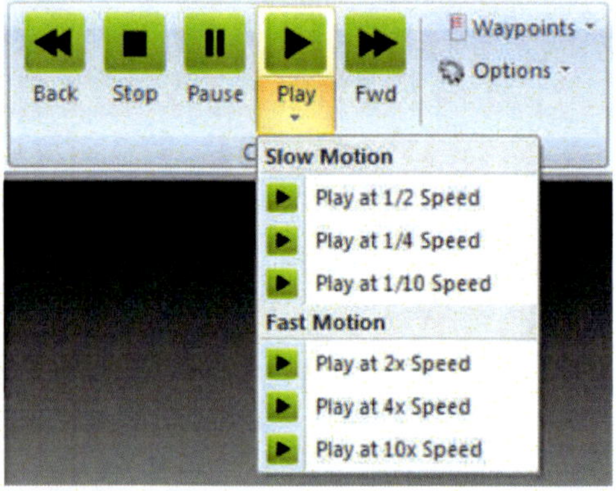

Fig. 13 The controls to replay a camera flight allow for simple speed adjustments

5 Summary and Outlook

Using a scene graph system as a basis for visualization in structural biology has already been proven to be helpful, as the management and network features which are part of OpenSG made it very simple to port the BioSG to a CAVE-like environment. In the future this will simplify integration of interdisciplinary research areas.

The current GUI and the tools that have been implemented provide a useful demonstration system for research of structural biology.

Although using a scene graph system usually imposes higher memory usage, the memory consumption of BioSG is very much comparable to that of other programs. Additionally it would be possible to store data which is common for a lot of nodes in a special data construct called "core" in OpenSG. This feature is not used yet, but it would work very well for residues. Big molecules may contain hundreds of residues of the same type. Moving data storage for these from nodes to cores should decrease memory consumption.

A scene graph that displays biomolecules opens up a whole new field of possibilities. This includes, for example, the option to use modules already developed for OpenSG, like annotation modules. These can be combined with the network interface part of OpenSG to form a collaborative working environment.

Being a strict extension of OpenSG, the BioSG can easily be embedded into any program that already uses OpenSG. This allows for more interdisciplinary work, for example with medical or biological applications, or even with programs for material testing.

On the other hand, algorithms already working on scene graph structures can be modified to work with the BioSG. Programs solving complex problems could use the visualization parts to display intermediate results in high quality.

References

1. Vinay Siddavanahalli, Chandrajit Bajaj, Peter Djeu, and Anthony Thane. TexMol - interactive visual exploration of large flexible multi-component molecular complexes. *IEEE Visualization*, 2004.
2. Forrester Consulting. The microsoft office fluent user interface: Information worker perception of productivity, training, and support requirements. User study, Forrester Research, Inc., 400 Technology Square, Cambridge, MA, USA, August 2007.
3. Mario Botsch, Christian Sigg, Tim Weyrich, and Markus Gross. GPU-based ray-casting of quadric surfaces. *Eurographics Symposium on Point-Based Graphics*, 2006.
4. Tolga Can, Yujun Wang, Yuan-Fang Wang, and Jianwen Su. FPV: fast protein visualization using Java 3D. In *Proceedings of the 2003 ACM symposium on applied computing*, 88–95. ACM press, 2003. http://www.ceng.metu.edu.tr/~tcan/fpv/.
5. W. L. DeLano. The PyMOL molecular graphics system. http://www.pymol.org/.
6. T. Exner, G. Moeckel, M. Keil, and J. Brickmann. Molecular modeling information transfer with VRML: from small molecules to large systems in bioscience. In *Pacific Symposium on Biocomputing*, 327–338. World Scientific, 1998.
7. Andreas Halm, Lars Offen, and Dieter Fellner. Visualization of complex molecular ribbon structures at interactive rates. In *IV '04: Proceedings of the Information Visualisation, Eighth International Conference on (IV'04)*, 737–744, Washington, DC, USA, 2004. IEEE Computer Society.
8. Andreas Halm, Lars Offen, and Dieter W. Fellner. BioBrowser: A framework for fast protein visualization. In *EuroVis05: Joint Eurographics - IEEE VGTC Symposium on Visualization, Leeds, United Kingdom, 1-3 June 2005*, 287–294. Eurographics Association, 2005.
9. C. W. V. Hogue. Cn3D: a new generation of three-dimensional molecular structure viewer. *Trends in Biochemical Sciences*, 314–316, 1997. ftp://ftp.ncbi.nih.gov/cn3d/.
10. Jmol: an open-source java viewer for chemical structures in 3D. http://www.jmol.org/.
11. M. E. Kabay. Arrogance or efficiency? A discussion of the microsoft office fluent user interface. *Ubiquity*, 1–4, March 2008.
12. M. Krone, K. Bidmon, and T. Ertl. GPU-based Visualisation of Protein Secondary Structure. In *Proceedings of TPCG 2008*, 115 – 122, 2008.
13. MFC Feature Pack for Visual C++ 2008. http://msdn.microsoft.com/en-us/library/bb982354.aspx.
14. Molekel. http://molekel.cscs.ch/wiki/pmwiki.php.
15. Lars Offen and Dieter Fellner. BioBrowser - visualization of and access to macro-molecular structures. In *Visualization in Medicine and Life Sciences*, 257–273. Springer Verlag, 2007.
16. OGRE - Object-Oriented Graphics Rendering Engine. http://www.ogre3d.org/.
17. OpenSceneGraph. http://www.openscenegraph.org/.
18. OpenSG. http://www.opensg.org/.
19. E. F. Pettersen, T. D. Goddard, C. C. Huang, G. S. Courch, D. M. Greenblatt, E. C. Meng, and T. E. Ferrin. UCSF Chimera – A visualization system for exploratory research and analysis. *Journal of computational chemistry*, 1605–1612, 2004. http://www.cgl.ucsf.edu/chimera/.
20. Dirk Reiners, Gerrit Voß, and Johannes Behr. OpenSG: Basic concepts. In *1. OpenSG Symposium OpenSG*, 2002.
21. Nvidia SceniX scene management engine. http://developer.nvidia.com/object/scenix-home.html.
22. J. L. Sussman, D. Lin, J. Jiang, N. O. Manning, J. Prilusky, O. Ritter, and E. E. Abola. Protein data bank (PDB): database of threedimensional structural information of biological macromolecules. *Acta Crysallogr.*, 1078–1084, 1998. http://www.pdb.org/.
23. H. Steen and M. Mann. The abc's (and xyz's) of peptide sequencing. *Nature Reviews Molecular Cell Biology*, 699–711, 2004.
24. R. Sayle and E. J. Milner-White. RasMol: Biomolecular graphics for all. *Trends in Biochemical Sciences*, page 374, 1995. http://www.rasmol.org/.

25. G. Voß, J. Behr, D. Reiners, and M. Roth. A multi-thread safe foundation for scene graphs and its extension to clusters. In *EGPGV '02: Proceedings of the Fourth Eurographics Workshop on Parallel Graphics and Visualization*, 33–37, Aire-la-Ville, Switzerland, Switzerland, 2002. Eurographics Association.
26. Andrew Dalke, William Humphrey, and Klaus Schulten. VMD – visual molecular dynamics. *Journal of Molecular Graphics*, 14:33–38, 1996. http://www.ks.uiuc.edu/Research/vmd/.

Linking Advanced Visualization and MATLAB for the Analysis of 3D Gene Expression Data

Oliver Rübel, Soile V. E. Keränen, Mark Biggin, David W. Knowles, Gunther H. Weber, Hans Hagen, Bernd Hamann, and E. Wes Bethel

Abstract Three-dimensional gene expression PointCloud data generated by the Berkeley *Drosophila* Transcription Network Project (BDTNP) provides quantitative information about the spatial and temporal expression of genes in early *Drosophila*

O. Rübel (✉) · B. Hamann
Computational Research Division, Lawrence Berkeley National Laboratory (LBNL), One Cyclotron Road, Berkeley, CA, 94720, USA
Institute for Data Analysis and Visualization (IDAV), Department of Computer Science, University of California, Davis, One Shields Avenue, Davis, CA, 95616, USA
International Research Training Group 1131, Technische Universität Kaiserslautern, Erwin Schrödinger-Straße, 67653, Kaiserslautern, Germany
e-mail: oruebel@lbl.gov; bhamann@ucdavis.edu

S.V.E. Keränen · M. Biggin
Genomics Division, LBNL, One Cyclotron Road, Berkeley, CA, 94720, USA
e-mail: svekeranen@lbl.gov; mdbiggin@lbl.gov

D.W. Knowles
Life Sciences Division, LBNL, One Cyclotron Road, Berkeley, CA, 94720, USA
e-mail: dwknowles@lbl.gov

G.H. Weber
Computational Research Division, LBNL, One cyclotron Road, Berkeley, CA, 94720, USA
e-mail: ghweber@lbl.gov

H. Hagen
International Research Training Group 1131, Technische Universität Kaiserslautern, Erwin Schrödinger-Straße, 67653, Kaiserslautern, Germany
e-mail: hagen@informatik.uni-kl.de

E.W. Bethel
Computational Research Division, Lawrence Berkeley National Laboratory (LBNL), One Cyclotron Road, Berkeley, CA, 94720, USA
e-mail: ewbethel@lbl.gov

L. Linsen et al. (eds.), *Visualization in Medicine and Life Sciences II*, Mathematics and Visualization, DOI 10.1007/978-3-642-21608-4_15,

embryos at cellular resolution. The BDTNP team visualizes and analyzes Point-Cloud data using the software application PointCloudXplore (PCX). To maximize the impact of novel, complex data sets, such as PointClouds, the data needs to be accessible to biologists and comprehensible to developers of analysis functions. We address this challenge by linking PCX and Matlab®[1] via a dedicated interface, thereby providing biologists seamless access to advanced data analysis functions and giving bioinformatics researchers the opportunity to integrate their analysis directly into the visualization application. To demonstrate the usefulness of this approach, we computationally model parts of the expression pattern of the gene *even skipped* using a genetic algorithm implemented in Matlab and integrated into PCX via our Matlab interface.

1 Introduction

The BDTNP has developed a novel type of spatial and temporal gene expression data called *PointCloud*. Single PointClouds are obtained via segmentation of two-photon microscopy images of *Drosophila* embryos and provide a quantitative representation of spatial gene expression levels of the *Drosophila* blastoderm at cellular resolution [19]. Multiple PointClouds representing a variety of genes at multiple developmental time intervals are registered into a single *Atlas PointCloud* describing the expression of many genes at multiple points in time [7].

PointCloudXplore (PCX) is the standard tool of the BDTNP for visualization and analysis of PointCloud data [28]. PCX supports various 2D and 3D physical embryo views and abstract data visualizations and provides a suite of analysis tools, e.g., methods for cell selection and clustering [20]. Matlab —a commercial, high-level programming language and analysis environment— is widely used throughout the BDTNP, e.g., for PointCloud creation and embryo registration, and provides with its rich feature-set an ideal platform for development of custom analysis functionality. Despite its versatility, we anticipate that increasing numbers of researchers will benefit from being able to quickly develop and integrate custom analysis capabilities with PCX.

To address this challenge we developed an interface between PCX and Matlab, enabling researchers to easily integrate their analysis with the visualization and providing biologists faster and more convenient access to advanced analysis functions (Sect. 3). Via the PCX-Matlab interface, one can call functions implemented in Matlab directly from PCX, while PCX automatically handles all necessary communication, including, start/close of Matlab, data transfer to and from Matlab, and initiation of function calls. No Matlab knowledge is required to access Matlab

[1]MATLAB is a registered trademark of The MathWorks Inc., 3 Apple Hill Drive Natick, MA 01760-2098, USA. Online at: http://www.mathworks.com/

functions via PCX. The interface also hides PCX's internal architecture from the Matlab developer, and very low effort is needed to make a Matlab function accessible to PCX. The interface supports fast prototyping and testing of new ideas and facilitates communication between bioinformatics researchers and experimental biologists.

We chose computational modeling of genetic regulatory networks as example to demonstrate the usefulness of the PCX-Matlab interface (Sect. 4). Expression regulatory models —e.g., by Janssens et al. [15] and Sanchez et al. [26]— often depend on extensive system-wide knowledge based on years of experimental work on mutants and transgenic constructs and specialized sets of equations and programs. With increasing number of components (i.e., genes), the number of potential interactions that need to be analyzed experimentally increases exponentially. Thus, computational methods are needed to identify probable candidate genes for experimental verification.

We describe a genetic algorithm for finding potential genetic regulatory interactions via optimization of a linear network model. We implemented this algorithm in Matlab and integrated it into PCX via our system interface. By integrating the modeling with the visualization, we can define the necessary input to the analysis quicker and more accurately, and are able to effectively validate the input and output of the analysis. We discuss the modeling of the expression pattern of the gene *even skipped* (*eve*) to illustrate the advantages and disadvantages of the employed modeling approach.

2 Related Work

PCX uses the established concept of *linked multiple views* [2] for visualization of high-dimensional 3D gene expression data. In the WEAVE system, a combination of physical and information visualization views is used for exploration of cardiac simulation and measurement data [9]. Henze [12] developed a multiple-view-based system for exploration of time-varying computational fluid dynamics data. Advanced queries are supported by this system via selection of data subsets in each view (here termed portraits). The concept of using abstract views to define data queries was formalized by Doleisch et al. [6]. Interfaces to programming and analysis languages and systems are often used in the context of visualization to support scripting and to ease integration of the visualization with simulation [16] or data analysis, e.g., R and GGobi [18].

Someren et al. [27], De Jong [4], and D'haeseleer et al. [5] provide an overview of genetic network modeling. Janssens et al. [15] used simulated annealing to describe a predictive model of transcriptional control of the gene *even skipped* (*eve*) based on spatial gene expression, regulatory sequence data, and FlyEx 2D cellular resolution quantitative protein expression data of *Drosophila* embryos, which has also been used for modeling of *Drosophila* anterior-posterior gap gene system [14]. Fowlkes et al. [7] described first results on expression modeling based on the BDTNP quantitative 3D expression atlas using linear regression to identify for each target pattern six likeliest candidate regulators. In this work, we describe the implementation of a genetic algorithm for finding genetic network models based on the BDTNP expression atlas. Genetic algorithms were surveyed by Whitley et al. [29] and Davis et al. [3]. The goal of this paper is not to develop a more sophisticated regulatory model but to provide a flexible environment to ease development of advanced analysis functions and to make them more accessible to the user.

3 The PointCloudXplore and Matlab Interface

The main goal of our research is to provide fast and easy access to new analysis functions for three-dimensional gene expression data, thus facilitating communication between biologists and bioinformatics researchers. To accomplish this goal, we added a Matlab interface (Fig. 1) to PointCloudXplore (PCX), making PCX more easily extensible by bioinformatics researchers. From a biologist's perspective, this interface is transparent with respect to which functionality is implemented via Matlab functions. No Matlab knowledge is necessary to use these functions.

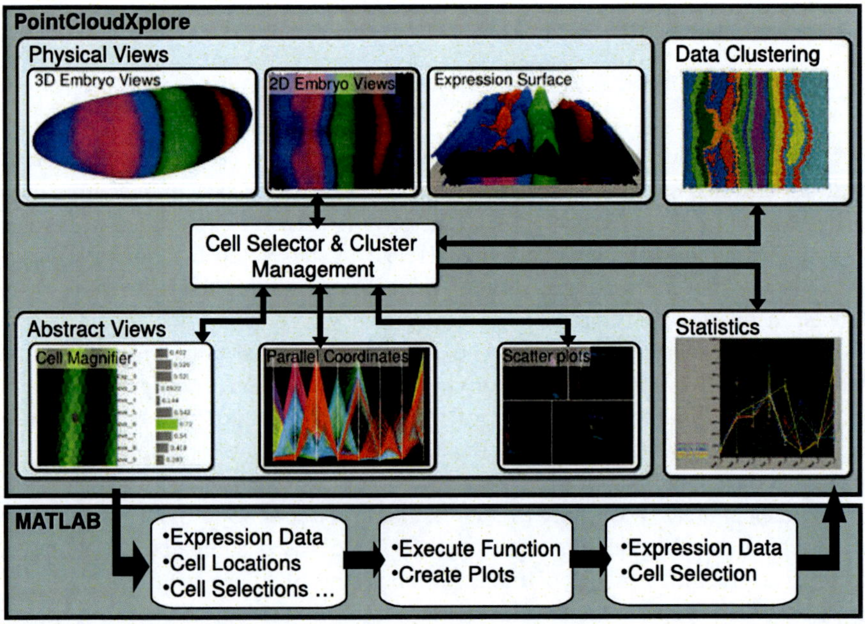

Fig. 1 Overview of PointCloudXplore and the interface to Matlab

From a bioinformatics researcher's perspective, the interface hides the internal PCX architecture and requires minimal effort to make a Matlab function accessible to PCX.

The PCX-Matlab interface provides means to initiate Matlab function calls from PCX's graphical user interface (GUI), while automatically handling all necessary inter-system communication. Data classes exported from PCX to Matlab include multiple user-defined lists of gene expressions and/or cell selections and their names, 2D/3D cell locations, cell neighbors, and additional user-definable function parameters (scalar double, Boolean, integer, or string). Information imported from Matlab to PCX includes multiple derived gene expression channels and/or cell selections including optional names.

In the following we describe the PCX-Matlab interface (Sect. 3.1), discuss the interface from a biologist's and bioinformatics researcher's view (Sect. 3.2), and review various example applications (Sect. 3.3).

3.1 Interface Design

The definition of a Matlab function for PCX consists of an *M*-file with the function implementation and a *PCXM* header file, describing the function for PCX (Fig. 3). This header file must specify:

- A function category (TYPE) used to group functions in the menu.
- A name for the function (NAME) to appear in the menu.
- The location of the *M*-file, if different from the *PCXM* file (DIR).
- The Matlab command to invoke the function (CALL).

The TYPE and NAME parameter may define an arbitrary string, allowing the user to group and name functions according to personal preferences.

Using strict name conventions for function call input and output parameters, PCX can automatically identify what data needs be transferred to and from Matlab — e.g., gene expressions, cell locations and neighbors, cell selections, and user-definable double, integer, Boolean, or string parameters— based on the function call itself. All data transfers are implemented via a series of function calls to the Matlab C++ API, allowing us to directly transfer vector/matrix data of varying type between the two systems.

For some applications it makes sense to pass input genes and names in multiple groups. For each list of genes or cell selections, one may optionally specify the expected size of each list (IN_GENES, IN_BRUSHES). Furthermore, one can define an optional URL for a HTML help page (HELP_URL). Thus, a PCXM header file provides PCX with information about: (a) what data needs to be exchanged between Matlab and PCX, (b) what parameters need to be exposed to the user, (c) how the Matlab function is invoked, and (d) how the function and its parameters should be represented within the GUI.

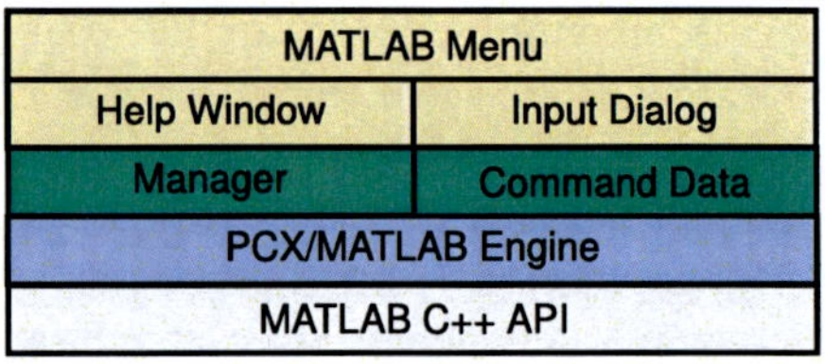

Fig. 2 Design of the interface between PointCloudXplore and Matlab

The PCX-Matlab interface consists of three main components: a communication layer (blue), a function management layer (green), and the GUI (tan) (see Fig. 2). The communication layer is based on Matlab's C++ API (gray) and provides higher-level functions for inter-system communication.

The function management layer manages all Matlab functions available to PCX. The *Manager* class parses the PCXM files, stores and provides access to information about the Matlab functions, and initiates the execution of Matlab functions. For each Matlab function, the Manager creates a *CommandData* object containing all information about the corresponding function.

The GUI portion of PCX's Matlab interface consists of: the *Matlab Menu*, a parameter *Input Dialog*, and the *Help Window*. The Matlab Menu is part of PCX's main menu bar and lists all available Matlab functions (NAME) grouped by their logical category (TYPE). Furthermore, the Matlab Menu provides access to the Help Window and various management options, such as updating the function list. Using the Help Window, one can conveniently access the HTML help pages of all Matlab functions. When selecting a Matlab function in the Matlab Menu, PCX dynamically creates an Input Dialog at runtime listing all user-definable input parameters of the corresponding function.

The basic work flow of a use-case involving both PCX and Matlab is as follows (Fig. 1): Upon start-up, PCX parses all available PCXM files and initializes the Matlab Menu and Help Window. To initiate the execution of a Matlab function, one selects the desired function in the Matlab Menu. Subsequently, PCX creates an Input Dialog for the function and starts Matlab if it is not already running. After one has specified and confirmed the function input in the Input Dialog, PCX transfers all the necessary data to Matlab and invokes the Matlab function. Finally, PCX imports any output expression channels or cell selections, once the Matlab function has terminated.

3.2 Perspective of the Biologist and Bioinformatics Researcher

When calling a Matlab function via PCX, the user interacts only directly with PCX. PCX automatically handles all communication with Matlab. In terms of access to the function it is irrelevant from a biologist's perspective whether a function is implemented in Matlab or directly in PCX. However, unlike Matlab, the interface of PCX is much more user-friendly and does not require any command-line input.

To simplify repeated analysis and test-runs, PCX supports saving and restoring gene input lists, and the integration of Matlab and PCX enables one to refine input parameters quickly between individual Matlab command executions by making it possible to quickly investigate input and output expression patterns. The interface between Matlab and PCX provides the user faster and more convenient access to advanced analysis functions.

From a bioinformatics researcher's perspective, the effort needed to make an existing Matlab function (defined in an M file) accessible in PCX is very low and limited to creating a simple PCXM header file. No recompilation of source code or knowledge of PCX design or implementation details is required. Matlab functions can be updated at runtime of PCX, making development and debugging more convenient and less time-consuming. In addition, PCX provides a convenient GUI for the Matlab functions and parses PointCloud data, i.e., a developer does not need to know how to read PointCloud data. PCX's advanced data display and selection mechanisms in combination with the simplicity of the Matlab interface simplify the development of usable scripts for processing PointCloud data. The PCX-Matlab interface provides an easy way to make new analysis functions quickly available to biologists. The interface supports fast prototyping and testing of new ideas and facilitates communication between bioinformatics researchers and experimental biologists.

Figure 3 shows a simple example function that computes the cell-by-cell difference between two expression patterns, illustrating the effort required for deploying a Matlab function in PCX. The function takes two genes and their names as input

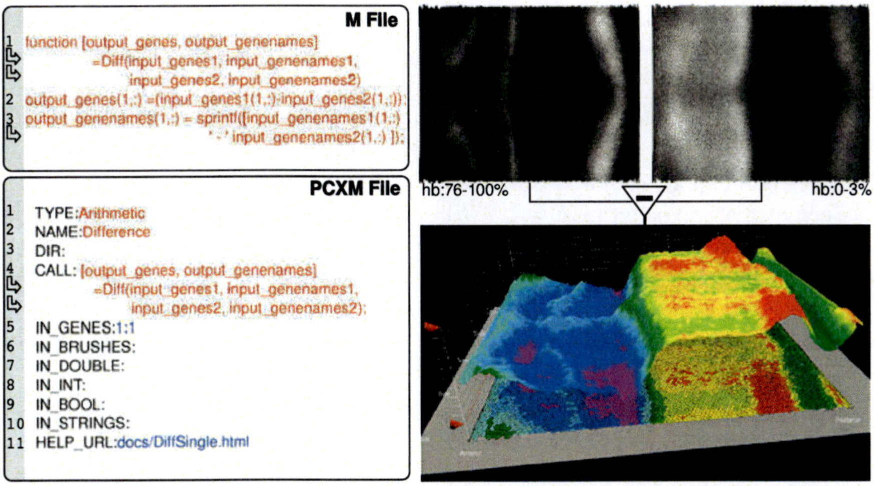

Fig. 3 M-file (top left) and PCXM header file (bottom left) of an example function for the computation of the cell-by-cell difference between two expression patterns. Color of text is used to illustrate which parts are mandatory (red), optional (blue), or provided in a template file (black). Right: Example showing the use of the function to compute the difference between late- and early-stage *hunchback* expression

and returns a new expression channel and its name as output. The PCXM header file specifies the TYPE, NAME, and function call to be issued (CALL). In addition, the developer defined a HTML help page (HELP_URL) and specified that both input gene lists should contain only a single gene (IN_GENES). Difference patterns are useful to analyze, e.g., the temporal variation of an expression pattern (Fig. 3, right side) or compare the pattern of different genes. In our example, we consider a PointCloud atlas file containing late (76-100%) and early (0-3%) blastoderm *hunchback* (*hb*) patterns. By applying the difference function to these genes, we can explore the temporal expression pattern change. As this example illustrates, the Matlab interface allows us to easily and quickly extend PCX with additional features.

3.3 Applications

The combination of PCX and Matlab has a wide range of applications ranging from plotting and cell selection to expression filtering.

Often, one needs to quickly create a specific plot, e.g., for a presentation. Matlab provides a rich set of plots that can be deployed quickly via PCX's interface to Matlab. Plots created in Matlab are displayed in an external Matlab plot window, supporting direct interaction with the plot. We implemented, e.g., a line-out plot in Matlab for comparison of the expression profiles of multiple genes along a line defined on the embryo surface (not shown).

A second application of PCX's Matlab interface is cell selection. This category of applications includes, e.g., functions for importing/exporting cell selections, segmenting gene expression patterns [13], clustering cells [20], and detecting features.

Expression filtering functions select expression patterns or create new derived "expression" patterns. Simple filtering functions include functions for expression data import or export, arithmetic functions for computing the cell-by-cell difference between expression patterns ($c = a - b$) or the cooperative pattern of genes (e.g., $c = a * b$). Other applications of expression filtering include smoothing and post-processing of expression patterns, masking of expression patterns, or dimension reduction (e.g., using principal component analysis). Advanced analysis functions, such as the gene expression modeling described in Sect. 4, often bridge various of the application categories mentioned above.

4 A Predictive Model of Expression Control

The BDTNP gene expression atlas provides us with information about the spatio-temporal expression pattern of genes, i.e., the input and output of the complex genetic networks responsible for the regulation of gene expression patterns. Based on this information one can attempt to infer potential regulatory interactions leading to the formation of a selected target pattern.

To explore and better understand gene pattern formation we first define a basic model for a genetic regulatory network. We use a linear model, i.e., the output of a model network is defined through a linear combination of its weighted inputs (Sect. 4.1). Given the input, output, and basic network model, we need to find a specific network model that defines the target well. To find a good network model we use a genetic algorithm [3, 29], which we implemented using Matlab (Sect. 4.2).

In the described network model, the output directly depends on the input. Validation and understanding of the input patterns is, therefore, crucial for the interpretation of a predicted model. Furthermore, we need to be able to compare the output of a network with the target pattern. PCX supports validation and comparison of expression patterns via a combination of physical and abstract visualizations and statistical analysis tools, making PCX ideal for these tasks. Using the PCX-Matlab interface we integrated the Matlab optimization function with PCX making the analysis easily accessible for the target user (Sect. 4.3).

In Sect. 4.4 we describe the example application of our system to model the expression pattern of the gene *eve*. It is beyond the scope of this paper to provide a conclusive model for the *eve* regulation. The goal of the research covered here is rather to investigate the advantages and disadvantages of the employed direct network modeling approach.

4.1 Expression Model

Next, we describe a continuous, linear expression model. Based on the expression values of the measured expression patterns of a set of input regulators $e_{regs} = \{e_1, e_2, ..., e_k\}$ we define the model expression of the cell c as:

$$v(w, e_{regs}, c) = \sum_{i=0}^{k} w_i e_i(c), \tag{1}$$

with $w = \{w_1, w_2, ..., w_k\} \in [-1, 1]^k$ being the weights of the different regulators e_{regs}. A weight of $+1$ indicates strong activation and a weight of -1 indicates strong inhibition of expression. In practice, a gene can only show positive expression, i.e., further inhibition of a non-expressed gene has no effect on the output. To account for this behavior, we define the model expression of a cell c as:

$$m(w, e_{regs}, c) = max(v(w, e_{regs}, c), 0). \tag{2}$$

The expression pattern defined by a network model w is defined as:

$$m(w, e_{regs}) = \{m(w, e, 1), m(w, e, 2), ..., m(w, e, n)\}. \tag{3}$$

A gene can also be ubiquitously expressed at some basic level even if no spatially regulated activator is present. To account for this behavior, a user can include a constant expression $e_{const} = 1$ as input factor with the corresponding weight $w_{e_{const}}$ defining the degree of basal expression. By including cooperative patterns — e.g., of the form $e_{ij} = e_i e_j$ — as input regulators, a user can also account for simple non-linear effects of cooperative regulation, i.e., two genes acting in cooperation to form a pattern.

Expression modeling — using binary as well as continuous models — has proven to be a powerful tool for the study of genetic regulatory networks [4, 26, 27]. The main limitation of linear expression models, such as the one described here, is that they are not able to account for non-linear regulatory effects — such as heterodimerization of two transcription factors for a specific regulatory effect [22] — and may not be able to predict all targets correctly.

4.2 Optimization Algorithm

The goal of the optimization is to find a model w for which $m(w, e)$ fits the target expression pattern e_{target} well. To define the similarity between two expression patterns e_i and e_j we use the correlation measure

$$corr(e_i, e_j) = \frac{\sum_{c=0}^{n}(e_i(c) - \overline{e_i})(e_j(c) - \overline{e_j})}{(n-1)s_{e_i}s_{e_j}} \tag{4}$$

as distance function, with s_{e_i} and s_{e_j} being the sample standard deviation of e_i and e_j, respectively. The correlation is a scale-invariant measure — i.e., $corr(e_{target}, m(w, e_{regs})) = corr(e_{target}, t * m(w, e)) \ \forall\ t > 0$ — allowing us to find models that capture the relative structure of the target pattern rather than its absolute expression. The correlation measure further has the advantage that it is easy to compute. The main disadvantage of the correlation measure is that it describes only the global fitness of a possible solution but does not consider the local structure of expression patterns in physical space.

The optimization problem we have to solve is:

$$\max_{w \in [-1,1]^k} (corr(e_{target}, m(w, e_{regs})). \tag{5}$$

The goal of the optimization is to find a weight vector w so that the correlation between e_{target} and $m(w, e_{regs})$ is maximal.

Finding the optimal weight vector w is an NP-hard problem. To find a possibly near-optimal solution we use the following genetic algorithm:

```
function [m, w, f]=GeneticNetworkOpt( e_target, e_regs, s_pop, p_cross, p_mutate, l_iter, l_corr )
  - pop_current : Create s_pop initial random solution vectors
  - for i=1:l_iter
      - f_pop : Compute the correlation between e_target and each network in pop_current
      - Sort pop_current according to f_pop
      - f = max(f_pop) : Find the best network in the current set of solutions
      - w = pop_current(1) : Copy the best weight vector to the output
      - m = m(w,e_regs) : Copy the expression pattern defined by w to the output
      - if(f > l_corr) then stop : Stop if desired minimum correlation has been reached
      - pop_new : Copy the two best networks pop_current(1) and pop_current(2) to pop_new
      - for j=1:(s_pop - 2)/2
          - Select two networks randomly from the ranked population pop_current
```

```
          - With probability p_cross perform random crossing of the selected networks
```

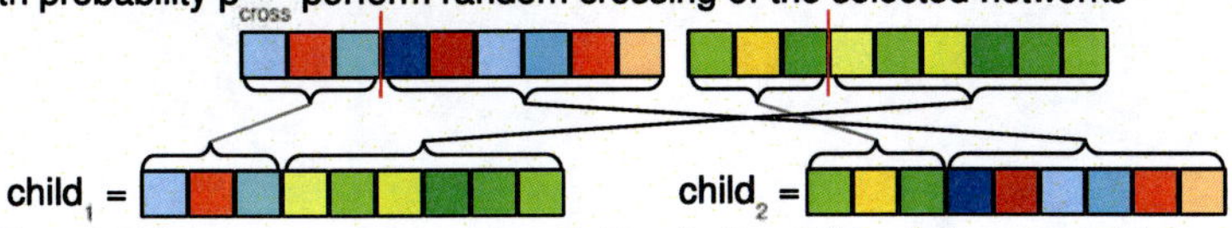


```
          - With probablity p_mutate mutate each weight w_i in the child solutions by adding a
            small, random value in the range of [-0.1 , 0.1] while ensuring that w_i ∈[-1,1]
```

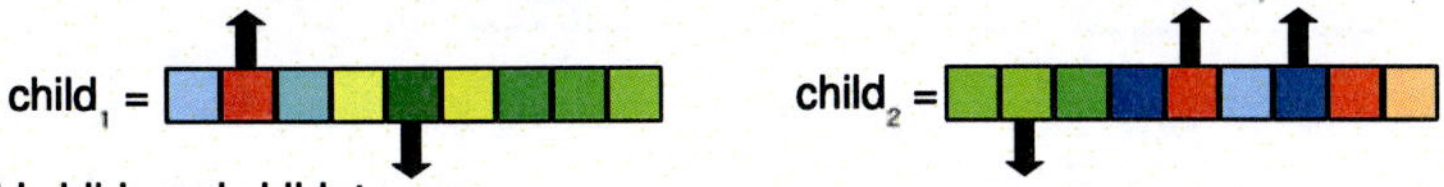


```
          - Add child1 and child2 to pop_new
      - Copy pop_new to pop_current
```

The algorithm is based on the canonical design of a genetic algorithm [29] and finds a single network w that defines the model expression pattern m with correlation f to the target pattern e_{target}. Besides the target pattern e_{target} and the input regulators e_{regs}, the algorithm has the following additional parameters: (a) s_{pop}, size of the population of possible solutions, (b) p_{cross}, probability of crossing between two selected solutions and (c) p_{mutate}, probability of mutating a particular entry w_i of a solution vector by a small random value. The parameters l_{iter} and l_{corr} serve as termination conditions defining the maximum number of iterations and the desired correlation score, respectively. Optionally, we also allow the user to specify: (a) whether the two best solutions should always be preserved, (b) whether mutations should only be accepted if they improve the solution, (c) whether the population should be initialized randomly or with all zero values, and (d) which types of plots should be shown during the optimization procedure.

Genetic regulatory networks are highly dynamic and often contain redundancies, i.e., in practice various good solutions may exists to the same problem. We repeat the optimization several times to investigate the diversity of good solutions and to increase the probability of finding a near-optimal solution.

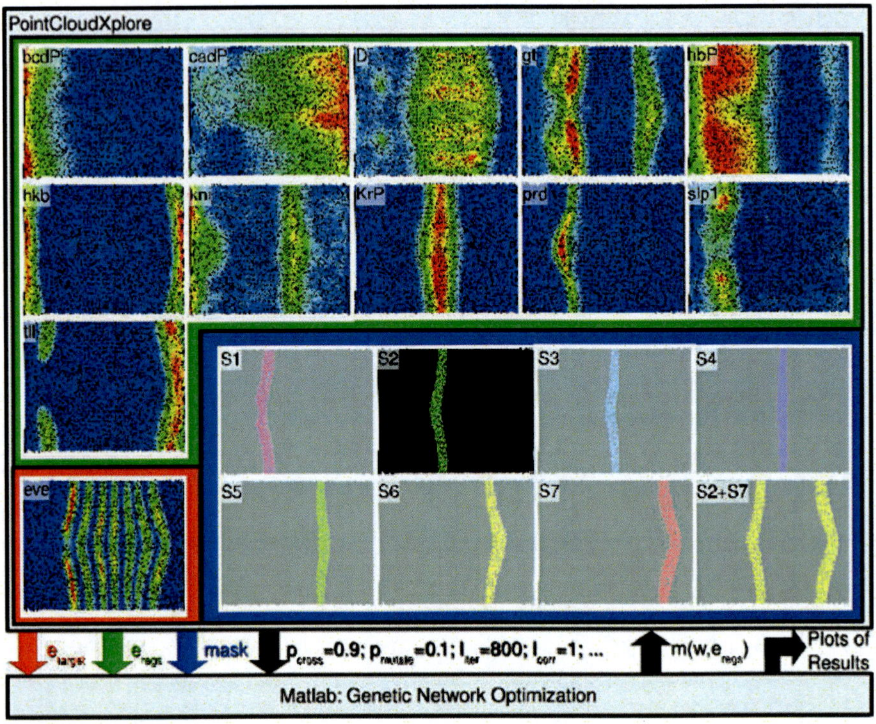

Fig. 4 Overview of the design of a network modeling experiment. Using PCX, the user specifies the necessary input of the optimization function, e.g.: (**a**) the set of input regulators e_{regs} (green box), (**b**) the target e_{target} (red box), (**c**) an optional mask specifying the area of interest (blue box), and (**d**) additional input parameters, such as p_{cross}. After the user has confirmed all settings, PCX sends all data to Matlab, calls the optimization function, and afterwards imports the model expression pattern $m(w, e_{regs})$. The optimization function also creates a set of plots with an overview of the analysis results

4.3 PointCloudXplore and Matlab for Network Modeling

We implemented the described optimization algorithm in Matlab and integrated the function with PCX via our Matlab interface. Figure 4 illustrates the design of a network modeling experiment using PCX and Matlab. Using the visualization and analysis capabilities of PCX, the user first identifies the target pattern of interest and a set of potential regulators, allowing the user to effectively validate the input expression patterns. A gene is often expressed in multiple distinct spatial domains — e.g., in seven stripes in the case of *eve* — which may be regulated by different *cis-regulatory elements* (CREs). Using PCX, a user can select the different expression domains of a gene. These cell selections can then be used to mask the expression pattern of the target to simulate the corresponding CRE pattern. Alternatively, one could use the cell masks directly in the optimization to compute the fitness of the model only in the area of interest to find a local model. Finally, the user is asked to specify the input parameters of the optimization algorithm, i.e., s_{pop}, p_{cross}, p_{mutate}, l_{iter} or l_{corr}, in the Input Dialog created automatically by PCX.

After the user has confirmed the input, PCX automatically starts Matlab, transfers the input data to Matlab, and initiates the execution of the optimization function. During the optimization process, the user is presented a set of plots visualizing the progress of the analysis. After completion of the optimization process, PCX imports the model expression pattern(s) $m(w, e_{regs})$, the relevant data is saved in a user-defined *.mat* Matlab state file, and a set of plots are shown to provide an overview of the results produced.

As this example illustrates, the close integration of PCX and Matlab allows advanced analysis functions to be deployed quickly to the biology user community while improving usability and accuracy of the analysis.

4.4 Modeling the eve Expression Pattern

In the following, we describe the application of these methods to the modeling of the expression pattern of the gene *even skipped* (*eve*). The *eve* pattern is characterized by seven stripes, which are usually presumed to result from the action of five CREs. While some stripes appear to have their own CRE, other stripes may share a module, and stripe 7 (the posteriormost stripe) has been speculated to be under redundant regulation. We here focus on modeling stripe 2 and 7 of the *eve* pattern. The goal of this experiment is to demonstrate the strengths and weaknesses of expression-based network modeling.

Figure 4 illustrates the basic setup of the experiment. In order to identify how the pattern of *eve* is initially formed we use the *eve* pattern at stage 5:9-25% as target, while applying selective masks to isolate individual stripes of the pattern. For the potential regulators (see Fig. 4) we used protein patterns at stage 5:4-8% because of the transcriptional delay between the regulator input and the output patterns. In cases where only mRNA expression data was available, we selected the patterns from stage 5:0-3% because of the translational delay from mRNA into a protein.

Figure 5 summarizes the results for the modeling of *eve* stripe 2. The algorithm consistently found similarly high-scoring models with $corr > 97\%$. The expression pattern of the predicted model fits the target well and even resembles the D/V variation of the target pattern. Commonly, the input regulators used are only considered as regulators along the A/P body-axis. The shown results suggest that the input regulators are also able to function as regulators along the D/V body-axis (Fig. 5d). These results are consistent with earlier modeling results based on data clustering [20]. The modeling suggests that *D*, hb, and possibly *prd* function as main activators of *eve* stripe 2 while *slp1*, Kr and gt function as inhibitors. The absolute weights of the remaining input regulators (i.e., *kni*, bcd, *hkb*, *tll*, and cad) are much lower, indicating that the function of these factors is not well characterized by the analysis.

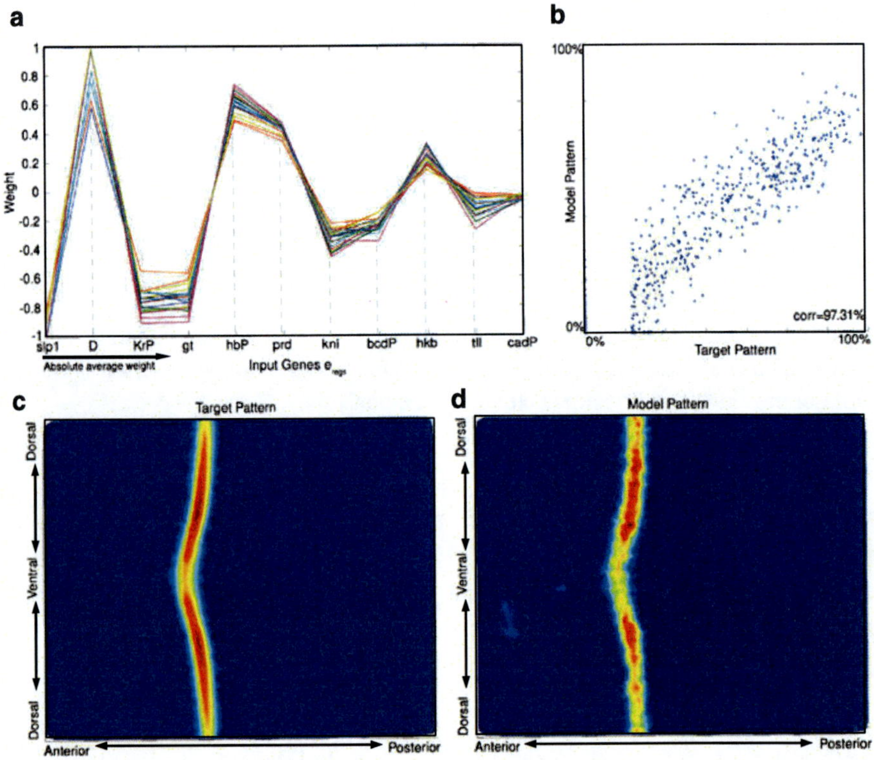

Fig. 5 Overview of the modeling results for *eve* stripe 2. We repeated the genetic algorithm 20 times while figures b-d show the model with the highest correlation. All experiments achieved a correlation of $> 97.x\%$. (**a**) Plot of the 20 models with each curve representing one model. The genes are ranked according to their absolute average weight. (**b**) Scatter plot of the target pattern and the best model pattern. (**c, d**) Unrolled view of the target and the model pattern, respectively. Color shows relative expression with blue = low and red = high expression. We can see that the model fits the target well and even resembles the D/V variation of the target stripe

Experimental data on *eve*-stripe regulation has shown that *eve* stripe 2 is activated by bcd [25] and hb [25] and potentially by D [21], and inhibited by gt [25], Kr [25] and weakly by slp1/2 [1]. prd is known to be able to activate late *eve* stripe expression via late enhancers [8], but is not essential for *eve* stripe 2 early upregulation. Notably, the modeling is able to find similarly high-scoring models even when removing *prd* from the input, indicating that *prd* may not be essential for defining *eve* stripe 2 (not shown). The reason why bcd was predicted as a weak inhibitor instead of an activator may be because (a) the highest bcd levels appear in a relatively wide gradient in the anterior of the embryo where *eve* stripe 2 is not located, and/or because (b) bcd activates also inhibitors of *eve* stripe 2, such as slp [10] and gt [17].

We also modeled *eve* stripe 7 using the same set of input regulators (Fig. 6a–c). The models we computed for stripe 7 score in general lower than the stripe 2 models

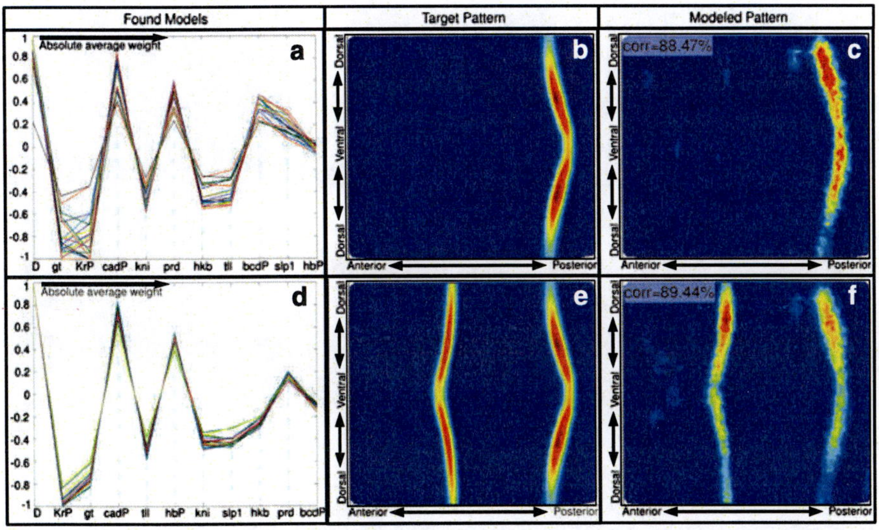

Fig. 6 Overview of the modeling results for *eve* stripes 7 (**a**–**c**) and 2 + 7 (**d**–**f**). We repeated the genetic algorithm 20 times while figure c and f show the result for the best model for stripe 7 and 2 + 7, respectively. All runs achieved a correlation of $87.x\%$ for stripe 7 alone and $> 88.9\%$ for stripe 2 + 7. (**a**, **d**) Curve plot of the 20 models. (**b**,**c**,**e**,**f**) Unrolled view of the target patterns and the respective best model pattern. Color shows relative expression with blue = low and red = high expression

($corr \approx 88\%$) most likely due to a strong bias resulting in lower "expression" on one side of the embryo. The input regulator cad — which functions as a main activator in the stripe 7 model — shows a similar bias due to measurement or data normalization errors, explaining the observed behavior. The boundaries of *eve* stripe 7 are, however, well-defined in the generated model. The model suggests *D*, cad, and *prd* as main activators and gt, Kr, *kni*, *hkb*, and *tll* as main inhibitors of *eve* stripe 7.

It is commonly presumed that *eve* stripe 7 arises from the action of *eve* stripe 3+7 CRE. With the standard *eve* 3 + 7 CRE, inhibition by kni has been experimentally verified, but mutation in tll actually abolished stripe 7 [24], and the demonstrated clear negative *hb* regulation was not found. Moreover, mutations in gt, Kr, and hkb have been reported not to affect the output of the 3 + 7 stripe element [24], though gt and Kr have been here predicted to be strong inhibitors. Likewise, experimental data does not support D as an activator of stripe 7 [21], and prd is not essential for early *eve*-stripes 7 upregulation [8]. However, Yan et al. [30] predicted *eve* stripe 7 to be activated by *Stat92E*, which was not available in the input dataset. The differences between *eve* stripe 3 + 7 CRE and other experimental data and our results suggest that modeling may be able to closely mimic the target pattern via a different network, even when important regulators are missing in the input.

While *eve* stripe 7 has usually been described as part of an output of the *eve* stripes 3 + 7 CRE, Janssens et al. [15] have proposed that *eve* stripe 2 CRE might also participate in regulation of *eve* stripe 7. The *eve* stripe 2 element is known to

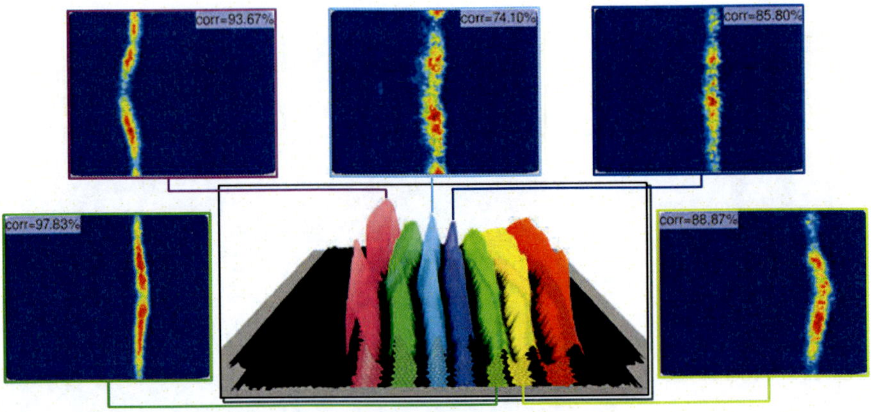

Fig. 7 Overview of modeling results for eve stripes 1, 3, 4, 5, and 6

often produce a weak *eve* stripe 7 expression and the same is observed in sepsid *eve* stripe 2 enhancers [11, 15, 23], while the stripe 7 produced by *eve* 3 + 7 CRE can be weaker than stripe 3 [24]. Therefore, we modeled *eve* stripe 2 and 7 at the same time (Fig. 6d–f). Qualitatively the model for *eve* stripe 7 does not change significantly while *eve* stripe 2 shows the same D/V bias as stripe 7, probably due to the higher-weighted *cad* pattern. The boundaries of *eve* stripe 2 are, however, still well-defined. The Janssens et al. [15] model supports our prediction of cad as an activator and gt and tll as repressors of *eve* stripe 7(+2) (Fig. 6a,d). These results suggest that *eve* stripes 2 and 7 could, indeed, be formed by a common network model and may be co-regulated. However, our model does not account for potential regulation by multiple or diffuse CREs, which may cause distortion due to too high a contribution of stripe 7 to a particular CRE model. To eliminate potential problems due to interference between CREs one would ideally measure the actual output of individual CREs, e.g., from transgenic constructs.

Figure 7 shows preliminary results for the modeling of *eve* stripes 1, 3, 4, 5, and 6. For *eve* stripe 1 and 5 we found high-scoring models ($corr > 93\%$) which — similar to the *eve* stripe 2 model — also reproduce the main variations in expression along the D/V axis. While the models for *eve* stripes 3 and 4 show expression in the expected location, the stripe-borders are not well-defined. Interestingly the algorithm is able to model stripe 3 well when including the cooperative factor $e_{hbP,KrP} = e_{hbP}e_{KrP}$ as input regulator, achieving much higher scores of $corr \approx 90\%$. While cooperative regulation of stripe 3 by *Kr* and *hb* may not be real [24], this behavior nevertheless indicates that — besides the fact that important input regulators may be missing — non-linear regulatory effects not captured by the used linear model may be responsible for the correct formation of at least *eve* stripe 3.

The fact that the modeling was able to predict a large range of regulators correctly for *eve* stripe 2 shows that modeling can provide interesting insights into, or at least hints of, possible regulatory interactions. Missing regulators (see stripe 7), noise (see stripes 7 and 2 + 7), and limitations of the employed computational model (see stripe 3), however, directly affect the quality of the predicted model and may also lead to false negatives (missing regulators) and false positives (misidentified regulators) in predictions (see stripe 7). Modeling results should, therefore, always be validated experimentally.

5 Conclusions

To fully exploit the collaborative research potential of teams of biologists, computational biologists, and computer scientists, it is essential to overcome true and perceived obstacles for collaboration. Biologists rarely do computation and computer scientists rarely do biology. To maximize the impact of novel, complex, high-dimensional data sets acquired via modern imaging or computational methods — such as the BDTNP 3D gene expression atlas data — the data needs to be accessible to biologists and comprehensible to developers of analysis and visualization software.

We addressed these challenges by linking the visualization system PointCloud-Xplore (PCX) and Matlab via a dedicated interface, providing biologists seamless access to advanced data analysis functions and enabling bioinformatics researchers to integrate their analysis directly into the visualization. By being able to test new analysis functions during development, biologists are able to provide feedback early, facilitating communication between the developer and the user. By utilizing PCX and Matlab, a developer can develop new functions more efficiently without having to know anything about the PointCloud data format or the architecture of PCX.

One potential focus for future research are methods for linking other types of data, such as *in vitro* and *in vivo* binding data, with the gene expression data analysis. Being able to incorporate different types of data is essential, e.g., for the development of advanced predictive models of gene expression.

Acknowledgments This work was supported by the National Institutes of Health through grant GM70444 and by Lawrence Berkeley National Laboratory (LBNL) through the Laboratory Directed Research Development (LDRD) program. Research performed at LBNL was also supported by the Department of Energy under contract DE-AC02-05CH11231. In addition, we gratefully acknowledge the support of an International Research Training Group (IRTG 1131) grant provided by the German Research Foundation (DFG), awarded to the University of Kaiserslautern, Germany. We thank the members of IDAV at UC Davis, the BDTNP at LBNL, the IRTG, and LBNL's Visualization Group.

References

1. L. P. Andrioli, V. Vasisht, E. Theodosopoulou, A. Oberstein, and S. Small. Anterior repression of a *Drosophila* stripe enhancer requires three position-specific mechanisms. *Development*, 129(21):4931–4940, November 2002.

2. M. Q. W. Baldonado, A. Woodruff, and A. Kuchinsky. Guidelines for using multiple views in information visualization. In *AVI '00: Proceedings of the working conference on Advanced visual interfaces*, 110–119, New York, NY, USA, 2000. ACM Press.
3. L. D. Davis and M. Mitchell. Handbook of genetic algorithms. *Van Nostrand Reinhold*, 1991.
4. H. De Jong. Modeling and simulation of genetic regulatory systems: A literature review. *Journal of Computational Biology*, 9(1):67–103, 2002.
5. P. D'haeseleer, S. Liang, and R. Somogyi. Genetic network inference: from co-expression clustering to reverse engineeing. *Bioinformatics*, 16(8):707–726, 2000.
6. H. Doleisch, M. Gasser, and H. Hauser. Interactive feature specification for focus + context visualization of complexsimulation data. In G.-P. Bonneau, S. Hahmann, and C. D. Hansen, editors, *Data Visualization 2003 (Proceedings of the Eurographics/IEEE TCVG Symposimon Visualization)*, 2003.
7. C. C. Fowlkes, C. L. L. Hendriks, S. V. E. Keränen, G. H. Weber, O. Rübel, M.-Y. Huang, S. Chatoor, A. H. DePace, L. Simirenko, C. Henriquez, A. Beaton, R. Weiszmann, S. Celniker, B. Hamann, D. W. Knowles, M. D. Biggin, M. B. Eisen, and J. Malik. A quantitative spatiotemporal atlas of gene expression in the *Drosophila* blastoderm. *Cell*, 133:364–374, April 2008.
8. M. Fujioka, P. Miskiewicz, L. Raj, A. A. Gulledge, M. Weir, and T. Goto. *Drosophila* paired regulates late *even-skipped* expression through a composite binding site for the paired domain and the homeodomain. *Development*, 122(9):2697–2707, Sept. 1996.
9. D. L. Gresh, B. E. Rogowitz, R. L. Winslow, D. F. Scollan, and C. K. Yung. WEAVE: A system for visually linking 3-d and statistical visualizations,applied to cardiac simulation and measurement data. In T. Ertl, B. Hamann, and A. Varshney, editors, *Proceedings IEEE Visualization 2000*, 489–492, Los Alamitos, CA, USA, 2000. IEEE Computer Society Press.
10. U. Grossniklaus, K. M. Cadigan, and W. J. Gehring. Three maternal coordinate systems cooperate in the patterning of the *Drosophila* head. *Development*, 120(11):3155–3171, 1994.
11. E. E. Hare, B. K. Perterson, V. N. Iyer, R. Meier, and M. B. Eisen. Sepsid *even-skipped* enhancers are functionally conserved in *Drosophila* despite lack of sequence conservation. *PLoS Genetics*, 4(6):e1000106., 2008.
12. C. Henze. Feature detection in linked derived spaces. In D. Ebert, H. Rushmeier, and H. Hagen, editors, *Proceedings IEEE Visualization '98*, 87–94, Los Alamitos, CA, USA, 1998. IEEE Computer Society Press.
13. M.-Y. Huang, O. Rübel, G. H. Weber, C. L. Luengo Hendriks, M. D. Biggin, H. Hagen, and B. Hamann. *Segmenting Gene Expression Patterns of Early-stage Drosophila Embryos*, 313–327. Springer Verlag, Germany, Jan 2008.
14. J. Jaeger, M. Blagov, D. Kosman, K. N. Kozlov, Manu, E. Myasnikova, S. Surkova, C. E. Vanario-Alonso, M. Samsonova, D. H. Sharp, and J. Reinitz. Dynamical analysis of regulatory interactions in the Gap gene system of *Drosophila melanogaster*. *Genetics*, 167(4):1721–1757, April 2004.
15. H. Janssens, S. Hou, J. Jaeger, A.-R. Kim, E. Myasnikova, D. Sharp, and J. Reinitz. Quantitative and predictive model of transcriptional control of the *Drosophila melanogaster even skipped gene*. *Nature Genetics*, 38, Sept. 2006.
16. C. Johnson, S. G. Parker, C. Hansen, G. L. Kindlmann, and Y. Livnat. Interactive simulation and visualization. *Computer*, 32(12):59–65, December 1999.
17. R. Kraut and M. Levine. Spatial regulation of the gap gene *giant* during *Drosophila* development. *Development*, 111(2):601–609, 1991.
18. D. T. Lang and D. F. Swayne. GGobi meets R: an extensible environment for interactive dynamic data visualization. In *Proceedings of the 2nd International Workshop on Distributed Statistical Computing*, Mar. 2001.
19. C. L. Luengo Hendriks, S. V. E. Keränen, C. C. Fowlkes, L. Simirenko, G. H. Weber, A. H. DePace, C. Henriquez, D. W. Kaszuba, B. Hamann, M. B. Eisen, J. Malik, D. Sudar, M. D. Biggin, and D. W. Knowles. Three-dimensional morphology and gene expression in the *Drosophila* blastoderm at cellular resolution I: Data acquisition pipeline. *Genome Biology*, 7(12):R123, 2006.

20. O. Rübel, G. H. Weber, M.-Y. Huang, E. W. Bethel, M. D. Biggin, C. C. Fowlkes, C. L. Hendriks, S. V. E. Keränen, M. B. Eisen, D. W. Knowles, J. Malik, H. Hagen, and B. Hamann. Integrating data clustering and visualization for the analysis of 3d gene expression data. *IEEE Transactions on Computational Biology and Bioinformatics*, 2008.
21. S. R. Russell, N. Sanchez-Soriano, C. Wright, and M. Ashburner. The *Dichaete* gene of *Drosophila melanogaster* encodes a SOX-domain protein required for embryonic segmentation. *Development*, 122(11):3669–3676, 1996.
22. F. Sauer and H. Jäckle. Heterodimeric *Drosophila* gap gene protein complexes acting as transcriptional repressors. *EMBO Journal*, 14(19):4773–4780, 1995.
23. S. Small, A. Blair, and M. Levine. Regulation of *even-skipped* stripe 2 in the *Drosophila* embryo. *The EMBO journal*, 11(11):4047–4057, Nov. 1992.
24. S. Small, A. Blair, and M. Levine. Regulation of two pair-rule stripes by a single enhancer in the *Drosophila* embryo. *Developmental Biology*, 175(2):314–324, May 1996.
25. S. Small, R. Kraut, T. Hoey, R. Warrior, and M. Levine. Transcriptional regulation of a pair-rule stripe in *Drosophila*. *Genes Dev*, 5(5):827–839, May 1991.
26. D. Thieffry and L. Sanchez. Dynamical modeling of pattern formation during embryonic development. *Current opinion in genetics & development*, 13(4):326–330, 2003.
27. E. P. van Someren, L. F. A. Wessels, E. Backer, and M. J. T. Reinders. Genetic network modeling. *Pharmacogenomics*, 3(4):507–525, 2002.
28. G. H. Weber, O. Rübel, M.-Y. Huang, A. DePace, C. C. Fowlkes, S. V. Keränen, C. L. Luengo Hendriks, H. Hagen, D. W. Knowles, J. Malik, M. D. Biggin, and B. Hamann. Visual exploration of three-dimensional gene expression using physical views and linked abstract views. *IEEE Transactions on Computational Biology and Bioinformatics*, 6(2):296–309, April-June 2009.
29. D. Whitley. A genetic algorithm tutorial. *Statistics and Computing*, 4(2):65–85, June 1994.
30. R. Yan, S. Small, C. Desplan, C. R. Dearolf, and J. E. Darnell Jr. Identification of a *Stat* gene that functions in *Drosophila* development. *Cell*, 84(3):421–430, Feb 1996.

Index

L. Linsen et al. (eds.), *Visualization in Medicine and Life Sciences II*, Mathematics and Visualization, DOI 10.1007/978-3-642-21608-4,